ANALYSIS
A Gateway to Understanding Mathematics

ANALYSIS

A Gateway to Understanding Mathematics

Seán Dineen

University College Dublin, Ireland

 **World Scientific**

NEW JERSEY · LONDON · SINGAPORE · BEIJING · SHANGHAI · HONG KONG · TAIPEI · CHENNAI

Published by

World Scientific Publishing Co. Pte. Ltd.

5 Toh Tuck Link, Singapore 596224

USA office: 27 Warren Street, Suite 401-402, Hackensack, NJ 07601

UK office: 57 Shelton Street, Covent Garden, London WC2H 9HE

British Library Cataloguing-in-Publication Data
A catalogue record for this book is available from the British Library.

ANALYSIS
A Gateway to Understanding Mathematics

ISBN-13 978-981-4401-38-8
ISBN-10 981-4401-38-2

Printed in Singapore by World Scientific Printers.

To

Traolach, Donal, Maura, Eithne and Emer.

Contents

PREFACE

It is not the strongest of the species that
survives, nor the most intelligent, it is the
one that is most adaptable to change.
Charles Darwin, 1809-1882

Searching for proof is one thing, but
searching for understanding is much
more important and they are not the
same thing.
Michael Atiyah, 2005

This book is based on an introductory first year course given by the author over a number of years to students of *Economics and Finance* at *University College Dublin*. This year long course was initially called, *One Variable Differential and Integral Calculus*, but was later retitled, *Introduction to Analysis*. These business students did not come to university to study mathematics but had knowingly enrolled in a program that required sophisticated mathematics during each year of their undergraduate studies. As the students were familiar with the technical aspects of the calculus I was motivated to help them understand what they knew and I was aware that today's students, particularly those with expectations of a lengthy career in any discipline that routinely applies mathematics, will need the ability to approach unknown areas of mathematics. This skill requires the confidence that accompanies understanding. In this book we show that it is possible to search for understanding while still meeting traditional expectations. Moreover, we found that mathematics as a language with clearly defined concepts and a reliance on deductive arguments had other long term practical benefits for our students.

The road to understanding is quite different to the technique oriented route. It is necessary to start at the very beginning, to make no assumptions, and to carefully and thoroughly examine one or two topics. The choice of topics is unimportant.[1] The students responded reluctantly to this unfamiliar approach but, as their involvement grew, they cautiously put aside previous attitudes and began to develop a confidence that accepts uncertainty. When this was in place, and it was neither immediate nor universal, technical skills were rapidly absorbed. As I put together handwritten notes to explain some chaotic lectures, I realized I was not addressing the usual audience for mathematical textbooks, *reluctant teenagers*, and a concise new title suggested itself: *Introducing Mathematics to Consenting Adults*. For a few years typed notes, with this title, were distributed to the students. Here, I have opted for a title that highlights a different aspect of this book. *Analysis* means separating a thing into its component parts and, by extension, discovering the general principles underlying individual phenomena. The founders and developers of the calculus used this word in their publications[2], acknowledging their analysis of infinite processes. Since then the word has been used in combinations such as, *Mathematical Analysis, Real Analysis, Complex Analysis, Numerical Analysis* and *Functional Analysis*, to loosely indicate areas of mathematics that rely heavily on limiting processes and the real number system. This applies in particular to the foundations of the differential and integral calculus.

My first title led me to ponder the extent that mathematics had seeped into the popular imagination. Many who protest that they know no mathe-

---

[1]The first chapter could have been devoted to any of the following: the Binomial Theorem, Pythagoras' Theorem, $\sqrt{2}$, solutions of linear equations in 3 unknowns.

[2]For example, *De Analysi per Aequationes numero terminorum infinitus,* by Newton in 1669, and *Introductio in analysin infinitum,* by Euler in 1748.

matics can read maps, and graphs, they can talk about speed, acceleration, inflation, betting odds, average temperatures, interest rates, and percentage increases, and when it suits they can follow logical arguments, while only vaguely associating these concepts with mathematics. But, for many, mathematical development ceases at the gates of academia. This is unfortunate as today there is an acute need for an educated public with a confident and non-trivial understanding of mathematics.

Mathematics has always played a role in civilized society and most are aware of its benign contributions. Unfortunately, over the last quarter century partially educated pseudo-experts, often in managerial positions, have realized that it is easy to associate numbers[3] with different human activities and to derive conclusions with little supporting evidence. Ironically, these developments are possible because mathematicians played a prominent role in developing three areas which facilitated the numerical analysis and interpretation of large quantities of data: *statistics, computer science* and the *internet*. The conclusions of these present day alchemists are rarely challenged by a unquestioning public and, unfortunately, the consequences may be serious. This is not a modern phenomenon. Throughout history, knowledge and ignorance have both been employed as a means of control. Priests in ancient Egypt controlled the calendar because it gave them power and the greatest promotor of statistics for purposes of political control in the modern era was Josef Stalin. Today's pseudo-experts rely on the mathematical ignorance of the masses and mathematicians should be alarmed, rather than flattered, by the widespread belief that everything can be measured and thus reduced to a number.

That mathematicians occupied over the centuries influential positions in politics, the church, the law, medicine, business, engineering, science, and the military is a tribute to the benefit of an education in clear think-

---

[3]We merely mention *biblio-metrics, citation indices, impact factors, key performance indictors, work-load models,* etc. One can almost assume that when the word *benchmark* is being used that mathematics is being abused. Many of these have impressive titles that become even more impressive and less intelligible when they evolve into catchy acronyms.

ing and abstract reasoning, a benefit recognized thousands of years ago by Socrates, Plato and Aristotle. Mathematical ideas evolved over considerable periods of time and the student of mathematics often follows, perhaps subconsciously, the path followed by those who developed the subject and may well struggle with the same difficulties that initially surrounded the introduction of new ideas. These were some of the thoughts that motivated our introduction of historical material into this text, later we used the opportunity to provide a background for mathematics. Nevertheless, no matter how much we read and no matter how strong our imagination we can never fully appreciate the overall atmosphere of a bygone age, but we can glimpse how things were, and we can be inspired. In the following paragraph we give one brief example.

The ten years, 1789-1799, was a turbulent period in French history. At the beginning of the decade the French monarchy of Louis XVI was in place. This was followed by the French revolution and by the end of the decade Napoleon was the uncrowned emperor of France. The revolution, usually portrayed only as a brutal class war, aimed at fundamentally changing society. During this chaotic period three mathematical legends, Lagrange, Laplace and Legendre, were at the height of their powers and lived in Paris while the future legends Fourier and Poisson were students in the same city. These mathematicians subscribed to the revolutionary philosophy of building a modern egalitarian state using science and reasoning. They maintained and developed, in unbelievably uncertain circumstances, their mathematical interests and became involved at a highest level in the revolution. All survived, and during that decade the three established mathematicians served on the committee that established the metric system of weights and measures while publishing important mathematics. The National Convention enacted a bill on October 30, 1794, setting up the École Normale to train a new type of teacher and each district in France was to send a small number of its most talented citizens to Paris to hear the experts lecture on their subject. Laplace, after diplomatically avoiding the call to be a talented citizen, was nominated as an expert and gave the inaugural lecture in mathematics to seven hundred mature students on January 20, 1795. His approach was revolutionary, he aimed at *showing the most important discoveries, their principles, the circumstances that led to their birth, the most direct route to them and the procedures for making new discoveries.* The ten lectures given by Laplace, were too advanced for his audience even though he used no equations in his earlier lectures and tried to explain abstract concepts in the language of his audience. They were published later and set the standard for public education in mathematics at second and third levels in France.

This book is devoted to the different aspects of one variable calculus. There are overlaps between the different chapters but each has its own dominant theme. The traditional differential calculus is developed in Chapters 10 and 11 and the integral calculus in Chapters 13 and 14. All other chapters are devoted to the background necessary to understand these chapters. Chapters 1 and 2 are an informal and intuitive introduction to mathematics, number systems are examined in Chapters 5 and 12, sets in Chapter 6, functions in Chapters 3 and 4, continuous functions in Chapter 9, and sequences and series in Chapters 7 and 8.

The features that distinguish this book from the many other books on the calculus are, hopefully, apparent to even casual readers as they proceed. We aim at a comprehensive, efficient, and rigorous treatment and introduce all the concepts that this necessitates. Functions, increasing (and decreasing) sequences, and bounds (upper and lower) play an important role in our study. Functions were often essential in allowing us to express ourselves clearly, the intuitive notion of increasing (and decreasing) convergent sequence helped us present a rigorous treatment of arbitrary sequences and series without resorting to the $\epsilon - \delta$ approach of Weierstrass. Increasing sequences of dyadic rationals allowed us to construct the real numbers without the use of either Dedekind sections or Cauchy sequences. By concentrating on polynomials, the exponential function, and power series, the fundamental concepts became more transparent. Moreover, students who persisted with our streamlined approach had little difficulty absorbing, and using in later courses, the traditional material omitted here.

There are diverse audiences for this book and, unlike the generic twentieth century encyclopedic tomes on The Calculus, this offering is not a catchall that can be recommended ubiquitously. The serious study of mathematics takes time, as does the study of any discipline, this is a fact and it should be anticipated. A positive predisposition to mathematics and a persistent commitment to mastering the material are the main prerequisites required of the reader. Those who have left the world of structured mathematics courses behind but who now, as adults, wish to study and understand mathematics may be interested in this book. The level of abstraction varies considerably between chapters and the nature of the discipline suggests that the lone reader, while following the *structure* of a formal course, should choose the *pace* that they find suitable. I recommend that these readers skim over Chapters 5, 6 and 12 on a first, and perhaps a second reading, and dip back into the material in these chapters when they feel ready to do so. The 200+ exercises, most of which were developed and tested in tutorials or as homework, were chosen because they forced understanding and are an *essential part* of this book. Some are routine, but many are not, and although most relate to earlier material this is not always the case, and to help the reader appreciate the unexpected a few have been placed strategically in unorthodox locations. Mathematics should be read backwards and forward. Each attempt at an exercise should lead to a focused perusal of at least one chapter and result in the understanding of at least one fact that was read, or merely passed over, a dozen times previously. This is how we make progress in mathematics. We have provided solutions to some exercises, partial solutions to others, and no solutions to a number.

This book may be used as a text in a standard one variable calculus or analysis course. But we only recommend it as the main text when the teacher has first experienced it as a supplementary text, perhaps to a small group. It is not our intention to appear elitist with this comment. We are quite serious. A casual reading may create a false impression about the pace and difficulty of the material. In particular, it may well hide the degree of involvement required of the teacher and the level of participation demanded of the student, both of which are essential if the approach we advocate is to succeed. Students with an aptitude, perhaps latent, for mathematics, intelligent students who have convinced themselves that mathematics is not for them, and weaker students with a sympathetic and experienced guide, appear to benefit most immediately from this book. The modestly successful average student, who has learned to cope with rote learning, may require additional motivation.

I have made strenuous efforts to eliminate errors in this book but experience tells me that some remain and I welcome information on these and any comments that readers wish to make.

It is a pleasure to acknowledge the help and support I received from many people in bringing this project to fruition. However, five people played very different, but equally important, crucial roles and I wish to thank them very explicitly. Michael Mackey created all the diagrams, helped with the layout, and his considerable patience was more helpful than he realizes. Domingo García was a source of practical encouragement and sensible advice when it was needed and it made a difference. I thank Guido Zapata for the many useful and insightful conversations that we had over many years about the teaching of mathematics. Terence Corish provided the questioning challenge that I needed in recent years while Livia Henderson helped with the design of the cover.

I would like to acknowledge the influence of my own students of Economics and Finance on the shape of this book and thank them for their involvement and interest over a number of years. Dan Golden conceived and developed the Economics and Finance program and gave me complete freedom to follow my own teaching instincts. Without his help and support I would never have had the opportunity to write this book. Marius Ghergu, who currently oversees the mathematical content of this program, has always been supportive while Mary Hanley was helpful in the final stages of proofreading. Lai Fun Kwong from World Scientific was the perfect editor.

Finally, I thank my family, especially my wife Carol, for patiently listening for years to my reasons and excuses for getting back to The Book.

Sean.Dineen@ucd.ie
School of Mathematical Sciences,
University College Dublin,
Belfield, Dublin 4,
Ireland.

Chapter 1

QUADRATIC EQUATIONS

And thus do we of wisdom, and of reach,
With windlasses, and with assays of bias,
By indirections find directions out.

William Shakespeare, 1600
Hamlet, Act 2, Scene 1.

Summary

Using quadratic equations as an example, we see how one might approach an unknown formula in mathematics.

1.1 Quadratic Equations

We begin our journey by considering an elementary result that is probably well known to most readers. We examine it as an example of how one might approach *any* mathematical result. Mathematics is a unified discipline and a thorough examination of any one part, even such a simple topic as quadratic equations, soon leads to fundamental questions that concern all of mathematics and to which we return, time and time again, as we proceed. Our approach allows us to wander along tangents into peripheral topics. These add to our overall understanding and help us develop our

1

own context for mathematics.

We consider the following result.

Theorem 1.1. *If*

$$ax^2 + bx + c = 0 \quad then \quad x = \frac{-b \pm \sqrt{b^2 - 4ac}}{2a}. \tag{1.1}$$

Although this result has been known[1] for over 4000 years, let us pretend it is something we have never seen before. How does one approach the unknown in mathematics?

It is always safe to begin by asking questions. What do the symbols mean? What does the theorem[2] claim to say? Does it make sense? Does it tell us something new? Does it agree with what we already know? Why is it written in this particular way? Where did it come from? Why would anyone consider such a theorem? How do we go about proving it? These are all natural questions and worth pursuing. We will, to some extent, answer them, but not immediately.

First note the formal format of the presentation. This is now standard within mathematics and, although it can be intimidating to the uninitiated, it is efficient. We have *two* statements that, for simplicity, we call **P** and **Q**; where

$$\textbf{P} \text{ is the statement} : \quad ax^2 + bx + c = 0$$

and

$$\textbf{Q} \text{ is the statement} : \quad x = \frac{-b \pm \sqrt{b^2 - 4ac}}{2a}$$

---

[1]Mathematical results were often discovered independently by different civilizations. Solutions to quadratic equations were known to the Babylonions who lived in Mesopotamia. They wrote with the slanted edge of a wedge shaped instrument called a *stylus* on tablets of unbaked clay and their arrow shaped writing is called *cuneiform*. These tablets, when baked, were easily preserved and hundred of thousands have survived to provide a penetrating insight into life thousands of years ago.

[2]Nowadays, the terms *proposition* and *theorem* are often used interchangeably but, in mathematical logic, a proposition is a proposal which has not yet been proved and indeed it may or may not be true. When a proof is provided, it becomes a theorem. See Section 12.2.

and these are connected by the assertion:

If   **P**   then   **Q**.

The expectation is that if **P** is accepted as true then eventually **Q** will also be similarly accepted. The process of convincing the reader to accept **Q** is called a proof. We discuss three principal methods[3] of proof in this book. The *synthetic* proof, used almost exclusively in classical Greek presentations, consists of a sequence of facts, acceptable to the reader, starting with **P** and cumulating in **Q**. It is not natural and hides the experimentation that invariably precedes the discovery of mathematical truths. The *analytic* proof begins by assuming what is sought and argues back to a known truth. This is a process of discovery and a careful examination of the completed analysis usually leads to a synthetic proof.[4] We discuss *proof by contradiction* later.

It is natural to ask:

what facts are acceptable?

and this innocent and highly non-trivial question is at the heart of mathematics. You will find your own answer in this book. Our aim in this chapter is to make Theorem 1.1 an acceptable fact.

Looking again at the presentation, we see a certain lack of symmetry between the letters a, b, c and x in both statements **P** and **Q**. There is almost a suggestion that we should treat them differently. Why not write statement **P** as

$$ad^2 + bd + c = 0$$

---

[3] As one dictionary dramatically and accurately puts it: *a proof consists in subduing the mind with evidence.* Thales of Miletus, c.624-548 BC, traveled, as a merchant, to Egypt and learned there geometrical facts derived by the Egyptians from repeated practical experiences. While inquiring into the validity of these facts, he, it is reputed, introduced the concept of proof into classical Greek mathematics. Around 430 BC, Hippocrates of Chios introduced the idea of arranging theorems so that later results could be proved on the basis of earlier ones. This approach was refined in Plato's Academy in Athens and culminated around 300 BC in *The Elements of Euclid.* The Hippocratic oath taken by members of the medical profession is due to Hippocrates of Cos.

[4] A synthesis is a *putting together to form a whole* while analysis implies *separating something into its component parts.* The natural order is to first separate and afterwards to put together.

or

$$xa^2 + ya + z = 0?$$

There is no mathematical or, indeed, no logical reason why the theorem could not have been written in either of these ways and if, for instance, we used the second of these, then statement **Q** becomes

$$a = \frac{-y \pm \sqrt{y^2 - 4xz}}{2x}.$$

However, there is a historical reason for the above *notation*. The original problems that gave rise to the above theorem were specific taxation and land division problems from Mesopotamia. Typical examples, in modern notation, are

$$x^2 - 5x + 6 = 0$$

and

$$10x^2 - 11x - 72 = 0$$

where all entries, except x, were known quantities. The problem was to find x. In Mesopotamia, such problems were written out in a very elaborate fashion. A major difference between our formulation and that of the Babylonions, separated by over 4000 years, lies in the use of symbols as mathematical shorthand. The Babylonians and other civilizations developing mathematics, began articulating mathematical concepts in the language they used every day.[5] Over the centuries, the introduction of new symbols

---

[5] The Babylonians did not have the concept of proof as we understand it but, by trial and error, had arrived at step-by-step procedures, now called algorithms, which led to a solution. They divided the collection of quadratic equations into nine different types. A typical school problem might read as follows: The length of a field exceeds the width by ten. If the area is six hundred what are the length and breath? For this example the following was the prescribed algorithm:

(1) Take half the difference of the length and width (the half difference); 5

(2) Square the half difference; 25

(3) Add the area; 625

(4) Take the square root; 25

(5) Length is square root + half difference; 30

and conventions led to clearer and more efficient presentations. For example, a pair of parallel lines ' $=$ ' was introduced to replace the Latin *aequalis* by Robert Recorde who stated '*bicause noe 2 thynges can be moare equalle*'. The $+$ symbol evolved from the Latin word for *and* (see Figure 1.1).[6] The

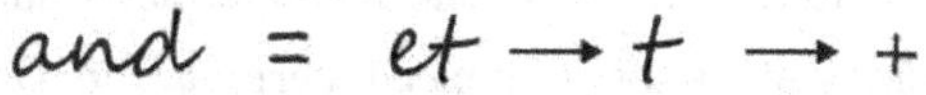

$$and \ = \ et \longrightarrow t \longrightarrow +$$

Fig. 1.1

simplicity of these changes hides the very long periods that were necessary for their germination and the amazing advances that they facilitated. This efficiency led to more and more information being compressed into less and less space and, as a result, to more demands on the reader. It may even lead to the erroneous conclusion that mathematics is exceptionally difficult, a common perception resulting from misunderstanding how mathematics has evolved as a *language* and from the failure to appreciate the amount of information being transmitted by mathematical statements.

In 1592 the French mathematician, François Viète, introduced the systematic use of letters in mathematical equations: using vowels for unknown

---

(6) Width is square root − half difference; 20.

Today we would use the following terminology. Let x and y denote the length and width of a field respectively. If $x - y = 10$ and $xy = 600$ find x and y. On substituting we obtain the quadratic equation

$$x^2 - 10x - 600 = 0$$

with solutions 30 and −20. The Babylonians only accepted positive solutions, hence $x = 30$ and $y = 30 - 10 = 20$.

[6]Robert Recorde, 1510-1558, from Tenby, Wales, was educated and may have taught mathematics at both Oxford and Cambridge. He practised medicine and was for a period controller of the royal mints in Bristol and Dublin. Latin, the language of scholars during his lifetime, was inaccessible to many and prompted him to write the basic mathematics texts *The Ground of Artes* (1543) and *The Whetstone of Witte* (1557) in English. His presentation was in the form of a Socratic dialogue and while he did not provide proofs he did supply motivation and explanation. Political intrigues landed him in jail and he died in prison.

quantities and consonants for known quantities. This allowed him to treat general, rather than specific, examples and to consider formulae rather than algorithms. Viéte[7] lived during the final period of the Renaissance. During this epoch, Europe rediscovered its classical Greek heritage. The ancient Greeks did not, as a matter of course, display their methods and intuitions and Viète was motivated to discover the secret of their success. His investigations and writings strongly influenced Fermat and Descartes in their discovery of *coordinate geometry* and Newton and Leibniz in their development of the *differential and integral calculus*. Prior to Viète, letters had been used sporadically as symbols for quantities, for instance, by Jordanus Nemorarius in the 13th century.

Fifty years after Viète, René Descartes revised this convention[8] by using the letters at the *beginning* of the alphabet for *known quantities* and the letters at the *end* to denote *unknown quantities*. This greatly simplified the presentation and has been in use ever since and, indeed, extended to the notation we use for functions (see Chapter 2).

Thus, in writing down

$$ax^2 + bx + c = 0,$$

we are indicating that we have been given three known quantities a, b and c and a relationship between them involving an unknown number x and it is our wish to find x. This unlocks the mystery surrounding the purpose of the theorem since we can now see clearly that $\mathbf{Q}$ gives us x directly.

---

[7]François Viète, 1540-1603, from Fontenay-le-Comte in western France, is regarded as the greatest French mathematician of the 16th century. He trained as a lawyer and his professional reputation led to a call to Paris to serve in the royal court. He was successful as advisor, negotiator, code-breaker, and privy councillor to Charles IX, Henry III and Henry IV. Political and religious intrigues led to his banishment from Paris for a five year period that he devoted entirely to mathematical studies.

[8]A tacit agreement not based on principle but sanctioned by custom and usage. Clearly, it is convenient and useful and nowadays necessary to have agreement on which side people should pass one another when travelling in opposite directions but there is no absolute reason to choose one option over the other. As a result, the world was, until recently, more or less evenly divided between the two *conventions* and, today, roughly two thirds of the world's population drive on the right while the remaining third drive on the left.

Thus, Theorem 1.1 turns an *implicit relationship* involving x into an *explicit formula* for x. Of course, you may still ask: why, if a, b and c are known, we do not write them as numbers like 2, 4 or 20? If we did, it would mean that for *every* triple of numbers, e.g. $\{2, 13, 4\}$, $\{122, 31, -74\}$, etc., we would have a *different* theorem. By using $\{a, b, c\}$ we have *one theorem* for *any* given triple of known quantities. This we call *abstraction*.

To proceed we need to be more precise about statements **P** and **Q**. As presented, they contain symbols, a, b, c and x about which we have been somewhat vague. They could mean anything.[9] Indeed, there are many meaningful possible choices. Are we being asked to consider positive integers, rational numbers, real numbers, complex numbers or matrices? For all these **P** makes sense. It is necessary to agree and settle on *one* of these and to stay with it. This will prevent misunderstandings and help us arrive at agreed conclusions. Let us suppose that all the letters represent real numbers. Now, of course, you may well ask:

What is a real number?

Once more we have encountered a highly non-trivial question. Most readers will have been using, manipulating and making statements about real numbers for years and accepted that the real numbers have always just been there, somehow given to us gratis and without explanation by nature.[10] But nothing is given in mathematics. Everything, including the real numbers, has to be agreed upon so that, in discussing mathematical concepts, results, and techniques there are no ambiguities. Without such precision we are operating intuitively and, although this may be quite effective, even for an extended period, it will, eventually, as the history of mathematics

---

[9]George Bernard Shaw: *Not a word was said to us about the meaning or utility of mathematics: we were simply asked to explain how an equilateral triangle could be constructed by the intersection of two circles, and to do sums in a, b, and x instead of in pence and shillings, leaving me so ignorant that I concluded that a and b must mean eggs and cheese and x nothing, with the result that I rejected algebra as nonsense, and never changed my opinion until in my advanced twenties Graham Wallas and Karl Pearson convinced me that instead of being taught mathematics I had been made a fool of.*

[10]In using the differential calculus to find maxima and minima, one implicitly assumes the existence of a logically sound construction for the real numbers.

shows, lead to contradictions that can only be resolved by clarifying basic concepts. Experience shows that a deeper understanding usually follows a successful resolution of such difficulties and, as a result, a richer and more secure theory emerges. Well, why not resolve these problems right now? Basically, because we are not ready and could easily, at this stage, be overwhelmed by technical details, that our intuition would not appreciate. Later we will be able to cope. To make progress we need to compromise and this usually takes the following form:

begin with an intuitive definition,

develop it and, when ready,

become formal and rigorous.

For the real numbers, we operate with the same intuitive geometric definition employed in classical Greek mathematics (Figure 1.2):

there is a one to one correspondence between the real numbers and the points on a line.

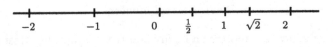

$$-2 \qquad -1 \qquad 0 \quad \tfrac{1}{2} \quad 1 \quad \sqrt{2} \quad 2$$

Fig. 1.2

As mathematicians were able to operate with this concept of real numbers for almost 3000 years, we can do the same for a much shorter period. This approach is standard and effective in all areas of mathematics. Students of mathematics, and indeed professional mathematicians, operate using a mixture of intuitive and well defined mathematical concepts. The unstated understanding always being that, at some stage, intuitive concepts will be replaced by mathematical definitions and rigorous proofs.

With real numbers, we can perform certain operations, we can *add, subtract, multiply, divide by non-zero numbers,* take *square roots of non-negative numbers* and, moreover, we can compute the *distance* between real numbers and also verify if one real number is *greater than, less than,* or *equal to* another real number.

Turning to statement **Q**, we see immediately that modifications are necessary. It contains a division by a, which may be 0, and a square root of $b^2 - 4ac$, which may be negative. Both of these operations are not allowed when dealing with real numbers.[11] Thus we must exclude these possibilities. We can modify either **P** or **Q** or both. For instance, we can replace **P** with the following statement that we call **P′**;

$$\mathbf{P'} : ax^2 + bx + c = 0, \quad a \neq 0, \quad b^2 - 4ac \geq 0$$

and, instead of attempting to prove Theorem 1.1, we shall seek to show that **P′** implies **Q**. That is we replace Theorem 1.1 with the following theorem.

Theorem 1.2. *If* $ax^2 + bx + c = 0$ *where* a, b, c *and* x *are real numbers,* $a \neq 0$ *and* $b^2 - 4ac \geq 0$, *then*

$$x = \frac{-b \pm \sqrt{b^2 - 4ac}}{2a}.$$

Now that we know what this piece of mathematics claims to say, do we believe it? Let us look at an example that we can verify. If

$$(x - 2)(x - 3) = 0$$

then

$$x = 2 \text{ or } x = 3.$$

This is based on an important property of 0:

if A *and* B *are real numbers and* $A \cdot B = 0$ *then either* $A = 0$ *or* $B = 0$.

Hence if $(x - 2)(x - 3) = 0$ then either $x - 2 = 0$ or $x - 3 = 0$, that is either $x = 2$ or $x = 3$. Now

$$(x - 2)(x - 3) = x^2 - 5x + 6$$

so our example says

$$\text{if} \quad x^2 - 5x + 6 = 0 \quad \text{then} \quad x = 2 \text{ or } x = 3.$$

---

[11] A real number x is *positive* if $x > 0$, it is non-negative if $x \geq 0$, and it is negative if $x < 0$. If we had assumed that a, b, c and x were complex numbers then we could take square roots. Thus, it is important to decide and stay with a choice of meanings from the very beginning.

To test the general formula, we see that $a = 1$, $b = -5$ and $c = 6$. Then $a \neq 0$ and $b^2 - 4ac = 25 - 4(1)(6) = 25 - 24 = 1 > 0$ so that the required conditions are satisfied. Now

$$\frac{-b \pm \sqrt{b^2 - 4ac}}{2a} = \frac{-(-5) \pm \sqrt{1}}{2} = \frac{5 \pm 1}{2} = 3 \text{ or } 2.$$

Thus, the theorem provides the correct solution for our test example.[12]

Now, given that we have some understanding of what the theorem says and some confidence in its truth, how do we go about proving it? A standard approach is to begin by considering particular or special cases. This may suggest a general proof or uncover the difficulties that have to be overcome. The result is supposed to work for all $\{a, b, c\}$ and x subject to $a \neq 0$ and $b^2 - 4ac \geq 0$. Let us consider the simple case $a = 1$ and $b = 0$. Statement **P′** now says that $b^2 - 4ac = -4c \geq 0$ and $x^2 + c = 0$. This means

$$x^2 = -c \text{ and } -c \geq 0. \tag{1.2}$$

The left- and right-hand sides of Equation (1.2) are consistent, that is they do not contradict one another as we have the following elementary but useful property of the real numbers: *if x is any real number then $x^2 \geq 0$, that is x^2 is non-negative*. We have already noted that we may take the square root of any non-negative real number. Taking square roots in (1.2) we obtain $x = \pm\sqrt{-c}$. We now verify that this agrees with statement **Q**. According to the theorem, the solutions should be

$$\frac{-0 \pm \sqrt{0^2 - 4c}}{2} = \frac{\pm\sqrt{-4c}}{2} = \frac{\pm\sqrt{4}\sqrt{-c}}{2} = \frac{\pm 2\sqrt{-c}}{2} = \pm\sqrt{-c}$$

and we have a proof in this special case. In the course of the above calculation we have used another property of the real numbers: if a and b are positive real numbers then

$$\sqrt{ab} = \sqrt{a} \cdot \sqrt{b}.$$

We return to this *law of indices* in Chapter 7.

---

[12]This also shows that we do in fact have examples. Just because we can write down a set of conditions does not guarantee that any examples exist. For instance, if we consider all real numbers x such that $x^2 < 0$ then clearly there are no examples. Other situations may not be so obvious. For example, consider all positive integers x, y and z such that $x^3 + y^3 = z^3$.

The next simplest particular case we look at is $a \neq 0$ and $b = 0$. This could be left as an exercise but, instead, we will go through it carefully as, in the process of doing so, we will use further properties[13] of the real numbers that we encounter again and again. We have

$$b^2 - 4ac = -4ac \geq 0 \quad \text{and} \quad ax^2 + c = 0. \tag{1.3}$$

Since $a \neq 0$ we have $a^2 > 0$ and hence $(1/a^2) > 0$. This implies

$$\frac{1}{4} \cdot \frac{1}{a^2} \cdot (-4ac) = \frac{-4ac}{4a^2} = -\frac{c}{a} \geq 0.$$

Combining this with the right-hand side of (1.3) we obtain

$$-\frac{c}{a} \geq 0 \text{ and } x^2 = -\frac{c}{a}$$

and this case reduces to that considered in (1.2). However, instead of using (1.2) we proceed directly and obtain further experience in manipulating symbolic formulae. We have obtained independently the solutions

$$x = \pm \sqrt{-\frac{c}{a}}$$

while the solutions according to the theorem are also

$$\frac{-0 \pm \sqrt{-4ac}}{2a} = \frac{\pm \sqrt{4} \cdot \sqrt{-ac}}{2\sqrt{a^2}} = \frac{\pm 2}{2} \sqrt{\frac{-ac}{a^2}} = \pm \sqrt{-\frac{c}{a}}.$$

Here we have used another *law of indices*:

$$\frac{\sqrt{A}}{\sqrt{B}} = \sqrt{\frac{A}{B}}$$

for any positive real number, B, and any non-negative real number A.

Moving on, we next consider the case $a \neq 0$ and $c = 0$, and encounter the equation

$$ax^2 + bx = 0 \tag{1.4}$$

and the condition $b^2 - 4ac = b^2 \geq 0$, which is clearly satisfied. Since (1.4) can be rewritten as $x(ax + b) = 0$ we obtain, on applying the rule we recently met about numbers whose product is zero, $x = 0$ or $ax + b = 0$ and $x = -b/a$. The theorem gives the solutions

$$x = \frac{-b \pm \sqrt{b^2 - 0}}{2a} = \frac{-b \pm \sqrt{b^2}}{2a} = \frac{-b \pm b}{2a} = \quad 0 \text{ or } \frac{-b}{a}$$

---

[13]If $x \neq 0$ we call $1/x$ the *reciprocal* of x. The product of positive numbers is again positive and the reciprocal of a positive number is also positive.

and this agrees with our expectations.

It is time to examine the above solutions and to see if they suggest a possible approach to a general proof. First note that, as $a \neq 0$, we can divide across to obtain the equation

$$x^2 + \frac{b}{a}x + \frac{c}{a} = 0$$

and if we let $B = b/a$ and $C = c/a$ we obtain

$$x^2 + Bx + C = 0 \tag{1.5}$$

and this might be simpler to work with, after all, it only involves three symbols, B, C and x in place of the original four symbols a, b, c and x. This is not a rule or a certainty but just an intuitive feeling that is, usually, worth following. We have also seen that the equation $x^2 = -c$ was rather easy to solve because the right-hand side was written as a square. Thus, if we could turn $x^2 + Bx + C$ into a square we could apply our earlier method. Since this expression contains x^2 let us look at $(x+\alpha)^2$ and compare it with $x^2 + Bx + C$. We have

$$(x + \alpha)^2 = x^2 + 2\alpha x + \alpha^2 \tag{1.6}$$

and, on comparing coefficients of x in (1.5) and (1.6), we should let $B = 2\alpha$, that is $\alpha = B/2$. We then have

$$\left(x + \frac{B}{2}\right)^2 = x^2 + Bx + \frac{B^2}{4}$$

and hence

$$x^2 + Bx + C = x^2 + Bx + \frac{B^2}{4} + C - \frac{B^2}{4} = \left(x + \frac{B}{2}\right)^2 + C - \frac{B^2}{4} = 0$$

that is

$$\left(x + \frac{B}{2}\right)^2 = \frac{B^2}{4} - \frac{4C}{4} = \frac{B^2 - 4C}{4}.$$

This is now in the form we were seeking. We have

$$x + \frac{B}{2} = \pm\sqrt{\frac{B^2 - 4C}{4}} = \frac{\pm\sqrt{B^2 - 4C}}{2}$$

and obtain the solutions

$$x = -\frac{B}{2} \pm \frac{\sqrt{B^2 - 4C}}{2} = \frac{-B \pm \sqrt{B^2 - 4C}}{2}.$$

On making the substitutions $B = b/a$ and $C = c/a$ we obtain

$$x = \frac{-\frac{b}{a} \pm \sqrt{\frac{b^2}{a^2} - \frac{4c}{a}}}{2}$$

$$= \frac{-\frac{b}{a} \pm \sqrt{\frac{b^2}{a^2} - \frac{4ac}{a^2}}}{2}$$

$$= \frac{-\frac{b}{a} \pm \sqrt{\frac{b^2 - 4ac}{a^2}}}{2}$$

$$= \frac{-\frac{b}{a} \pm \frac{\sqrt{b^2 - 4ac}}{a}}{2}$$

$$= \frac{-b \pm \sqrt{b^2 - 4ac}}{2a}.$$

Now that we have proved the formula under certain conditions, let us look to see if these conditions are really necessary. We first note that statement **P**, that is $ax^2 + bx + c = 0$, makes sense regardless of whether $a = 0$ or $b^2 - 4ac < 0$. Perhaps we can find an explicit formula for x even if these conditions are not satisfied. If $a = 0$ then **P** becomes $bx + c = 0$. Hence if $b \neq 0$ then $bx = -c$ and $x = -c/b$. In this case we have only one solution while, previously, we had *two* solutions if $b^2 - 4ac > 0$ and *one* solution if $b^2 - 4ac = 0$.

What if the condition $b^2 - 4ac \geq 0$ is not satisfied. We then have $b^2 - 4ac < 0$ and this means that $a \neq 0$. Otherwise, that is if $a = 0$, we would have $b^2 < 0$ and as we have already noted no real number has this property. Returning to our previous calculations and using the notation $B = b/a$ and $C = c/a$, we see that

$$\left(x + \frac{b}{2a}\right)^2 = \left(x + \frac{B}{2}\right)^2 = \frac{B^2 - 4C}{4} = \frac{b^2 - 4ac}{4a^2}. \tag{1.7}$$

The left-hand side of (1.7) is non-negative while the[14] right-hand side is negative. This cannot occur. So we have no real number x satisfying $ax^2 + bx + c = 0$ when $b^2 - 4ac < 0$. We summarize, in the form of a theorem, all that we have learned about solutions to the equation $ax^2 + bx + c = 0$.

Theorem 1.3. *If a, b, c and x are real numbers and $ax^2 + bx + c = 0$ then*

---

[14]We use here the property that the product of a positive and a negative number is negative.

(a) *if $a \neq 0$ and $b^2 - 4ac > 0$ we have two solutions*
$$\frac{-b + \sqrt{b^2 - 4ac}}{2a} \; , \; \frac{-b - \sqrt{b^2 - 4ac}}{2a}$$

(b) *if $a \neq 0$ and $b^2 - 4ac = 0$ we have a unique solution*
$$-\frac{b}{2a}$$

(c) *if $a = 0$ and $b \neq 0$ we have a unique solution*
$$-\frac{c}{b}$$

(d) *if $b^2 - 4ac < 0$ there are no solutions.*

1.2   Further Consequences

If we so wished, this could be the end of the story, particularly, as we have completely answered the problem we originally considered. However, at this stage it is as fruitful to ask questions about the result and how we found it as it was initially to ask questions about the result when we knew little about it. Are there any obvious consequences of the result we have just proved that are worth recording? Does the result suggest any similar results that it might be worth trying to prove? Did we use any logical arguments that may be used again? Did we develop any useful techniques? Can we find another proof? At a deeper level, one may reasonably ask: where are all these rules coming from? How do we know which are the legitimate rules? Is there no end to them? In this section we address some of these questions.

One technique that we employed is so useful that it has been given a special name: *completing squares*. Given $ax^2 + bx + c$ with $a \neq 0$ we have

$$ax^2 + bx + c = a\left(x^2 + \frac{b}{a}x + \frac{b^2}{4a^2}\right) + c - \frac{ab^2}{4a^2} = a\left(x + \frac{b}{2a}\right)^2 + c - \frac{b^2}{4a}.$$

Its usefulness is often combined with two simple facts that we have already used: $x^2 \geq 0$ for any real number x and any non-negative real number has a square root.

In proving Theorem 1.3 we used, without advertising the fact, an argument that is occasionally used outside mathematics. In mathematics it is so useful that it has been given a special name, *proof by contradiction*, and is also known by its Latin name *reductio ad absurdum*. It was, apparently, first used formally by Zeno of Elea, c.490-430 BC, and introduced into mathematics by Hippocrates of Chios. The historian Plutarch, 45-120, relates in his life of Pericles, that Zeno had perfected a technique of cross-examination which enabled him to corner his opponent by question and answer. He reports that Timon of Phlius described Zeno as *an assailer of all things, whose tongue like a double-edged weapon argued on either side with an irresistible fury.*[15]

We recall a simple result that we previously used:

$$\text{if } b^2 - 4ac < 0 \text{ then } a \neq 0.$$

We supposed, assumed or hypothesized, that the result is not true, that is we supposed that $a = 0$. This implies $b^2 - 4ac = b^2 < 0$ and, as we already noted, no real number has this property. We have arrived at a contradiction and as our analysis was correct we must conclude that one of our hypotheses or assumptions was false. But we only made one assumption, that $a = 0$. This must be false and hence we must have $a \neq 0$. This is what we required.

As this method of argument is so useful, we discuss it formally. The acceptance that all mathematical statements are either true or false[16] led to this method of proof. The basic idea is as follows. Suppose we wish to prove that a certain statement, call it A, is true. We begin by posing as a *hypothesis*, that A is false. From this assumption we draw a number of consequences which eventually lead to a conclusion that *we know* to be false. The only possible flaw in our argument is our hypothesis, and we conclude that this hypothesis is false. Hence the statement 'A is false' is

---

[15]Zeno was noted for the *paradoxes* he constructed. These often involved the *infinite* and have continued to fascinate philosophers, novelists and mathematicians through the ages. Mathematical explanations of Zeno's paradoxes required satisfactory definitions of *limits, infinite sets* and *continuity*. The Argentinian novelist, Jorge Luis Borges, 1899-1986, made frequent use of Zeno's ideas, e.g. the infinite divisibility of time and space. We refer to his short stories and, in particular, to the collection *Ficciones* for details.

[16]This is known as the *Law of the Excluded Middle*. It is accepted as a *rule of inference* or as a *Law of Thought*; that is as a logical principle that justifies deriving one truth from another, by most, but not all, mathematicians. See Section 12.3.

incorrect, and A must be true. This is what we required.

Useful or noteworthy observations that follow immediately or almost immediately from a theorem are often displayed, especially if required later, as *corollaries*. An inspection of Theorem 1.3 shows, immediately, that for any real numbers, a, b and c the equation $ax^2 + bx + c = 0$ has at most two solutions. In theory, at least, it may be possible to obtain this directly, that is without using Theorem 1.3. Sometimes this is possible and of interest and at other times it may be as difficult as proving the main theorem. Even so, it is often worth spending a little time to see how one might proceed directly. We state this result as a corollary and prove it independently of Theorem 1.3, by contradiction.

Corollary 1.4. *If a, b and c are real numbers, at least one of which is non-zero, then there are at most two real numbers x which satisfy the equation*
$$ax^2 + bx + c = 0.$$

Proof. The result is clearly true if $a = 0$ so we suppose that $a \neq 0$. Suppose the corollary is not true. Then for some real numbers a, b and c, where $a \neq 0$, there are 3 distinct real numbers x satisfying
$$ax^2 + bx + c = 0.$$
Since we will be referring to these number frequently, we should give them names. Names or, if you wish, notation are as important in mathematics as in any language. They should be distinct and recognizable and should provide us with a context and if possible with information. Since they are related to x and are three in number, it is natural to denote them by x_1, x_2 and x_3.[17] Our assumption can now be rewritten as

$$ax_1^2 + bx_1 + c = 0 \qquad (1.8)$$

$$ax_2^2 + bx_2 + c = 0 \qquad (1.9)$$

$$ax_3^2 + bx_3 + c = 0 \qquad (1.10)$$

---

[17]One may consider what they have in common, x, as the *family name* and that we distinguish between them by their *given names*, the subscripts $1, 2$ and 3. If we let $A = \{x_1, x_2, x_3\}$ then A is the set of solutions. We say that the elements in A are indexed by the set $\{1, 2, 3\}$. The set A is written in different ways: $A = \{x_i\}_{i=1,2,3}$, $A = \{x_i\}_{i \in \{1,2,3\}}$, or $A = \{x_i : i = 1, 2, 3\}$. A large set requires a large indexing set.

where x_1, x_2 and x_3 are three *distinct* real numbers. On subtracting (1.9) from (1.8) and using the identity $x_1^2 - x_2^2 = (x_1 - x_2) \cdot (x_1 + x_2)$ we obtain

$$a(x_1^2 - x_2^2) + b(x_1 - x_2) = a(x_1 - x_2) \cdot (x_1 + x_2) + b(x_1 - x_2)$$
$$= (x_1 - x_2)\big[a(x_1 + x_2) + b\big]$$
$$= 0.$$

Since $x_1 \neq x_2$, $x_1 - x_2 \neq 0$ and we may divide by $x_1 - x_2$ to obtain

$$a(x_1 + x_2) + b = 0. \tag{1.11}$$

By following the same procedure, but using (1.10) and (1.9) in place of (1.9) and (1.8), we obtain

$$a(x_2 + x_3) + b = 0 \tag{1.12}$$

and, on subtracting (1.12) from (1.11), we obtain

$$a(x_1 - x_3) = 0.$$

Since $x_1 \neq x_3$ this implies $a = 0$. We have arrived at a contradiction and completed the[18] proof of Corollary 1.4. $\qquad\square$

We have concentrated on the *quadratic equation*

$$ax^2 + bx + c = 0$$

but clearly one might also look at equations like

$$4x^3 + 10x^2 - 7x + 8 = 0 \tag{1.13}$$

or even more generally at

$$a_n x^n + a_{n-1} x^{n-1} + \cdots + a_1 x + a_0 = 0. \tag{1.14}$$

Equation (1.13) is an example of a *cubic equation*. The Babylonians used $n^2 + n^3$ tables to solve certain cubic equations. The Persian mathematician and poet Omar Kkayyam, c.1044-1123, obtained a procedure for obtaining geometrically the positive roots of any cubic and, around 1535, Nicolo

---

[18]The expression *completed the proof* refers to the proof under construction. There may be a *first proof*, or even a first correct proof, given generally when a result was discovered. Many results have more than one known proof (see Section 12.3).

Fontana, c.1477-1557, also known as Tartaglia, obtained a formula for the solution of cubics in which all three solutions are real. Mathematical formulae were sometimes reserved in the middle ages as secret weapons for public mathematical contests and a serious row erupted in 1545 when Giralamo Cardano, 1501-1576, who obtained the formula in the strictest confidence from Tartaglia, published it in his book *Ars Magna* as his own discovery.

The 16th century approach was hampered by suspicions about negative numbers and negative square roots and the complete treatment of cubic equations using complex numbers was only completed by Euler in 1732. An explicit formula for the solutions of all cubics, similar to that given in Theorem 1.2 for quadratic equations, follows from Exercises 1.3 and 11.28.[19] In the second half of the 17th century, Isaac Newton's study of cubic equations led him to manipulate power series, in particular binomial series with non-integer exponents, as if they were polynomials. This experimentation contributed to the later development of the calculus.

There are practical geometrical reasons for solving cubic equations but no immediately obvious reasons why one should consider the general *polynomial equation of degree n* given in (1.14). However, mathematicians thrive on curiosity and the history of pure and applied mathematics shows that many abstract results obtained with no practical purposes in mind are often found later to have serious applications. Nevertheless, we may adapt the proof of Corollary 1.4 to show that any polynomial of degree n has at most n real solutions (Exercise 1.2). This shows why alternative proofs may be useful and how a corollary may lead to new information in cases where the original result cannot be generalized. In Example 9.9 we will see, from a different perspective, that every polynomial equation of *odd degree* has at least one *real* solution.[20]

While all polynomial equations up to degree 4 have explicit formulae for their solutions, the Norwegian mathematician, Niels Henrik Abel, showed

---

[19]Ludovico Ferrari, 1522-1565, a student of Cardano obtained a method for solving quartic equations and thus finding the roots of polynomials of degree 4.
[20]That the polynomial equation $P(z) = 0$ has a complex solution z is one version of a result known as the *Fundamental Theorem of Algebra*.

that there are equations of any higher degree which do not. This provided
the initial motivation for the development of an important area within
modern algebra, now called *Galois Theory*, after the French mathematician Évariste Galois.[21] Both Abel and Galois exploited the fact that the
coefficients $a_i, i \leq n$, in the expression $P(x) := a_n x^n + \cdots + a_1 x + a_0$ are
symmetric functions of the solutions of the equation $P(x) = 0$.

1.3   Exercises

(1.1) By completing squares show that $x^8 - 3x^4 - 5x^2 - 4x + 17$ is positive
for all real numbers x.

(1.2) Show that the equation $a_n x^n + a_{n-1} x^{n-1} + \cdots + a_1 x + a_0 = 0$ has
at most n solutions among the real numbers. Give examples where it
has precisely n solutions and other examples were it has strictly less
than n real solutions.

(1.3) Let $P(x) = ax^3 + bx^2 + cx + d = 0$ where a is non-zero. Divide
across by a to obtain $x^3 + Bx^2 + Cx + D = 0$. If $x = y - \left(\frac{B}{3}\right)$, show
that $y^3 + py + q = 0$ where $p = C - \frac{B^2}{3}$ and $q = D - \frac{CB}{3} + \frac{2B^3}{27}$.
Solve $y^3 + q = 0$. If $p < 0$ solve $y^3 + py = 0$. Find all solutions of
$x^3 + 3x^2 + x - 1 = 0$.[22]

(1.4) Find conditions on the real numbers A, B and C which are equivalent
to $Q(x, y) = Ax^2 + Bxy + Cy^2 > 0$ for all real numbers x and y,
$(x, y) \neq (0, 0)$.

(1.5) If $a, b,$ and c are real numbers, $a \neq 0$ and $b^2 - 4ac \geq 0$ show, by substitution, that the numbers $(-b \pm \sqrt{b^2 - 4ac})/2a$ satisfy the equation
$ax^2 + bx + c = 0$. State this result using the statements **P** and **Q**
from Theorem 1.1.

(1.6) Find all real solutions of the equation $2x^4 - 10x^2 - 3 = 0$.

(1.7) Let $f(x, y) = x^4 + y^4 - 2x^2 - y^2 - 4x + 2y + 9$. By completing squares
show that f is always positive.

---

[21] Both Abel, 1802-1829, and Galois, 1811-1832, died in their twenties, Abel of poverty
at 26 and Galois in a duel at 20.
[22]The number of solutions of the equation $y^3 + py + q = 0$ is examined in Example 11.6.

(1.8) If x, y and z are real numbers and $2x^2 + 2y^2 + 4z^2 - 2xy - 4yz + 2x = 15$ show, by first completing squares, that $-5 \leq x \leq 3, -9 \leq y \leq 7$ and $-6.5 \leq z \leq 5.5$. Find all solutions of the above equation which have the form $(3, y, z)$.

(1.9) Show that $x^4 - 4x^3 + 8x^2 - 2x + 10 > 0$ for any real number x.

(1.10) Discuss the equation $ax^2 + bx + c = 0$ where a, b, c and x are 2×2 matrices.

(1.11) Why is the equation $ax^2 + bx + c = 0$ called *quadratic*? Why do we say x-squared for x^2?

References [7; 8; 9; 11; 17; 18; 22; 24; 27; 30; 31; 32; 35]

Chapter 2

DIAGRAMS

Dubito, ergo cogito, ergo sum.
I doubt, therefore I think,
therefore I am.
René Descartes

It is desirable in order to aid the
concept of an equation to let the
two unknown magnitudes form
an angle which we would
assume to be a right angle.
Pierre Fermat

Summary

To further develop an intuitive approach, prior to a formal investigation, we sketch and modify the graphs of some basic functions.

2.1 Graphs

We take an elementary approach to *graph sketching* in this chapter and use graphs to gain insight into the results in Chapter 1 and to prepare the

21

way for the rigorous definitions in Chapter 3. It is important to consider examples before and after the introduction of abstract concepts. Sketching diagrams and graphs[1] to understand and illustrate mathematics does not come naturally to everybody but, to most people, it is a skill that can, with patience, be developed.[2] A notable exception was the French mathematician, Siméon Denis Poisson, 1781-1840, who made fundamental contributions to almost all areas of applied mathematics under investigation during his lifetime. He had a deep theoretical insight into physical phenomena such as the movement of planets, mechanics, electricity, magnetism, heat, vibrations, etc., and was a very clear expositor. He was also very clumsy, and this affected his career in a number of ways: he had to abandon his first apprenticeship in medicine because of lack of coordination, he could not apply for a position in the civil service because of his inability to draw diagrams and, for the same reason, he could not pursue, to any great extent, his study of geometry. Through the efforts of Pierre Simon Laplace, 1749-1827, the most influential French mathematician of that period, he obtained a position in a physics institute and, wisely, did not attempt to verify his theoretical conclusions by experiment.

Classical Greek geometry is based on the straight line and circle. In this setting, complete and accurate diagrams may be constructed and are

---

[1]In mathematics and computer science, *graph theory*, a branch of network analysis, is the study of collections of vertices and the edges joining them. This theory dates back to a paper by Leonhard Euler in 1736, *The Seven Bridges of Königsberg*. It has many applications, for instance, in distribution theory, traffic problems, and the analysis of molecular structure. The same article of Euler played an important role in the development of another area of mathematics, called the *geometry of position* by Gottfried Leibniz but now known as *topology*.

[2]After having developed and used the material in this chapter for a number of years I came across the book: *Functions and Graphs* by I.M. Gelfand, E.G. Glagoleva and E.E. Shnol, Birkhäuser, 1990. This wonderful book, one in a series for the *Mathematical School by Correspondence*, was written during the 1960's to enable high school students in the Soviet Union, who might not have access to high school teachers of mathematics, to study mathematics independently. It contains an excellent development of ideas, touched on in this chapter. Israel M. Gelfand, 1913-2009, was one of the most renowned and creative mathematicians of the 20th century and it is worth noting the attention he paid to the so-called elementary topics.

acceptable as a legitimate component in a rigorous proof. However, the diverse nature of differentiable functions, defined and studied in Chapter 10, cannot be captured with the same precision and, in developing the calculus, diagrams are no longer included in acceptable proofs. Nevertheless, diagrams may be used to sharpen our intuition and to lead us to and guide us through proofs. They can be used to illustrate what we already know and to help us find what we are seeking. Moreover, the functions that illustrate our diagrams will re-appear in different contexts throughout this book and provide the concrete examples by which we come to understand the significance of abstract results. Archimedes, who reputedly drew diagrams in the cold ashes of the dawn, appreciated help provided by non-orthodox and non-rigorous sources.

Bishop Nicolas Oresme,[3] 1323-1382, one of the great intellectuals of the 14th century, used coordinate systems for particular mathematical problems and made, in a century relatively devoid of intellectual development, important contributions to mathematics, physics, economics, philosophy and musicology. He often used ideas from one discipline as inspiration in another. In economics he expounded a surprisingly modern theory of *money and inflation*, in physics he studied bodies moving under constant acceleration and discovered results that we now obtain using integration theory. His musicology influenced his approach to mathematics where he developed fractional exponents and showed that the *harmonic series* $\sum_{n=1}^{\infty}(1/n)$ diverged. His philosophical-mystical approach to *infinity*, both to the infinitely large and infinitely small, freed the concept from the inflexible role to which it had been consigned by the ancient Greeks, and gave mathematicians in the 16th and 17th centuries the confidence to speculate and experiment. This proved an important preparatory step in the development of the differential and integral calculus. Oresme's thinking was too advanced for his contemporaries and as he died over 50 years before the invention of the printing press the potential, inherent in his ideas, was not

---

[3]Oresme, from the French city of Caen, was councillor and confessor to Charles V of France for most of his adult life and became Bishop of Lisieux in 1377. Arithmetic, geometry, astronomy, and music were grouped together in medieval European universities.

realized and his contributions overlooked.

The independent discovery and development of *coordinate geometry*[4] in 1637 by two French mathematicians Pierre Fermat and René Descartes uncovered the universal role of graphs and led to their systematic use. Fermat, 1601-1665, from Beaumont-de-Lomagne in the south of France, lived most of his life in the Toulouse region. He trained as a lawyer, studied mathematics under students of Viéte, and was a member of the local regional parliament. His mathematics appeared in informal notes and in his voluminous correspondence and, although a serious mathematician, this casual approach allowed him to set his own rules of engagement. As a result, he often only hinted at proofs, he sometimes gave outlines, and often promised, but usually did not give proofs later. He is regarded as the founder of modern number theory and he developed methods, for finding maxima, minima, tangents and areas, that were important milestones on the road to the calculus. Additionally he contributed to the foundations of probability theory.

Descartes, 1596-1650, from La Haye near Tours in France, had a varied life as a young man in which he studied, traveled and experienced the world as a soldier of fortune. At the age of thirty two, he settled in Holland and devoted himself to the development of a philosophy which could be applied to discover the truth about anything and everything. This he based on simple uncontroversial ideas and logic. His essay on geometry, together with essays on optics and meteorology, were written to provide examples of his philosophical *method*.

The possibilities opened up by the notational advances of Viète and their interest in discovering the methods of classical Greek mathematics led

---

[4]In the history of mathematics we have other well known examples where an important area was discovered independently and almost simultaneously by two or more people; Newton and Leibniz discovered the *differential and integral calculus*, while Abel and Galois investigated the solvability of *polynomial equations*. It is interesting to speculate why this is so and to compare and contrast different approaches and their subsequent influences. It appears that, often, the way has been subconsciously prepared but yet it takes deep insight and concentrated effort to bring forth new material. Coordinate Geometry is also known as *Analytic Geometry*.

them to their discoveries. The informal notes of Fermat, *Ad locos planos et solidos isagoge*, were circulating in Paris while Descartes awaited the galley proof of his masterpiece, *Discours de la méthode pour bien conduire sa raison et chercher la vérité dans les sciences.*[5] In comparing our two heroes, we note that Fermat started with equations and looked for the curve, while Descartes always began with a geometric curve, to him the real object of interest, and looked for the equations. Fermat was easy to read and had a surprisingly modern style, but yet was unpublished. He wrote in Latin which, until the 19th century, was the language of science. Descartes wrote in French and was, according to his own admission, deliberately obscure in order to deter others from claiming his ideas. His work was translated into Latin in 1649 and its originality of content and style were immediately recognized. Descartes and Fermat influenced Newton and Leibniz. Both gave the same basic technique of relating algebra and geometry. Fermat attempted to replace the geometric reasoning of the ancient Greeks by algebraic arguments. Using coordinates, he was able to determine the locus of any *quadratic equation* in two variables. Descartes, an ambitious philosopher, as the long but accurate title of his work implies, applied his method to solve geometrical problems using algebraic methods and equations.

Coordinate geometry provided a background that nurtured the concept of function. The differential calculus, which deals with the rate of change of functions, needed graphs, coordinate geometry, and an intuitive notion of function for its development.

There are other reasons, not immediately obvious, why we have taken a rather specific approach in this chapter. Mathematics is a deductive science whose foundations are constantly being examined and occasionally augmented. Moreover, the store of mathematical knowledge is constantly increasing and, without some rationalization, mathematicians would be overwhelmed by facts. Alternative approaches[6] frequently result in simplifications, amalgamations, refinements, and the presentation of material

---

[5] *Introduction to Plane and Solid Loci* (Fermat) and *Discourse on the Method for Rightly Directing One's Reason and Searching for Truth in the Sciences* (Descartes).

[6] No one doubts the advantage enjoyed by a footballer who kicks with *both* feet.

in more manageable proportions. The material in this chapter is our *first* approach to graph sketching, the differential calculus will provide us with a *second*. Moreover, the differential calculus, universally acknowledged as one of the great achievements of western civilization over the last two thousand years, can spin an illusion that leads to false expectations. The differential calculus may appear deceptively easy. A simple reading, familiarity with a few rules and a small number of examples are sufficient to convince some that they *know calculus*. The elementary technical skills of these novices may even impress those unfamiliar with the subject. However, it is rarely the case that mathematics can be so directly and immediately applied to practical problems, a deeper understanding is usually required. By the time you read this book, you will have observed the highly non-trivial nature of the subject, you will value alternative approaches to the same material, and see that a proper understanding of the differential calculus involves an appreciation of the foundations and thinking behind much more mathematics.

In sketching graphs it is necessary, initially, to plot points but once familiarity with a *few basic shapes* and their mathematical presentation has been established, simple techniques allow us to sketch a variety of graphs and obtain significant information by merely glancing at a graph. Moreover, many new functions may be obtained from these basic functions by applying the operations that we inherit from the real numbers, *addition*, *multiplication*, their inverse operations *subtraction* and *division*, and a new construction *composition of functions*.

Having considered the *equation* $ax^2 + bx + c = 0$ in Chapter 1, we now examine the *expression* $ax^2 + bx + c$ as x ranges over the real numbers $\mathbf{R}$. Since the constants a, b and c will not change in our discussion, we write, for convenience[7], $f(x)$, in place of $ax^2 + bx + c$, that is we let

$$f(x) = ax^2 + bx + c \tag{2.1}$$

for any real number x. We call f a function of the real variable x. As x changes so does $f(x)$ and different methods have been developed to uncover

---

[7]In Chapter 3 we will formally introduce functions. In the meantime we rely on an unspecified intuitive concept.

and display how $f(x)$ changes with x. One of these, in which we plot the points $(x, f(x))$ for different values of x, leads to a diagram, the *graph* of f (Figure 2.1). In graphs the horizontal line is called the x-axis and, as we often write y in place of $f(x)$, the vertical line is called the y-axis.

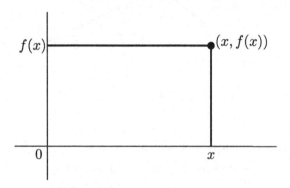

Fig. 2.1

If we first consider the simple case, $a = 0$ in (2.1), we obtain $f(x) = bx + c$, and on experimenting with different values of x we see that no matter what choice of b and c we use we always obtain a *straight line*. On examining the graphs for different values[8] of b we obtain when $b = 0$ a *horizontal line*, if $b > 0$ we obtain a line which slopes *upwards* to the right and say that such a line is *increasing*, if $b < 0$ we get a line which slopes *downwards* to the right and say that such a line is *decreasing* (see Figure 2.2). The line $f(x) = x$, given by $b = 1$ and $c = 0$, is of special interest. All points on its graph are equidistant from the x and y axes (see Figure 2.4).

If the point $(x, f(x))$ on the graph lies on the x-axis, then $f(x) = 0$. The fact that straight lines with $b \neq 0$ cut the x-axis once and once only is a geometric representation of the result in Theorem 1.4(c) that the equation $f(x) = bx + c = 0$ has precisely one solution. We use the term *linear function* from now on. No choice of b gives rise to a *vertical line* (the reason why becomes clear in Chapter 3).

We now turn to $ax^2 + bx + c$ with $a \neq 0$. The simplest such *quadratic*

---

[8]The constant b is called the slope of the line. It is often denoted by m.

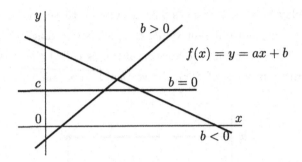

Fig. 2.2

function f occurs when $a = 1$ and $b = c = 0$, that is $f(x) = x^2$. By plotting the points corresponding to $x = 0$, $x = \pm 1$, $x = \pm 2$ and a few more, a distinctive shape soon emerges and we obtain the classical *parabola* (Figure 2.3).

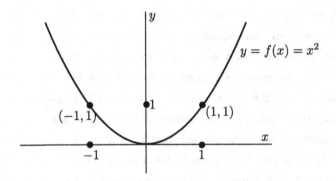

Fig. 2.3

If we sketch on the same diagram the graphs of $f(x) = x$ and $g(x) = x^2$, we obtain the following diagram (Figure 2.4). When $0 < x < 1$ we have $f(x) = x > x^2 = g(x)$ and for $x > 1$ we have $f(x) = x < x^2 = g(x)$. When the two graphs intersect, that is when $x = 1$ and $x = 0$, we have $f(x) = x = x^2 = g(x)$. If we had no graph and just used the result from the last chapter and considered the equation $x^2 = x$, we would have obtained

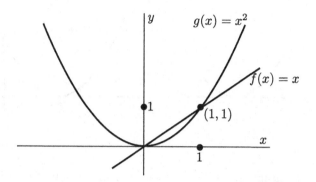

Fig. 2.4

the quadratic equation

$$x^2 - x = 0 = x(x - 1)$$

with solutions $x = 0$ and $x = 1$. This establishes a relationship between the solutions of equations and results that may be observed by inspecting graphs. As these remarks apply to *all graphs* and are a useful guide to our intuition, we list them formally. If the graphs of the functions f and g are sketched on the same diagram then the following hold:

(a) *the graphs of f and g intersect on the vertical line through the point x on the x-axis if and only if $f(x) = g(x)$,*

(b) *on the vertical line through the point x on the x-axis the graph of f lies above the graph of g if and only if $f(x) > g(x)$.*

To draw the graph of

$$g(x) := 2x^2 + 8x - 10 \tag{2.2}$$

we complete squares to get

$$g(x) = 2(x^2 + 4x - 5) = 2(x^2 + 4x + 4) - 8 - 10 = 2(x + 2)^2 - 18$$

and proceed by modifying the graph of $f(x) = x^2$. We first sketch the graph of x^2 and then obtain the graph of $(x + 2)^2$ by moving the graph of x^2 *two units to the left.* This is because the value of x^2 at, say, 6 is the same as the value of $(x + 2)^2$ at $6 - 2 = 4$ which is 2 units to the left (see Figure 2.5).

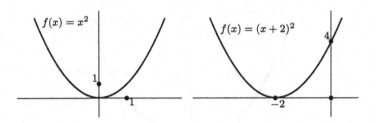

Fig. 2.5

Next, we graph $2(x+2)^2$ by modifying the graph of $(x+2)^2$. In this case 0 remains at 0 and all positive numbers are doubled so that the graph retains its basic shape but becomes narrower and rises faster. Finally we modify $2(x+2)^2$ to sketch $g(x) = 2(x+2)^2 - 18$ by *subtracting* 18, this is implemented by moving the graph *down* 18 units (see Figure 2.6).

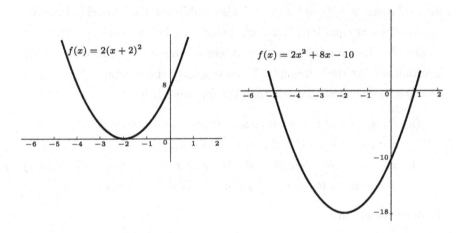

Fig. 2.6

The process of taking a simple function and changing it by a sequence of elementary operations will be seen later as *decomposing* a function, that is reversing the process of *composing* functions.

We next consider the graph of the modulus of g, h, where

$$h(x) = |2x^2 + 8x - 10| \qquad (2.3)$$

and $|x|$ denotes the (absolute) distance of the real number x to the origin.

Clearly g and h are related, if $g(x) \geq 0$ then $h(x) = g(x)$, while if $g(x) \leq 0$ then $h(x) = -g(x)$. Geometrically, this means that whenever $(x, g(x))$ lies *below* the x-axis we obtain $(x, h(x))$ by reflecting on the x-axis, otherwise $(x, g(x))$ and $(x, h(x))$ coincide. Thus a simple modification of the graph of g gives us the graph of h (see Figure 2.7).

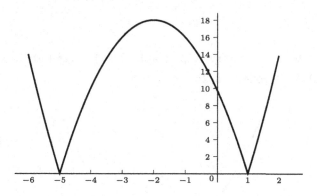

Fig. 2.7

For the general quadratic $f(x) = ax^2 + bx + c$ with $a \neq 0$ we have the following possibilities (see Figures 2.8 and 2.9). If $a > 0$, f has a minimum, that is it achieves a smallest value, and no maximum, that is it never achieves a largest value. If $a < 0$, f has a maximum and no minimum.[9]

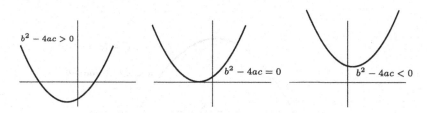

Fig. 2.8 $a > 0$

If we wish to incorporate the information we obtained in Theorem 1.3 into our graph we obtain the following diagrams (Figures 2.10 and 2.11). Note that the $\pm$ in the solution of the quadratic equation $ax^2 + bx + c = 0$

---

[9]The expressions absolute maximum and absolute minimum are also used.

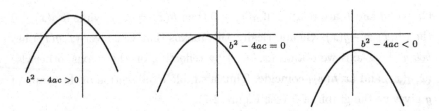

Fig. 2.9 $a < 0$

corresponds to the symmetry of the graph of $f(x) = ax^2 + bx + c = 0$ about the vertical line through $-b/2a$.

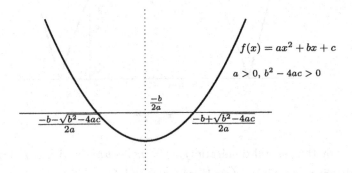

Fig. 2.10

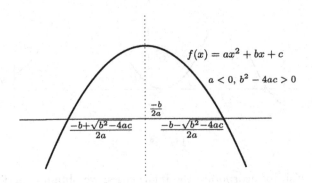

Fig. 2.11

By experimenting with the numbers ± 1, ± 10, ± 100, ± 1000 and $\pm 1/10$,

±1/100, ±1/1000, etc. one sees, helped by the following observations, how to modify the graph of a sufficiently regular function f on an *interval* to obtain the graphs of the functions $1/f$ and $-f$:

(a) *when $f(x)$ is large and positive, $1/f(x)$ is small and positive,*

(b) *when $f(x)$ is small and positive, $1/f(x)$ is large and positive,*

(c) *when $f(x)$ is large and negative, $1/f(x)$ is small and negative,*

(d) *when $f(x)$ is small and negative, $1/f(x)$ is large and negative,*

(e) *f is increasing and positive $\Longleftrightarrow 1/f$ is decreasing and positive,*

(f) *f is increasing and negative $\Longleftrightarrow 1/f$ is decreasing and negative,*

(g) *f is increasing $\Longleftrightarrow -f$ is decreasing,*

(h) *f has a local maximum[10] at $p \Longleftrightarrow -f$ has a local minimum at p,*

(i) *f has a non-zero local maximum at $p \Longleftrightarrow 1/f$ has a non-zero local minimum at p,*

(j) *if f has a local maximum at p and $f(p) = 0$ then $(1/f)(x) \longrightarrow -\infty$ as $x \longrightarrow p$,*

(k) *if f has a local minimum at p and $f(p) = 0$ then $(1/f)(x) \longrightarrow +\infty$ as $x \longrightarrow p$,*

(l) *if $f(p) = 0$ then $(1/f)(x) \longrightarrow \pm\infty$ as $x \longrightarrow p$.*

Note that (e) and (f) only apply to functions which are either always positive or always negative on the given interval. We now apply *graphing by modification* to the graph of $f(x) = x$ to obtain the graph of $k(x) = 1/x$, see Figure 2.12. Since $k(0)$ does not make sense the graph of k will not contain the point $(0, k(0))$. Note that we make use of properties (a), (b), (c), (d), (e), (f) and (l) above. Carrying this approach a step further we consider the graph of

$$l(x) = x + \frac{1}{x} = \frac{x^2 + 1}{x}. \tag{2.4}$$

We sketch the graphs of both f where $f(x) = x$ and k where $k(x) = 1/x$ on the *same diagram*, Figure 2.13(a), and then combine them to obtain

---

[10]If the restriction of f to the set of all points within some small but positive distance from p has a maximum (respectively a minimum) at p we say that f has a local maximum (respectively local minimum) at p.

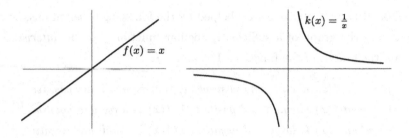

Fig. 2.12

the graph of $l(x) = f(x) + k(x)$. On each vertical line through $x \neq 0$ we add $f(x)$ and $k(x)$. Near 0, $k(x)$ is large while $f(x)$ is close to 0. Hence $l(x) \approx k(x) = 1/x$. As $x \longrightarrow \pm\infty$, $k(x) \longrightarrow 0$ and hence $l(x) \approx f(x) = x$. On sketching this information on a diagram we obtain Figure 2.13(b). As both f and k are *regular* away from 0 it seems reasonable to suppose that the missing parts of the graph of l take the form given in Figure 2.13(b).

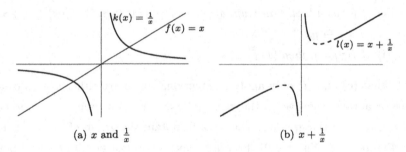

(a) x and $\frac{1}{x}$ (b) $x + \frac{1}{x}$

Fig. 2.13

This indicates that l has a *local minimum* for some positive value of x and a *local maximum* for some negative value of x. Further properties of l that may be observed in the graph are given in Exercises 2.1 and 2.3.

Our sketches of some basic shapes in this chapter, although simple, did lead to useful techniques. The methods, particularly those connected with increasing and decreasing functions and the limiting behavior of $f(x)$ as x tends to infinity touch on ideas that will come into focus when we discuss sequences and convergence. The examples will reappear as we proceed. Our approach has been informal and we have only considered two basic topics: *solving quadratic equations* and *graphing simple functions*. We have asked simple questions and any insight and understanding gained has resulted from attempting to answer these questions.

Example 2.1. We consider in this example the function

$$f(x) = \frac{x}{8 + x^3} = \frac{1}{\frac{8}{x} + x^2}.$$

In Figures 2.14 we sketch the graphs of $g(x) = 8/x$ and $h(x) = x^2$ and combine them in Figures 2.15 to get $k(x) = \frac{8}{x} + x^2$. From the graph,

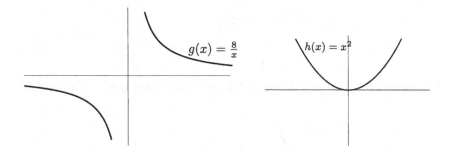

Fig. 2.14

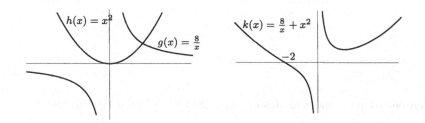

Fig. 2.15

we see that the equation $k(x) = 0$ has at least one negative solution and as $xk(x) = 8 + x^3$ we obtain the unique solution -2. We perform an inversion in Figure 2.16 to obtain f. Finally we examine the graph to see if the information it provides can be observed in the function. We see that $f(x) \approx \frac{1}{x^2} \longrightarrow 0$ as $x \longrightarrow \pm\infty$ and $f(x) \approx \infty$ as $x \approx -2$.

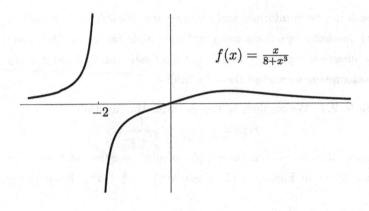

$$f(x) = \frac{x}{8+x^3}$$

Fig. 2.16

Example 2.2. The possibility of different approaches gives rise to oppor-

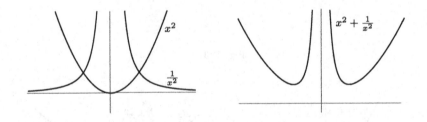

Fig. 2.17

tunities to experiment as this example shows. Consider the function

$$f(x) = \frac{x^2}{1+x^4} = \frac{1}{\frac{1}{x^2}+x^2} = \frac{1}{(x+\frac{1}{x})^2 - 2}.$$

In Figures 2.17 and 2.18 we give one approach. Alternatively, one can sketch the graph following the sequence $x+\frac{1}{x}$ (see Figure 2.13(b)), $(x+\frac{1}{x})^2$, $(x+\frac{1}{x})^2 - 2$ and then perform an inversion to obtain the graph of f.

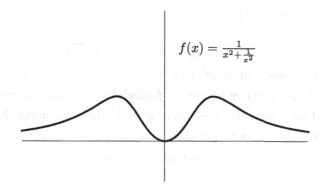

$$f(x) = \frac{1}{x^2 + \frac{1}{x^2}}$$

Fig. 2.18

2.2 Exercises

(2.1) Let $f(x) = x + (1/x)$ for all $x \neq 0$. Show that $f(x) \geq 1$ for all $x > 0$ and $f(x) \leq -1$ for all $x < 0$. For $y > -1$ and $x = 1 + y$ show that

$$f(x) = f(1+y) = 2 + \frac{y^2}{1+y}.$$

Hence show that f has a minimum of $+2$ over all positive real numbers and that it has a maximum of -2 over all negative real numbers.

(2.2) Sketch the graphs of the following functions;

$$f(x) = x^2, g(x) = x^2 + 2, h(x) = (x-1)^2 + 2, k(x) = x^2 - 2x + 4.$$

(2.3) A function f is *even* if $f(x) = f(-x)$ for all real numbers x and it is *odd* if $f(x) = -f(-x)$ for all real numbers x. Show that $f(x) = x^2$ is even and $g(x) = x + (1/x)$ is odd. If f is even show that the graph of f is got by reflecting the part for $x \geq 0$ on the y-axis while if g is odd show that the graph of g is obtained by reflecting the part for $x \geq 0$ first on the y-axis and then on the x-axis.

(2.4) Sketch the graphs of

$$\frac{1}{x}, \quad \frac{1}{x+1}, \quad \frac{1}{(x+1)^2}, \quad \frac{-1}{(x+1)^2}, \quad 1 - \frac{1}{(x+1)^2}.$$

(2.5) Find, by using a sketch or otherwise, all real numbers x such that

$$|x - 1| < |x - 5|.$$

(2.6) Sketch the graphs of

$$\frac{1}{x^2 - 2x + 3} \text{ and } \frac{1}{x^2 + 2x - 3}.$$

(2.7) Sketch a diagram of $ax^2 + bx + c$ when $0 < 2a < 2c < b$.

(2.8) Sketch the graph of a function f which has a local minimum at -2, a local maximum at $+3$, no other local maxima or minima, $f(x) \longrightarrow 0$ as $x \longrightarrow \pm\infty$ and $f(x) \longrightarrow \infty$ as $x \longrightarrow 0$. Sketch the graphs of $g := 1/f$ and $h := -f$ and describe the behavior of both at $0, -2, +3$ and $\pm\infty$.

(2.9) Show that $f : \mathbf{N} \times \mathbf{N} \longrightarrow \mathbf{R}$ defined by $f(n, m) = |(n/m)^2 - 2|$ has no minimum.

(2.10) Show that

$$f(x) := 5x^5 - 20x^3 + 8x + 2 = x[5(x^2 - 2)^2 - 12] + 2.$$

Sketch, in turn, the graphs of $x^2, x^2 - 2, (x^2 - 2)^2, 5(x^2 - 2)^2, 5(x^2 - 2)^2 - 12, x[5(x^2 - 2)^2 - 12]$, and f.

(2.11) If a is a positive constant, find the values of x for which $a(x^2 + 2x - 8)$ is negative. Find the value of a if this function has a minimum value of -27.

(2.12) Sketch the graph of

$$f : \mathbf{R}\backslash\{-1/2\} \longrightarrow \mathbf{R}, \quad f(x) = \frac{x^2 + 2}{2x + 1}.$$

(2.13) Sketch the graph of

$$f : \mathbf{R}_0^+ \longrightarrow \mathbf{R}_0^+, \quad f(x) = \frac{x}{2x^2 + 3}.$$

Find the minimum and maximum of f.

(2.14) Show that $x^4 - 20x^2 - 2x + 50 = 0$ has four real solutions.

(2.15) Sketch diagrams illustrating the observations (a), (b), ... ,(l) made prior to Example 2.1.

(2.16) If $x - x^{-1} = 2$ show that $x^3 - x^{-3} = 14$. Find $x^5 - x^{-5}$.

References [9; 23; 30]

Chapter 3

FUNCTIONS

*If some quantities so depend on other quantities
that if the latter are changed the former
undergoes change, then the former quantities
are called functions of the latter.*

Leonhard Euler, 1755

*Drawing graphs is one way to see formulae
and functions.*

Israel M. Gelfand, 1967

Summary

We define functions rigorously and discuss their basic properties.

3.1 Functions

As a language, mathematics is both *precise* and *efficient*. We have seen in Chapter 1 that expressions such as

$$ax^2 + bx + c \qquad (3.1)$$

are not precise unless each symbol is clearly defined and all the operations involved make sense. Thus, there is always background information that should be public knowledge. As regards efficiency, mathematics takes a minimalist approach and tries to build its structures on a foundation that is as small as possible. Over the centuries, indeed over the millennia, it has never ceased to search for stronger and simpler foundations. These twin properties can come into conflict as we proceed. Familiarity with the material, the desire for simplicity, and the wish to avoid repetition can tempt us to assume that certain information is understood and lead to imprecise thinking. We need to be careful.

The immediate practical and theoretical success of the differential and integral calculus towards the end of the 17th century convinced practitioners in the early 18th century that the calculus was consistent with the rest of mathematics. Even though Leibniz and Newton were often obscure and non-rigorous[1], it was generally felt that, with time, all would be satisfactorily explained. A new Zeno in the form of Bishop George Berkeley[2] appeared. In 1734 he published a book, *The Analyst: or a discourse addressed to an infidel mathematician*, which showed that the founders of the calculus had not given precise definitions of the concepts they employed and that their logic was faulty. His arguments could not be refuted and motivated prominent mathematicians to search for a rigorous foundation for the results of Newton and Leibniz. The situation was serious but not urgent and, as a result, it took over two hundred years to put the correct foundations in place. The subject which arose to justify the calculus has been called any one of the following: *Mathematical Analysis*, *Real Analysis*, or *Analysis*. It was developed during the 18th and 19th centuries and deals with concepts that had, for centuries, been widely and informally discussed such as *real numbers, infinity, limits* and *functions*. We discuss *all* of these

---

[1]Their standard of logic would not have impressed the ancient Greeks.

[2]George Berkeley, 1685-1753, from Kilkenny was educated and later became a faculty member at Trinity College Dublin. He was ordained an Anglican priest in 1708 and became bishop of Cloyne in 1734 but lived in England and Italy for extended periods. He was a noted 18th century philosopher.

in this book, beginning now with *functions*.

Functions, like many important concepts in mathematics, are the end result of ideas that evolved over a very extended period of time and it is still a matter of opinion when they were first clearly enunciated. The Greek mathematician, Menaechmus, 375-325 BC, a disciple of Eudoxus, discovered *conic sections* (see Figure 3.1) and gave us the classical *curves*, such as the straight line, the parabola, the ellipse, the circle and the hyperbola.

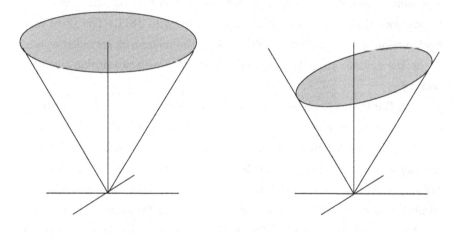

Fig. 3.1

With the development of algebra and coordinate geometry in the first half of the 17th century these curves were described by algebraic formulae. At the beginning of the 18th century, it was universally accepted that functions were just natural, that is non-artificial, formulae of some sort. Most mathematicians had a intuitive idea of what a function should be and, in their writings, gave reasonable definitions. Moreover, practical examples, even if based on different definitions, were usually acceptable to the community of mathematicians. Leonhard Euler, who is responsible for the notation $f(x)$, took a step forward by assuming that a function was any expression that had a *precise computational meaning* and, as he allowed the use of an infinite number of algebraic operations in computations, he considered convergent power series to be functions. During the 18th and early

19th centuries, anomalies arose that gave rise to debates on the concept of function. One controversy was generated by Fourier's memoir *Théorie Analytique de la Chaleur*, in which he claimed, erroneously, that *any* function could be expanded as a convergent trigonometric series, today we use the term Fourier series expansion. He did obtain such an expansion for certain discontinuous functions (see Chapter 9) but his research received mixed reviews when first presented to the academy in 1807. Fourier sent in supplementary notes in 1808 and 1809 to meet criticisms, and eventually he was awarded, with some reservations, a prize in 1811. The third and final version was published in 1822. Today the memoir is regarded as one of the key contributions to pure and applied mathematics during the 19th century.

Jean Baptiste Joseph Fourier, 1768-1830, from Auxerre in France lived in turbulent times and had a colorful life. During 1787-1789 he trained for the priesthood. In 1790 he read a paper on algebraic equations to the Royal Academy in Paris and became a teacher in his home town. In 1793 he joined the local revolutionary committee and was arrested on a number of occasions during one of which he feared for his life, he was only released in a general amnesty following Robespierre's execution. In 1795 he enrolled in the first cohort of students at the École Normale in Paris where his teachers included the eminent mathematicians Lagrange, Laplace and Monge and a few months later, when the school failed, he became a dean of discipline at the École Polytechnique. In 1797 he succeeded Lagrange as professor of analysis and mechanics and, in 1798, he published his first paper on applied mathematics and joined Napoleon's army in Egypt as a scientific officer and administrator. There he helped set up a scientific academy and led the archaeological exploration which found the Rosetta stone. Napoleon, always interested in science and mathematics, became one of the twelve members of the mathematics section of the Cairo academy. Fourier, returned in 1801 to a France where Napolean was emperor in all but name, and tried to resume his old position as professor of analysis at the École Polytechnique but, instead, was sent by Napoleon as prefect to

Isére, an area near Grenoble. He remained there until 1815 and, while carrying a heavy administrative burden, managing some large engineering projects, and reporting on the expedition to Egypt, he completed the first draft of his major contribution to mathematics. For the final period of his life, Fourier was able to concentrate on mathematics. On a personal level Fourier was recognized as an outstanding lecturer and as a noted political orator. He had excellent personal relations with practically everyone, even with his political enemies, and had the ability to maintain his integrity and friendships even when faced with great uncertainty. When Napolean escaped from Elba his return to Paris was via Grenoble where Fourier was still prefect. Fourier tried to entice the populace to oppose Napoleon and, as Napoleon arrived, Fourier fled. Napoleon was furious but soon after made him prefect of the Rhône. Fourier resigned after a short time rather than carry out a political purge of his administration but, nonetheless, must have maintained a good relationship with Napoleon as Napoleon awarded him a pension, which was, however, annulled by virtue of Napoleon's final defeat.

The criticism of Fourier prompted Cauchy, who was about to publish his own lecture notes on analysis, to give clear non-controversial definitions. This paved the way for Dirichlet to give essentially the definition we use. The controversy that arose later when Cantor developed set theory was mainly based on the perception that it moved functions away from the computational definition of Euler. The final definition presumes a rigorous notion of *sets*, a presumption only justified late in the 19th century. We assume here an informal notion of set. A set A consists of elements and we suppose that there is *no ambiguity* concerning which elements[3] belong to A. We write $x \in A$ if x belongs to A and use the notation $\{x : \ \}$ and $\{ \ , \ , \ , \ \}$ to describe sets. In the first case the set consists of all x which satisfy the conditions given after the colon : and in the second we list all the elements in the set. For example $\{1, 2, 3, 4\}$ denotes the set containing

---

[3]Real numbers can be identified with *points* on a line and as many sets that we consider are subsets of **R** we frequently follow popular custom and refer to elements in a set as *points*.

the elements 1, 2, 3 and 4 and

$$\{x : x \in \mathbf{R}, x \text{ is not a positive integer}\} = \{x : x \in \mathbf{R}, x \notin \mathbf{N}\}$$

is the set of all real numbers which are not positive integers. Certain sets occur frequently and it saves time if we give them special names or symbols. For example, we have already used $\mathbf{R}$ to denote the real numbers. When introducing a new symbol for a particular set we use the convention $A := \{\ \}$, the inclusion of : indicates that we are using this notation here for the *first time*. We let $\mathbf{R}^+ := \{x \in \mathbf{R}, x > 0\}$ denote the set of *positive*[4] real numbers, let $\mathbf{R}_0^+ := \{x \in \mathbf{R}, x \geq 0\}$ denote the set of *non-negative* real numbers and call $[a, b] := \{x \in \mathbf{R}, a \leq x \leq b\}$ a *closed interval*.

Definition 3.1. A function consists of a triple $\{A, B, f\}$ or $f : A \to B$ where A and B are sets and f is a rule assigning to each element in A a unique element in B. We call A the *domain* and B the *range* of the function.[5]

It is common to use the letter x when referring to an *arbitrary* element of A. Thus x *varies* over A and, as a result, we call x a *variable* and f a function of the variable $x \in A$. This slightly flawed use of language is quite helpful. It is important to note that two functions are equal if and only if they have the *same domain*, the *same range* and the *same rule*. One often comes across the expression: *the function f*, followed by the rule or formula for f, where it is taken as understood which x are being used in the expression $f(x)$. However, this may lead to ambiguities and misunderstanding later and it is always useful to keep in mind that, whenever we are discussing a function, there are both a domain and a range in the background. One can think of f as an operation which takes each point in A and changes it so that it becomes an element in B. *Each* point in A is transformed by f into a *unique* point in B and, as x varies over A, the transformed point,

---

[4]The notation $\mathbf{R}_{>0}$ is also used in place of $\mathbf{R}^+$ and $\mathbf{R}_{\geq 0}$ in place of $\mathbf{R}_0^+$.

[5]In the literature, functions are also called *mappings* and *transformations*, terms that refer to roles played by functions. The range is often called the *codomain* but we prefer to maintain a clear linguistic distinction between the two sets used to define functions. Later, we will need to consider the *image* of a function.

denoted by y and also by $f(x)$, varies within B. In analogy with language, one could say that elements in A and B correspond to *nouns* and f acts like a *verb* (see Figure 3.2).

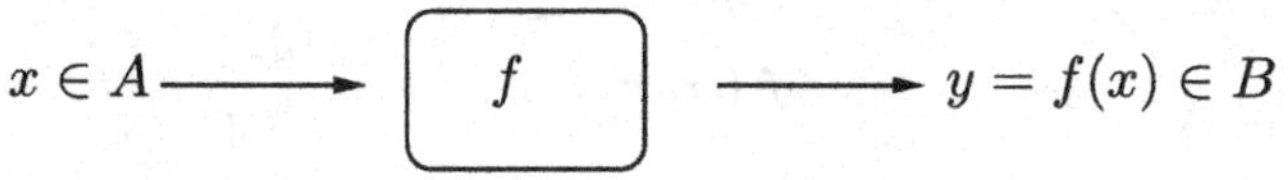

$$x \in A \longrightarrow \boxed{f} \longrightarrow y = f(x) \in B$$

Fig. 3.2

In the previous chapter we encountered the function $f : \mathbf{R} \to \mathbf{R}, f(x) = x^2$. One has to be careful as we could also consider the function $g : \mathbf{R} \to \mathbf{R}_0^+, g(x) = x^2$. These two functions are not equal; even though they have the same domain, the same rule, and the same graph, as they do not have the same range. We shall see later that they have different properties. A review of Chapter 2 shows that we have not been precise in describing functions and there is ambiguity regarding the function $f(x) = 1/x$. It could either be:

$$f : \{x : x \in \mathbf{R}, x \neq 0\} \to \mathbf{R}, \quad f(x) = 1/x$$

or

$$f : \{x : x \in \mathbf{R}, x \neq 0\} \to \{x : x \in \mathbf{R}, x \neq 0\}, \quad f(x) = 1/x.$$

Either would have been suitable for graph sketching purposes, our focus at the time, as both graphs would look the same but, in other situations, for example, in looking for an inverse function, one has to distinguish between them.

There are different ways of representing functions diagrammatically that improve understanding and intuition. We first look at *arrow diagrams* which are useful when A is small. Let $A = \{1, 2, 3, 4\}$, $B = \{a, b, c, d, e\}$, $f(1) = a, f(2) = a, f(3) = b$ and $f(4) = d$. Since we have a rule f which assigns a unique point in B to each point in A, we have defined a function, $f : A \to B$. In Figure 3.3 we sketch the arrow diagram for f.

We call the elements in A *sources* and elements in B *targets* and the lines joining the elements of A to elements of B are called *arrows*. The important

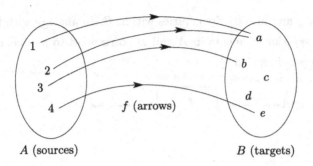

Fig. 3.3

features in this diagram are the essential properties of the definition of a function: there is *one and only one arrow* out of each source and *each arrow hits a target in B*. Without these we would not have a function and with these we do have a function. If we had two arrows coming out of the same source, say 1, we would be unsure of the value of $f(1)$. For instance, if one arrow hits a and the other hits b, is $f(1) = a$ or $f(1) = b$? If we had no arrow coming out of source 4 what meaning could we give to $f(4)$?

It is also worth observing the different roles played by A and B in defining a function. Note that target a is hit twice while no arrows hit targets c and e. Moreover, if we reverse the direction of the arrows we *do not* always obtain a function.

Next we consider *graphical representation*, a method already discussed for special functions in the previous chapter.[6] Given sets A and B, we let $A \times B$ denote the set of all pairs (x, y) where $x \in A$ and $y \in B$. Using our previous notation we have

$$A \times B := \{(x, y) : x \in A, y \in B\}.$$

We call $A \times B$ the *product* of the sets A and B (see Figure 3.4) and write A^2 in place of $A \times A$.

Mathematically the graph of the function $f : A \to B$ is the following subset of $A \times B$:

$$\text{Graph}(f) = \{(x, f(x)) : x \in A\} = \{(x, y) : x \in A, y \in B, y = f(x)\}.$$

[6]The second volume of Euler's *Introductio in analysin infinitorum* contains the earliest systematic graphical study of functions.

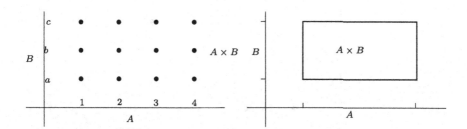

Fig. 3.4

We represent the points in A along a *horizontal line* and the points in B on a *vertical line*. Through each $x \in A$ we draw a vertical line until we reach the point opposite $f(x)$. This is a point on the graph and, when we have marked all these points, we have drawn the graph of f. In the case of the above function we obtain the following graph (Figure 3.5).

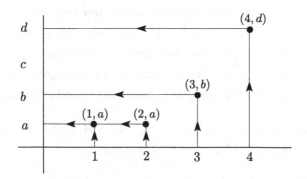

Fig. 3.5

In the above diagram we have labeled the points on the graph with a •. The graph contains all the information required to reconstruct the function but it has become a little crowded so we remove some of the non-essential features and obtain the graph in Figure 3.6. This consists of just a few dots and, mentally, we can put in the horizontal and vertical lines that we removed.

Clearly, such a diagram, in which we include all points, will only be useful when the number of points involved is small. We have already men-

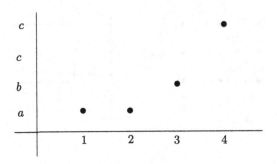

Fig. 3.6

tioned that mathematics is precise and efficient. It is also *selective* and experimenting over centuries by many people has led to the isolation of an extremely small number of highly useful particular functions, classes of functions and operations. What is surprising is how little we start with and how far we go. The particular functions we use over and over in our analysis are, not surprisingly, regular and well behaved and, when we start plotting points on their graphs, this regularity begins to appear and a full sketch is outlined from a few points. This was the case with the graphs displayed in the previous chapter. For example, consider the following functions.

Example 3.2. We sketch the graphs of $f : \mathbf{R} \to \mathbf{R}$, $f(x) = x^2$ and $g : \mathbf{R} \to \mathbf{R}$, $g(x) = |x|$ in Figures 3.7(a) and (b) respectively.

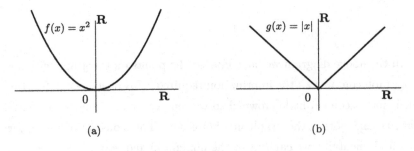

(a)   (b)

Fig. 3.7

It is possible to sketch many different curves. How does one recognize curves which are the graphs of functions? In such cases, is it possible to identify the *domain* and the *range* of the function? On translating the fact that out of each source there is precisely one arrow, we see that a curve represents a function with *domain A* if and only if every *vertical line* through A hits the curve *precisely once*. This allows us to identify the domain of the function. Thus, the curve in Figure 3.8(a) is not the graph of a function while Figure 3.8(b) is the graph of a function with domain $[-1, +1] = \{x \in \mathbf{R} : -1 \le x \le 1\}$.

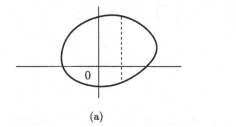   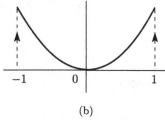

(a)   (b)

Fig. 3.8

We have previously noted that two different functions may have the same graph. From our comments above, it should be clear that functions with the same graph have the same domain and the same rule and hence the only difference must lie in their ranges. This means that we cannot always identify the precise range from the graph alone. However, we can still get some information. The range will always include the targets that are hit. This is the subset[7] of B that we call the *image* of f and denote by $im(f)$ or $f(A)$. More precisely we have the following definition.

Definition 3.3. For the function $f : A \to B$ we let

$$im(f) := f(A) := \{f(x) : x \in A\}.$$

The set $im(f)$ consists of all elements on the y-axis which lie on *horizontal lines* which hit the graph of f. For example, if $f : [-1, 2] \to \mathbf{R}, f(x) = x^2$

---

[7]A set A is a subset of the set B if every element in A is also an element of B. We write $A \subset B$ when A is a subset of B. Clearly, for sets A and B we have $A = B$ if and only if $A \subset B$ and $B \subset A$.

then, as in Figure 3.9, we have $im(f) = [0, 4]$. Using equations we get an-

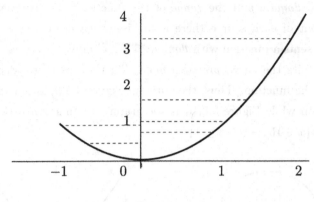

Fig. 3.9

other perspective.

Theorem 3.4. *If $f : A \to B$ is a function then, given $y \in B$, the equation*

$$f(x) = y$$

has a solution for some $x \in A$ if and only if $y \in im(f)$.

It is natural to ask: why consider both the image and range of a function? In many ways the image is a more intuitive concept and, in practical situations, it is usually more informative than any suitable but arbitrarily specified range. However, it may be difficult to explicitly describe the image, e.g. what is the image of the function $f : \mathbf{R} \longrightarrow \mathbf{R}$ where

$$f(x) = \frac{x^{10} - 6x^4 + 3x^2 - 67}{x^{18} + 4x^{12} + x^6 + 24}.$$

Using limits, the Intermediate Value Theorem and the fundamental existence theorem for maxima and minima, one can show that the image is a closed bounded interval (see Chapters 9 and 10) and, if necessary, we could probably find it, or at least approximate it, to any preassigned degree of accuracy. However, by using the image and not the range, it would be difficult to write down, explicitly, very many functions and composing functions would only rarely be possible (see Exercise 9.11).

Arrow diagrams and graphs are useful aids in understanding and developing an intuitive understanding of functions but it is reassuring to have the formal definition available as a final and definitive reference and referee.

Different mathematical concepts and results often arose in response to practical problems[8] in everyday life such as economics, physics, engineering, finance, etc. and conversely mathematical results which were discovered out of pure intellectual curiosity were often found later to have important practical applications. This symbiotic relationship has been of mutual benefit to both pure and applied mathematicians and for the student it is often useful to see the reach and motivation of mathematical results.

Example 3.5. We consider a typical function that might arise in a simple economic model. Let $C(x)$ denote the cost of producing x items of a given object, e.g. light-bulbs. In situations such as this we have two cultures converging, mathematics and economics, and both have their own rules, intuitions, habits and language. Both should be borne in mind and used to help one another. Useful conventions in economics include the suggestive naming of functions, such as C to denote the cost function[9], and following Descartes the writing, where possible, of fixed unknowns as positive numbers. For example, when economic analysis implies that an unknown constant is always positive, we might write it as $A > 0$ and, when it implies that it is always negative we write it as $-A$ with $A > 0$. The function

---

[8]A remarkably prescient document published by the National Research Council in 1987 *Mathematical Sciences: A Unifying and Dynamical Resource* contained the following paragraph: *There is no longer any question as to whether mathematical analysis will substantially influence discussions on public policy but only whether it will be used appropriately and effectively. It is essential that those making the decisions understand and influence the assumptions used to form the mathematical model and that mathematicians comprehend the applications sufficiently well that they address and solve the correct problem. In fields where mathematical models are not subject to experimental verification–such as those with the most drastic consequences–it is especially essential that the mathematics be critically scrutinized.*

[9]Traditionally C is used for the cost function, P for the price and, as it creates problems if the same letter is used with different meanings, the letter K has been used for *capital* and Π (the Greek P) for the profit function.

$C(x)$ is said to *model* the situation. Models are generally built up from the simplest and most basic assumptions, tested and then refined. In this way, quite complicated models may be constructed.

Before producing any light-bulbs there are fixed costs, the cost of the factory, the cost of equipment, etc. This is positive and we denote it by a. To produce one light-bulb a certain amount of material and labor is required. This will cost a certain positive amount and we denote it by b. Clearly, x light-bulbs will cost bx in material and labor. We now have a simple model. It will cost

$$C(x) = a + bx$$

to produce x light-bulbs and since one can only produce a positive integer number of light-bulbs we have $x \in \mathbf{N}$, where $\mathbf{N}$ is the set of positive integers, that is

$$\mathbf{N} := \{1, 2, 3, \ldots\}.$$

However, for mathematical reasons and, in particular, so that we can use the differential calculus, we suppose that x may be any positive real number. Moreover, since $C(x)$ may take any positive value, we let $C : \mathbf{R}^+ \to \mathbf{R}^+$ and we have our first simple model of the cost function. Now we start refining the model. Since the factory owner would have to buy the raw material and pay the workers each week or month and have to sell the light-bulbs to set up an income flow, it is important to consider the cost of producing x light-bulbs in a given time period so we suppose $C(x)$, with all the conditions specified above, denotes the cost of producing x light-bulbs in one month. In this case, there is a restriction on x since the factory has a certain capacity and thus there is a positive number number M such that $x \leq M$. We now have

$$C : [0, M] \to \mathbf{R}^+, \quad C(x) = a + bx.$$

Of course, it costs more per light-bulb to produce 1 rather than say 2000. This indicates a certain economy of scale which takes effect as x increases. This reduces the cost and we write it as $-cx^2$ where c is a positive constant. Our refined model now has the form

$$C : [0, M] \to \mathbf{R}^+, \quad C(x) = a + bx - cx^2.$$

Of course this is just a preliminary model that might be considered by someone interested in setting up a light-bulb factory. It would now be necessary to model revenue, sales and profit functions. We leave these aside and just consider one further simple point. Suppose the prospective factory owner had a certain amount of capital k and wished to know how much could be produced prior to any revenue from sales becoming available. This requires a solution to the *quadratic equation*

$$C(x) = a + bx - cx^2 = k$$

where $0 \leq x \leq M$ and a, b, c, k and M are positive constants. We can solve this problem using Theorem 1.3 once the constants are known and, by looking at different possible values for the constants, various scenarios can be examined. Moreover, we may have no solution, a unique solution or two solutions. In the case of two solutions, further information from the revenue or profit functions may determine a preference for one over the other. We can also rephrase the above in the terminology we have introduced in this chapter: does k belong to $im(C)$? does the horizontal line through k hit the graph of C?

3.2 Generating New Functions

The operations of addition, subtraction, multiplication and division may be applied to generate new functions using the simple examples of real–valued functions that we already have when the domains coincide and certain obvious necessary conditions are satisfied.

Example 3.6. Suppose $f : A \to \mathbf{R}$ and $g : A \to \mathbf{R}$ are functions.

(a) $f + g : A \to \mathbf{R}$ is defined by letting $(f + g)(x) = f(x) + g(x)$ for all $x \in A$,

(b) $f - g : A \to \mathbf{R}$ is defined by letting $(f - g)(x) = f(x) - g(x)$ for all $x \in A$,

(c) if $c \in \mathbf{R}$ then $cf : A \to \mathbf{R}$ is defined by letting $(cf)(x) = cf(x)$ for all $x \in A$,

(d) $f \cdot g : A \to \mathbf{R}$ is defined by letting $(f \cdot g)(x) = f(x) \cdot g(x)$ for all $x \in A$,

(e) if $g(x) \neq 0$ for all $x \in A$ then $f/g : A \to \mathbf{R}$ is defined by letting $\left(\frac{f}{g}\right)(x) = \frac{f(x)}{g(x)}$ for all $x \in A$.

We now define the composition of functions, this provides a rich supply of new functions.

Definition 3.7. If $f : A \to B$ and $g : C \to D$ are functions and $B \subset C$ then the composition $g \circ f : A \to C$ is defined by letting $(g \circ f)(x) = g(f(x))$ for all $x \in A$.

In terms of arrow diagrams we have Figure 3.10.

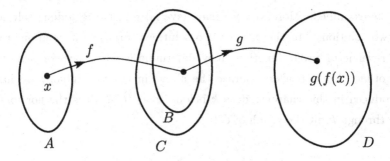

Fig. 3.10

If the composition $g \circ f$ is defined then $range(f) \subset domain(g)$. It is important to note that the functions f and g play different roles in the composition and, even when both $g \circ f$ and $f \circ g$ are defined we may, and usually do, have $f \circ g \neq g \circ f$.

Example 3.8. Let $f, g : \mathbf{R} \to \mathbf{R}, f(x) = x^2, g(x) = 1 + x$. If $x \in \mathbf{R}$ then

$$(g \circ f)(x) = g(f(x)) = g(x^2) = 1 + x^2.$$

In this case we can also define $g \circ f$ and we have, for $x \in \mathbf{R}$,

$$(f \circ g)(x) = f(g(x)) = f(1 + x) = (1 + x)^2 = 1 + 2x + x^2.$$

The functions $f \circ g$ and $g \circ f$ both have $\mathbf{R}$ as their domain and range and they will coincide if they have the same rule. They have the same rule if

$(f \circ g)(x) = (g \circ f)(x)$ for all $x \in \mathbf{R}$. Hence to show $f \circ g \neq g \circ f$ we must find x such that $(f \circ g)(x) \neq (g \circ f)(x)$. Let $x = 0$, then $(f \circ g)(0) = 1 = (g \circ f)(0)$. Let us try $x = 1$. Then $(f \circ g)(1) = 4 \neq 2 = (g \circ f)(x)$. Hence $f \circ g \neq g \circ f$. Note that it is not sufficient just to glance at the rule to determine that two functions are different, as functions may be presented in different ways. For example the functions h and k, both with $\mathbf{R}$ as domain and range and rules $h(x) = (x + 1)^2 - x^2$ and $k(x) = 2x + 1$, are equal.

To enlarge our set of examples we introduce, informally, the graphs of two important functions, the *exponential* and *log* functions (see Definition 7.16). We have (Figure 3.11)[10]

$$\exp : \mathbf{R} \longrightarrow \mathbf{R}^+, \, im(\exp) = \mathbf{R}^+,$$

and

$$\log : \mathbf{R}^+ \longrightarrow \mathbf{R}, \, im(\log) = \mathbf{R}.$$

We may be presented with functions in a rather complicated fashion and it can be useful to reverse the process of composing and to *decompose* the function into simpler functions.

Example 3.9. Suppose we wish to write $f : \mathbf{R} \to \mathbf{R}, f(x) = \exp{(4x^2 + 1)}$ as a composition of simpler functions. We first calculate f at an arbitrary

---

[10]Leonhard Euler, 1707-1783, from Basel in Switzerland was one of the most original and prolific mathematicians of all time. His professional career was divided between St. Petersburg, 1727-1741 and 1766-1783, and Berlin, 1741-1766. In mathematics, Euler contributed to the development of analysis and infinite series (showing, for instance, that $\sum_{n=1}^{\infty}(1/n^2) = \pi^2/6$), number theory (where he explained, supplied proofs, and extended the work of Fermat), differential equations, the calculus of variations, differential geometry, mechanics, astronomy, hydrostatics, etc. Additionally, his professional duties in St. Petersburg and at the court of Frederick the Great in Berlin necessitated his involvement in theoretical physiology, cartography, science education, shipbuilding, management of botanical gardens, advising on lotteries and pensions, publications of maps and calendars, etc. This is all the more remarkable as he had eye problems from his early twenties, lost the use of an eye around 1740 and became totally blind in 1771. We use many concepts due to Euler in this book and often follow his notation. He introduced the exponential function, the notation e for $\exp 1$ and Σ for summation, etc.

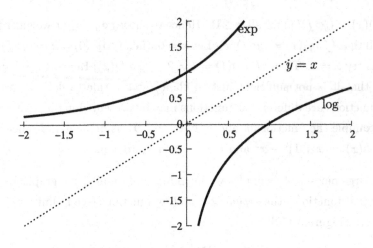

Fig. 3.11

specific point, say $x = 10$, and examine carefully the steps taken and the order in which they are taken to make this calculation. We take 10, *square it, multiply it by 4*, then *add 1* and finally *take the exponential*. Each of these steps represents an operation or function. We call these functions f_1, f_2, f_3 and f_4. We have

$$f_1(x) = x^2, \quad f_2(x) = 4x, \quad f_3(x) = x + 1, \quad f_4(x) = \exp(x),$$

and

$$f_4 \circ f_3 \circ f_2 \circ f_1(x) = f_4 \circ f_3 \circ f_2(x^2) = f_4 \circ f_3(4x^2) = f_4(4x^2 + 1) = f(x).$$

Hence $f = f_4 \circ f_3 \circ f_2 \circ f_1$. Note that the order in which we compose the functions is important.

When composing more than two functions there is *associativity*. Specifically, given functions $f : A \to B$, $g : C \to D$ and $h : E \to F$, where $B \subset C$ and $D \subset E$, then $(h \circ g) \circ f = h \circ (g \circ f)$. To show this it suffices to note that for $x \in A$ we have

$$((h \circ g) \circ f)(x) = (h \circ g)(f(x)) = h(g(f(x))) = h((g \circ f)(x)) = (h \circ (g \circ f))(x).$$

Hence $(h \circ g) \circ f = h \circ (g \circ f)$ and we may write, without ambiguity, the rule for the composition of these functions as

$$h \circ g \circ f,$$

that is we do not need to use brackets.

3.3 Exercises

(3.1) Find the largest set of real numbers that can form the domain of a function which maps into the real numbers with rule $f(x) = \frac{1}{\sqrt{x-5}}$. Find $im(f)$ and sketch the graph of f.

(3.2) If $f, g : \mathbf{R} \to \mathbf{R}$ are given by $f(x) = 2x^2 + 1$ and $g(x) = (x-1)/2$ find $f \circ g$ and $g \circ f$. Show that $f \circ g \neq g \circ f$.

(3.3) Let $f; \mathbf{R}^+ \longrightarrow \mathbf{R}$ be given by $f(x) = 1 - \frac{1}{x^2}$. Write f as $f_4 \circ f_3 \circ f_2 \circ f_1$. Find the image of each function. Sketch the graphs of $f_1, f_2 \circ f_1, f_3 \circ f_2 \circ f_1$ and f.

(3.4) Let $A := \{1, 2, 3, 4, 5\}$ and $B = \{1, 2, 3, \ldots, 10\}$. Sketch the graph and draw an arrow diagram for the function

$$f : A \longrightarrow B, \quad f(x) = 1 + |x - 3|.$$

From your diagrams, find all $y \in B$ such that the equation $f(x) = y$ has a unique solution in A. Find $im(f)$.

(3.5) If a is a real number, sketch the graph of the function

$$f(x) = e^{a+x} - e^x - e^a, \quad x \in \mathbf{R}.$$

Find all solutions of the equation

$$e^{a+x} - e^x - e^a = 0.$$

(3.6) Let $(A_i)_{i=1}^5$ denote 5 sets and let $f_i : A_i \to A_{i+1}$ denote 4 functions. Show that

$$(f_4 \circ f_3) \circ (f_2 \circ f_1) = f_4 \circ ((f_3 \circ f_2) \circ f_1) = ((f_4 \circ f_3) \circ f_2) \circ f_1.$$

References [9; 20; 30]

Chapter 4

INVERSE FUNCTIONS

If a variable y is so related to a variable x that
whenever a numerical value is assigned to x,
there is a rule according to which a unique
value of y is determined, then y is said to be
a function of the independent variable x.

Lejeune Dirichlet, 1837

In mathematics, exactness is not everything, but
without it there is nothing: a demonstration
which lacks exactness is nothing at all.

Henri Poincaré

Summary

We define and discuss special collections of functions: surjective, injective, bijective, inverse, strictly increasing, strictly decreasing, indicator and convex functions.

4.1   Injective and Surjective Functions

Although functions were introduced as part of the foundations for the differential and integral calculus, it soon became apparent that they provided an exceptionally flexible basic concept for all of mathematics. They extended mathematics as a language in a logically secure fashion and their adaptability led to rapid progress in different areas of mathematics. Variations of the concept began to appear everywhere; linear functions in matrix theory, continuous functions in analysis, analytic functions in complex analysis, measurable functions in integration theory, random variables in probability theory, homomorphisms in group theory, etc.[1] We will see that functions bring clarity and sharpness to many situations. For example, in defining infinite sets, embedding the natural numbers in the rational numbers and introducing sequences.

We begin by blending, as in Chapter 3, intuitive techniques and rigorous definitions. Our approach to examining the relationship between the *range* and *image* of f will be based on existence and uniqueness of solutions of the equation

$$f(x) = y.$$

At the same time we consider the problem of reversing operations. Since operations are implemented by functions, this amounts to looking for an *inverse function*.

Definition 4.1. A function $f : A \to B$ is *surjective* or *onto* if

$$im(f) = B.$$

From our previous discussion, we immediately have the following theorem which allows us to tell by inspection when arrow diagrams and graphs represent surjective functions.

---

[1]Towards the middle of the 20th century a new subject, based on the strong similarities between fundamental concepts and constructions in different areas of mathematics, emerged and today, *category theory*, plays a foundational role in both *mathematics* and *theoretical computer science*.

Theorem 4.2. *If $f : A \to B$ is a function, then the following are equivalent conditions:*

(1) f is surjective,

(2) $\operatorname{im}(f) = B$,

(3) all targets are hit,

(4) all horizontal lines through B hit the graph of f,

(5) for all $y \in B$ the equation

$$f(x) = y$$

has a solution $x \in A$.

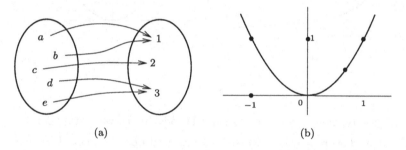

(a)   (b)

Fig. 4.1

The arrow diagram in Figure 4.1(a) portrays a surjective function while the function $f : [-1, +1] \to [-1, +1]$, drawn in Figure 4.1(b), is not surjective. We next consider uniqueness and obtain a similar result.

Definition 4.3. A function $f : A \to B$ is said to be *injective* or *one to one* if, for all $y \in B$, there exists, at most one $x \in A$ such that $f(x) = y$.

Our prior analysis immediately yields the following theorem.

Theorem 4.4. *If $f : A \to B$ is a function, then the following are equivalent conditions:*

(1) f is injective,

(2) no target is hit more than once,

(3) horizontal lines through B hit the graph of f at most once,

(4) if x_1 and x_2 belong to A and $x_1 \neq x_2$ then $f(x_1) \neq f(x_2)$,

(5) if x_1 and x_2 belong to A and $f(x_1) = f(x_2)$ then $x_1 = x_2$,

(6) the equation $f(x) = y$ has at most one solution $x \in A$ for each $y \in B$.

In Figure 4.2(a) we sketch an injective function while the function in Figure 4.2(b) is not injective.

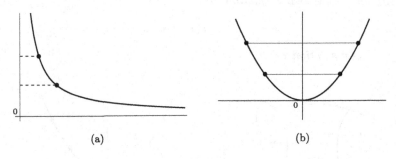

(a)   (b)

Fig. 4.2

To discuss the interesting logical relationship between statements (4) and (5) in Theorem 4.4 we make some general observations. We denote the *opposite or negation* of a statement **P** by **NP** and use the expression 'not **P**'. Thus, **NP** is true precisely when **P** is false and it is also clear that **N(NP) = P**. Statements (4) and (5) have the form *if . . . then*. Let **P** denote the statement: $x_1 \neq x_2$, and let **Q** denote the statement: $f(x_1) \neq f(x_2)$. Then **NP** is the statement $x_1 = x_2$ and **NQ** is the statement $f(x_1) = f(x_2)$. Condition (4) says:

$$\textit{if} \quad \mathbf{P} \quad \textit{then} \quad \mathbf{Q} \tag{4.1}$$

while condition (5) says

$$\textit{if} \quad \mathbf{NQ} \quad \textit{then} \quad \mathbf{NP}. \tag{4.2}$$

Conditions (4.1) and (4.2) are equivalent for *any* statements **P** and **Q** and this equivalence is known as *The Law of Contraposition,* it is often useful in proving results. Thus instead of starting with **P** and aiming to arrive at **Q** one can start with **NQ** and aim to arrive at **NP**.

We take this opportunity to simplify our notation on the application of logic to mathematics. George Boole[2] discovered, in the middle of the 19th century, that the rules of logic, for example Theorem 4.5, could be formulated using algebraic notation and, in this way, clarity and a saving of effort achieved. Combining, *if* **P** *then* **Q**, and *if* **Q** *then* **P**, we obtain **P** *if and only if* **Q**. In place of writing *if* **P** *then* **Q**, we write **P** $\Longrightarrow$ **Q**, and, in place of writing **P** *if and only if* **Q** we write **P** $\Longleftrightarrow$ **Q**. The following theorem summarizes the above.

Theorem 4.5.

$$\{P \Longrightarrow Q\} \Longleftrightarrow \{NQ \Longrightarrow NP\}.$$

4.2 Bijective Functions and Inverses

For a given function $f : A \to B$, we have seen in the previous section that the equation

$$f(x) = y \tag{4.3}$$

has a solution $x \in A$ for all $y \in B$ if f is *surjective* and, it has at most one solution if f is *injective*. Combining these conditions we obtain a unique solution for all $y \in B$.

Definition 4.6. A function $f : A \to B$ is bijective if it is injective and surjective.

---

[2]George Boole, 1815-1864, was a native of Lincoln and first professor of mathematics at Queen's College Cork. His most profound work, *An Investigation of the Laws of Thought*, introduced the subject *Mathematical or Symbolic Logic*. In this area effective use is made of the symbols $\forall$ (all), $\exists$ (some), $\vee$ (and/or), and $\wedge$ (and). For example given propositions **P** and **Q**, we have **N(P $\vee$ Q)** $\Longleftrightarrow$ **NP** $\wedge$ **NQ**. This symbolic language is powerful but so condensed that it is not suitable for technical use in an introductory text. It has led, in recent years, to the concept of formal proof and helped mathematics to secure its foundations (see Chapter 12) and has contributed to some surprising and interesting results in mathematics, philosophy, computer science and logic over the last hundred years. The English mathematician Charles Dodgson, 1832-1898, used the pseudo-name Lewis Carroll and employed mathematical logic while writing the children's story *Alice in Wonderland*.

From the properties of injective and surjective functions we immediately obtain the following theorem.

Theorem 4.7. *For the function $f : A \to B$, the following conditions are equivalent:*

(1) f is bijective,
(2) each target in B is hit precisely once,
(3) each horizontal line through B hits the graph of f precisely once,
(4) for each $y \in B$ the equation $f(x) = y$ has a unique solution $x \in A$.

Since Equation (4.3) has precisely one solution for all $y \in B$ whenever $f : A \to B$ is bijective, it is natural to try to find the unique solution x for a given y in terms of y. In fact, we can formulate a rule to achieve this: for each $y \in B$ we let $g(y)$ denote the unique point x in A satisfying $f(x) = y$. This defines a function $g : B \to A$. We can now find both compositions $g \circ f : A \to A$ and $f \circ g : B \to B$. Since $f(x) = y$ and $g(y) = x$ we have for all $x \in A$

$$g \circ f(x) = g(f(x)) = g(y) = x \tag{4.4}$$

and for all $y \in B$

$$f \circ g(y) = f(g(y)) = f(x) = y. \tag{4.5}$$

There is, in fact, only one function which satisfies (4.4) and (4.5) for if $h : B \to A$ satisfies $h \circ f(x) = x$ for all $x \in A$ and $f \circ h(y) = y$ for all $y \in B$ then, for all $y \in B$, we have

$$g(y) = g(f(x)) = x = h(f(x)) = h(y)$$

and $g : B \to A$ coincides with $h : B \to A$. We formalize this information in the following definition.

Definition 4.8. If $f : A \to B$ is a bijective function we call the unique function, $f^{-1} : B \to A$ satisfying $f^{-1} \circ f(x) = x$ for all $x \in A$ and $f \circ f^{-1}(y) = y$ for all $y \in B$ the *inverse function* of f.

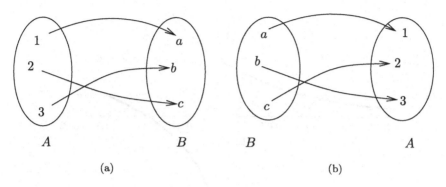

Fig. 4.3

Example 4.9. Consider the function $f : A \to B$ with arrow diagram given in Figure 4.3(a).

This function is easily seen to be bijective and its inverse has arrow diagram given in Figure 4.3(b). This is, in fact, the general situation since, if we start at the source $x \in A$ and apply the rule f, we hit the target $f(x) = y \in B$ and, if we start at the source $y \in B$ and apply the rule f^{-1}, we hit the target $x = f^{-1}(y)$. Thus, the arrow diagram for the function $f^{-1} : B \to A$ is obtained by *reversing* the direction of the arrows in the arrow diagram for $f : A \to B$.

Example 4.10. If $f : A \to B$ then

$$Graph(f) = \{(x,y) : x \in A, y \in B, y = f(x)\} \subset A \times B.$$

If f is bijective then, as $y = f(x)$ if and only if $x = f^{-1}(y)$, we have

$$Graph(f^{-1}) = \{(y,x) : y \in B, x \in A, x = f^{-1}(y)\} \subset B \times A$$
$$= \{(y,x) : y \in B, x \in A, y = f(x)\} \subset B \times A.$$

This shows that $Graph(f^{-1})$ is obtained from $Graph(f)$ by switching co-ordinates and amounts to reflecting $Graph(f)$ on the diagonal line $x = y$. For example log and exp are inverse functions and we see in the following diagram (Figure 4.4) how to obtain one graph from the other.

Reflecting on the line $y = x$ can also be accomplished by rotating to the left (anti-clockwise) through $90°$ (Figure 4.5(b)) and then reflecting on the

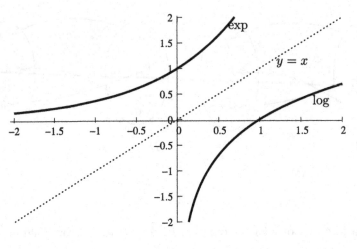

Fig. 4.4

vertical axis (Figure 4.5(c)). See also Figure 10.5. Analytically we have

$$(x, y) \overset{rotation}{\longrightarrow} (-y, x) \overset{reflection}{\longrightarrow} (y, x)$$

and geometrically we obtain the following diagram (Figure 4.5).

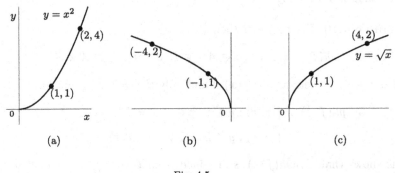

Fig. 4.5

If we consider the rule for a bijective function $f : A \to B$ as a process which transforms each element in A into an element of B, then the inverse function *reverses* or *undoes* this process. In a number of cases it is easily seen how this reverse operation works and we may write down the rule for the inverse. Following this approach we obtain[3] the following inverses.

(1) $f_1 : \mathbf{R} \to \mathbf{R}, f_1(x) = x + 1, \quad f_1^{-1} : \mathbf{R} \to \mathbf{R}, f_1^{-1}(x) = x - 1,$

(2) $f_2 : \mathbf{R} \to \mathbf{R}, f_1(x) = 4x, \quad f_2^{-1} : \mathbf{R} \to \mathbf{R}, f_1^{-1}(x) = x/4,$

(3) $f_3 : \mathbf{R}^+ \to \mathbf{R}^+, f_3(x) = x^2, \quad f_3^{-1} : \mathbf{R}^+ \to \mathbf{R}^+, f_3^{-1}(x) = \sqrt{x},$

(4) $f_4 : \mathbf{R}\backslash\{0\} \to \mathbf{R}\backslash\{0\}, f_4(x) = 1/x, \quad f_4^{-1} : \mathbf{R}\backslash\{0\} \to \mathbf{R}\backslash\{0\},$
$f_4^{-1}(x) = 1/x.$

Note that $f_4 = f_4^{-1}$. We had to take some care with f_3 since the function $f : \mathbf{R} \to \mathbf{R}, f(x) = x^2$ is not bijective; by restricting the domain, but keeping the same rule, we obtained an injective function on $\mathbf{R}^+$. Since the image of this function is also $\mathbf{R}^+$, we obtained the bijective function f_3. We call f_3^{-1} the *positive square root function*. We could have taken a different choice of domain and this would have led to a *different* square root function. The same way of thinking leads to the following theorem.

Theorem 4.11.

(a) *If $f : A \to B$ and $g : B \to C$ are injective then $g \circ f : A \to C$ is injective,*

(b) *If $f : A \to B$ and $g : B \to C$ are surjective then $g \circ f : A \to C$ is surjective,*

(c) *If $f : A \to B$ and $g : B \to C$ are bijective, then $g \circ f : A \to C$ is bijective and*

$$(g \circ f)^{-1} = f^{-1} \circ g^{-1}.$$

Proof. (a) If $x_1, x_2 \in A$, $x_1 \neq x_2$ then, since f is injective, $f(x_1) \neq f(x_2)$ by the implication (1) $\Longrightarrow$ (4) in Theorem 4.4 . By the same implication we have $g(f(x_1)) \neq g(f(x_2))$ since g is injective. Hence $(g \circ f)(x_1) \neq (g \circ f)(x_2)$ and by the reverse implication, (4) $\Longrightarrow$ (1) in Theorem 4.4, $g \circ f$ is injective.

(b) Let $z \in C$. Since g is surjective, the implication (1) $\Longrightarrow$ (5) in Theorem 4.2 shows that there exists a $y \in B$ such that $g(y) = z$. Since f is surjective, the same argument shows that there is an $x \in A$ such

---

[3] We use in this example the following standard notation: if A and B are both subsets of some set C then $A\backslash B := \{x \in C : x \in A, x \notin B\}$.

that $f(x) = y$. Hence $(g \circ f)(x) = g(f(x)) = f(y) = z$. The implication (5) $\Longrightarrow$ (1) in Theorem 4.2 shows that $g \circ f$ is surjective.

(c) By (a) and (b), $g \circ f$ is injective and surjective and hence, by definition, it is bijective. If $x \in A$ and $z \in C$ then, by associativity of functions,

$$(f^{-1} \circ g^{-1}) \circ (g \circ f)(x) = f^{-1} \circ (g^{-1} \circ g) \circ f(x)$$
$$= f^{-1} \circ (g^{-1} \circ g)(f(x))$$
$$= f^{-1}(f(x))$$
$$= x$$

and

$$(g \circ f) \circ (f^{-1} \circ g^{-1})(z) = g \circ (f \circ f^{-1}) \circ g^{-1}(z)$$
$$= g \circ (f \circ f^{-1})(g^{-1}(z))$$
$$= g(g^{-1}(z))$$
$$= z.$$

This completes the proof. $\square$

Note that in taking the inverse of a composition of functions, we take the composition of inverses in *reverse order*. This, after a moments reflection, is obvious. Think of the operations of putting on your socks and shoes in the morning, at night you reverse the process, the shoes come off first and then the socks.

Example 4.12. Theorem 4.11(c) allows us, by decomposing functions, to find the rule for rather complex functions. For example, consider the function

$$f : \mathbf{R}^+ \to \{x \in \mathbf{R} : x \geq \exp(1)\}, \quad f(x) = \exp\left(4x^2 + 1\right)$$

from the previous chapter. This function can be written as a composition of the four functions

$$f_1 : \mathbf{R}^+ \to \mathbf{R}^+, \quad f_1(x) = x^2,$$

$$f_2 : \mathbf{R}^+ \to \mathbf{R}^+, \quad f_2(x) = 4x,$$

$$f_3 : \mathbf{R}^+ \to \{x \in \mathbf{R}, x \geq 1\}, \quad f_3(x) = x + 1,$$

and

$$f_4 : \{x \in \mathbf{R}, x \geq 1\} \to \{x \in \mathbf{R}, x \geq \exp(1)\}, \quad f_4(x) = \exp x.$$

We will prove later that the exponential function is bijective and has the log function as inverse (see Theorem 7.18). We leave it as an exercise to show that the remaining functions are bijective and that

$$f = f_4 \circ f_3 \circ f_2 \circ f_1.$$

Hence

$$f^{-1} = f_1^{-1} \circ f_2^{-1} \circ f_3^{-1} \circ f_4^{-1}.$$

Since

$$f_1^{-1}(x) = \sqrt{x}, \quad f_2^{-1}(x) = x/4, \quad f_3^{-1}(x) = x - 1, \quad f_4^{-1}(x) = \log x$$

we have

$$
\begin{aligned}
f^{-1}(x) &= f_1^{-1} \circ f_2^{-1} \circ f_3^{-1} \circ f_4^{-1}(x) \\
&= f_1^{-1} \circ f_2^{-1} \circ f_3^{-1}(\log x) \\
&= f_1^{-1} \circ f_2^{-1}((\log x) - 1) \\
&= f_1^{-1}\left(\frac{((\log x) - 1)}{4}\right) \\
&= \sqrt{\frac{(\log x) - 1}{4}}.
\end{aligned}
$$

4.3 Increasing, Decreasing, Convex and Indicator Functions

The examples we have presented up to now have been very specific and, in each case, we have given an explicit formula for the rule. In this section we define collections of real-valued[4] functions defined on intervals in $\mathbf{R}$ whose rule satisfies a certain property. For real numbers a and b, $a \leq b$, we let

$$(a, b) := \{x \in \mathbf{R} : a < x < b\}, \quad (a, b] := \{x \in \mathbf{R} : a < x \leq b\}$$

---

[4]A function whose range is a subset of $\mathbf{R}$ is called a real-valued function.

and

$$[a, b) := \{x \in \mathbf{R} : a \le x < b\}$$

and call these sets and, the previously defined $[a, b] := \{x \in \mathbf{R} : a \le x \le b\}$, bounded intervals. The sets $(-\infty, b] := \{x \in \mathbf{R} : x \le b\}$, $(-\infty, b) := \{x \in \mathbf{R} : x < b\}$, $(a, +\infty) := \{x \in \mathbf{R} : a < x\}$, $[a, +\infty) := \{x \in \mathbf{R} : a \le x\}$ and $\mathbf{R}$ are the unbounded intervals. We call (a, b) an *open interval* and recall, from Chapter 3, that $[a, b]$ is called a *closed interval*. Using upper and lower bounds we discuss intervals abstractly in Chapter 6.

Definition 4.13. Let I denote an interval in $\mathbf{R}$. A function $f : I \to \mathbf{R}$ is said to be strictly increasing (respectively decreasing) if, for any $x, y \in I, x < y$ implies $f(x) < f(y)$ (respectively $f(x) > f(y)$).

The graphs of strictly increasing and strictly decreasing functions are easily recognized (Figure 4.6).

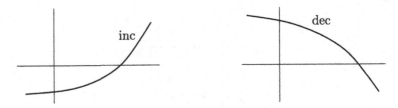

Fig. 4.6

Theorem 4.14. *A strictly[5] increasing or strictly decreasing function $f : I \to \mathbf{R}$, where I is an interval in $\mathbf{R}$, is injective.*

Proof. Suppose f is strictly increasing. If $x, y \in I, x \ne y$, then either $x < y$ and $f(x) < f(y)$ or $x > y$ and $f(x) > f(y)$. In either case we have $f(x) \ne f(y)$ and f is injective by Theorem 4.4. A similar proof can be constructed when f is strictly decreasing. □

---

[5]We could also define an increasing function on an interval I by requiring that $x \le y$ implies $f(x) \le f(y)$. These functions, and the corresponding decreasing functions are of interest, but we prefer not to confuse the reader by introducing too many different collections of functions. See Exercise 4.22.

Definition 4.15. A function $f : (a, b) \to \mathbf{R}$ is convex if the straight line joining any two points on the graph of f always lies above the graph.

From the definition, we can see by inspecting graphs whether or not, a given function is convex. For example, the functions $f : \mathbf{R} \to \mathbf{R}, f(x) = x^2$ and $g : \mathbf{R} \to \mathbf{R}, f(x) = |x|$ are easily seen to be convex (see Figure 3.7). As we may not always have a graph in place, it is useful to have an intrinsic description of convex functions. If we take any two points x and y in the interval (a, b) then any point between x and y has the form $tx + (1 - t)y$ where $0 \le t \le 1$ (if $t = 1$ we get the point x, if $t = 0$ we get the point y, and if $t = 1/2$ we get the point mid-way between x and y). From Definition 4.15 (see Figure 4.7) we obtain the following equivalent formal definition.

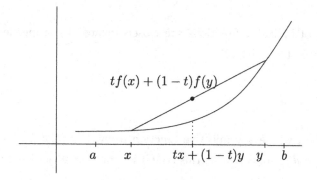

Fig. 4.7

Definition 4.16. The function $f : (a, b) \to \mathbf{R}$ is convex if and only if for all $x, y \in (a, b), x \le y$, and all $t, 0 \le t \le 1$, we have

$$f(tx + (1 - t)y) \le tf(x) + (1 - t)f(y). \tag{4.6}$$

From Definition 4.16 it is easily seen that the sum of convex functions is again convex and we obtain examples of convex functions, such as

$$f : \mathbf{R}^+ \to \mathbf{R}^+, f(x) = x^2 + \exp(x^3) - \log(x),$$

whose graphs are difficult to sketch. Definition 4.15 is intuitive and helpful in developing a preliminary understanding of convex functions. It may be considered a typical example of an 18th century approach. In contrast Definition 4.16 is mathematically precise and representative of 19th century thinking. These comparative interpretations illustrate a point we make on a number of occasions.

The final type of function we consider, the *indicator function*, takes only two values 1 and 0. It plays an important role in approximation theory, measure theory, probability theory and mathematical logic. If A is a *subset* of a larger set Ω, we define the indicator function of A in Ω as follows:

$$1_A : \Omega \longrightarrow \mathbf{R}$$

$$1_A(x) = \begin{cases} 1 & \text{if } x \in A \\ 0 & \text{if } x \notin A. \end{cases}$$

Properties of indicator functions are closely related to properties of sets (see Exercises 4.8 and 4.13).

4.4   Exercises

(4.1) Explain why every injective function from $A = \{1,2,3,4\}$ into $B = \{a,b,c,d\}$ is bijective. Is this true if we consider surjective functions?

(4.2) Let
$$f; \mathbf{R}^+ \longrightarrow \mathbf{R}^+, \quad f(x) = \log\left(x^2 + 1\right).$$
Show that f is bijective and find f^{-1}.

(4.3) Show that there are no surjective functions from the set $A = \{1,2,3,4\}$ onto $B = \{1,2,3,4,5\}$.

(4.4) Give examples of bijective functions which (a) map the positive integers onto the even positive integers and (b) map the positive integers onto the integers.

(4.5) If $f : A \to B$ is bijective show that $f^{-1} : B \longrightarrow A$ is bijective and that $(f^{-1})^{-1} = f$.

(4.6) Let $f : \mathbf{R} \to \mathbf{R}$ and $g : \mathbf{R} \to \mathbf{R}$ denote convex functions and let $c \in \mathbf{R}^+$. Show that $f + g$ and cf are convex.

(4.7) Show that the strictly increasing surjective function $f : (a, b) \rightarrow (c, d)$ is bijective and that $f^{-1} : (c, d) \rightarrow (a, b)$ is strictly increasing. State a similar result for strictly decreasing functions. Show, by example or otherwise, that the inverse of a bijective convex function need not be convex.

(4.8) If A and B are subsets of a set Ω show that $A \subset B$ if and only if $1_A \leq 1_B$. Show that

$$\min\{1_A, 1_B\} = 1_A \cdot 1_B = 1_{A \cap B}$$

and

$$\max\{1_A, 1_B\} = 1_A + 1_B - 1_{A \cap B} = 1_{A \cup B}.$$

(4.9) Let $A = \{1, 2, \ldots, 10\}$ and let $f; A \longrightarrow \mathbf{R}$ be given by

$$f = 1_{\{1,4\}} - 31_{\{2,4,6,8\}} + 41_{\{6,10\}}.$$

Draw an arrow diagram for f and sketch its graph.

(4.10) Let

$$f(x) = 1/x, \quad g(x) = \log(x^2), \quad h(x) = x^2$$

for all strictly negative real numbers. Sketch the graphs of f, g, and h. Which of these functions are increasing and which are convex?

(4.11) Write the statement '$a \cdot b = 0$ implies $a = 0$ or $b = 0$' in the form 'If **P** then **Q**' and afterwards as 'If **NQ** then **NP**'.

(4.12) If $p > 1$ is a positive integer prove that 6 divides p^2 if and only if 6 divides p. Is this true if 6 is replaced by 8? Justify your answers.

(4.13) If A, B and C are subsets of a set Ω show that

$$1_A + 1_B = 1_C$$

if and only if

$$A \cap B = \emptyset \quad \text{and} \quad A \cup B = C.$$

(4.14) If $|b| < 2$ show that

$$x^2 - bxy + y^2 > 0 \quad \text{for all} \quad (x, y) \neq (0, 0) \in \mathbf{R}^2.$$

(4.15) Let

$$f : \{x : x \geq -1\} \longrightarrow [2, \infty), \quad f(x) = \sqrt{(x+1)^2 + 4}.$$

Write f as a composition of bijective functions. Find f^{-1}. Check that $f^{-1} \circ f(0) = 0$ and that $f \circ f^{-1}(2) = 2$. Sketch the graphs of f and f^{-1}.

(4.16) Suppose f, g and h are the rules for bijective functions and the following hold:

$$f(x) = \exp(\sqrt{g((x-2)^2)}), \quad f^{-1}(x) = \left(h(\log x) + 1\right)^{1/2} + 2.$$

Find g and h.

(4.17) If A is a non-empty set and $f : A \longrightarrow B$ is injective for any set B and any mapping f prove that A has only one element. State similar results for surjective and bijective mappings.

(4.18) If $0 < t_i < 1$ for $i = 1, 2, \ldots, n$, $\sum_{i=1}^{n} t_i = 1$, and $\phi : (a, b) \longrightarrow \mathbf{R}$ is convex show that

$$\phi\left(\sum_{i=1}^{n} t_i x_i\right) \leq \sum_{i=1}^{n} t_i \phi(x_i)$$

for any $\{x_i\}_{i=1}^{n} \subset (a, b)$.

(4.19) If $f : \mathbf{R} \longrightarrow \mathbf{R}$ and $im(f)$ is a finite subset of $\mathbf{R}$ show that $f = \sum_{i=1}^{n} a_i 1_{A_i}$ where $a_i \neq a_j$ and $A_i \cap A_j = \emptyset$ when $i \neq j$. Is the converse true? Justify your answer. If

$$g : \mathbf{R} \longrightarrow \mathbf{R}, \quad g = 2 1_{[-1,3]} + 1_{(1,4)} - 3 1_{[3,5]}$$

sketch the graph of g. Write g in the form

$$g = \sum_{i=1}^{n} b_i 1_{B_i} \quad \text{where} \quad b_i \neq b_j \quad \text{and} \quad B_i \cap B_j = \emptyset$$

when $i \neq j$.

(4.20) If $f, g : \mathbf{R} \longrightarrow \mathbf{R}$ are strictly *decreasing* functions show that $g \circ f$ is strictly *increasing*.

(4.21) Show that the composition of a strictly increasing function and a strictly decreasing function is strictly decreasing. What can you say about the composition of n functions each of which is either strictly increasing or strictly decreasing?

(4.22) Let $f : \mathbf{R} \longrightarrow \mathbf{R}$ denote a function. Find the relationship between the statements,

$$\mathbf{P} : x \le y \Longrightarrow f(x) \le f(y)$$

and

$$\mathbf{Q} : x < y \Longrightarrow f(x) < f(y).$$

What happens if (a) f is injective and (b) f is not constant on any non-empty open interval. Prove your assertions.

(4.23) Show that the following hold:

$$(\mathbf{P} \Longleftrightarrow \mathbf{Q}) \Longleftrightarrow (\mathbf{NP} \Longleftrightarrow \mathbf{NQ}), \quad \mathbf{N}(\mathbf{NP} \vee \mathbf{NQ}) \Longleftrightarrow \mathbf{P} \wedge \mathbf{Q}.$$

(4.24) Find all real numbers p such that

$$f : \mathbf{R} \longrightarrow \mathbf{R}, \quad f(x) = x^3 + px$$

is injective. Prove your assertions and illustrate your results using sketches of

$$g, h : \mathbf{R} \longrightarrow \mathbf{R}, \quad g(x) = x^3, \quad h(x) = px$$

and f.

(4.25) Suppose the coefficients p and q in the equation $x^3 + px + q = 0$ satisfy

$$4p^3 + 27q^2 = 0.$$

If $q = 2\alpha^3$ find p in terms of α and hence show that α solves the equation. Find all solutions of the equations

$$x^3 + px + q = 0 \quad \text{and} \quad x^3 + px - q = 0.$$

Solve

$$27x^3 - 225x - 250 = 0.$$

References [9; 20]

Chapter 5

NUMBERS

We conclude therefore that there are no absurd, irrational, irregular, inexplicable, or deaf numbers; but there is in them such excellence and agreement that we have subject matter for meditation night and day in their admirable perfection.

Simon Stévin, 1548-1620

Numbers carry an aura of precision simply because they are numbers.

Mary Poovey, 2002

Summary

We examine the natural numbers **N**, the integers **Z** and the rational numbers **Q**. Properties of these collections suggest germs of ideas that will make later developments appear natural and almost inevitable.

5.1 The Positive Integers

While we may never know precisely how different civilizations arrived at the basic concept of number, we can speculate and propose a plausible sequence of events. Probably the most primitive notion was one of *size*, an important factor in primitive man's struggle for survival. The recognition of a very small group, a small group, groups of the same size, a large group, a very large group, etc. led, after some time, to more precise distinctions and eventually, to counting and to the *natural numbers* or *positive integers*, **N**. The first relationship between natural numbers was based on these comparisons and we write $n > m$ if n is bigger than m, $n = m$ if n and m are equal and $n < m$ if n is smaller than m. Between any two natural numbers precisely one of these relationships holds. We can quantify this relationship with the introduction of addition, thus $n > m$ if and only if $n = m + p$ for some natural number p and $n < m$ if and only if $n + q = m$ for some natural number q. We write $n \leq m$ if n is less than or equal to m and $n \geq m$ if n is greater than or equal to m. If $n \geq m$ and $n \leq m$ then $n = m$. Note that every positive integer n except 1 has a unique immediate predecessor.[1]

Next we can consider the smallest and largest elements contained in a set of natural numbers. Let A denote a non-empty subset of **N**. If $m \leq n$ for all $n \in A$ we call m a *lower bound* for A. Clearly 1 is a lower bound

---

[1]The *Principle of Mathematical Induction* is a useful logical procedure based on this property of **N**. Suppose we wish to prove a proposition P that involves the positive integer n. For instance, suppose we are interested in showing that $P(n) : \sum_{j=1}^{n} j^2 = \frac{n(n+1)(2n+1)}{6}$ holds for every positive integer n (see Exercise 11.1). It is natural to begin by proving $P(1)$ since intuitively this should be simpler than the general case and if this were not true then the general result would not be true. Next, one might consider proving $P(2)$, but now it is permissible to use the fact that $P(1)$ is known to be true. Proceeding in this way, one may be able to show generally that the truth of $P(n)$ implies the truth of $P(n+1)$. This completes the proof of P. This method of proof was called *Mathematical Induction* by De Morgan to distinguish it from the philosophically very different notion of *Induction*. Science proceeds by *Induction*; experiments are performed to check the validity of a scientific theory and each successful experiment gives the theory more credibility. Mathematics is not a science, it is deductive, not inductive, its truths are deduced from accepted axioms. There are mathematical propositions, such as the Riemann Hypothesis, which, on the basis of experiments, would be regarded as scientific results but which are not acceptable as mathematical truths. Symbolically the Principle of Mathematical Induction can be written as:

$$\mathbf{P} \iff \{\mathbf{P}(1) \wedge (\forall n \in \mathbf{N}, \mathbf{P}(n) \implies \mathbf{P}(n+1))\}.$$

Cantor developed a theory of transfinite (that is infinite) induction.

for every subset of $\mathbf{N}$. A lower bound for A which is greater than all other lower bounds is called the *greatest lower bound* of A. If $n_1 \in A$ then either $n_1 \leq n$ for all $n \in A$ or there exists $n_2 \in A$ such that $n_2 < n_1$. Thus, if $n_1 \in A$ is a lower bound for A, it is the greatest lower bound. We can repeat this process with n_2 and continue, but, as there only a finite number of positive integers less that n_1, we will eventually arrive at a greatest lower bound for A. This establishes the following principle.

Well Ordered Principle. *Any non-empty set of positive integers contains a greatest lower bound or smallest element.*

On applying the Well Ordered Principle to the set $A = \{n, m\}$ we recover the result that either $n = m$, $n > m$ or $n < m$ for any positive integers n and m. The claim that a set of natural numbers may not always have a largest element is reasonable and believable. We prove this assertion by producing an example, we show that $\mathbf{N}$ does not have a largest element. Suppose n were the largest element in $\mathbf{N}$. Then $n+1 \in \mathbf{N}$. Since $n+1 > n$ we have found an element in $\mathbf{N}$ larger than the largest element. This is impossible, so our assumption that $\mathbf{N}$ has a largest element is false. Hence $\mathbf{N}$ does not have a largest element.

It may be important to know when a set has a largest element. A non-empty set A of positive integers is *bounded above* if there exists a positive integer m such that $n \leq m$ for all $n \in A$. If

$$B := \{m \in \mathbf{N} : n \leq m \text{ for all } n \in A\}$$

then, B consists of all upper bounds for A. The set B is non-empty if and only if A is bounded above. Let n_0 denote the greatest lower bound for B. Since $n_0 \in B$, $n \leq n_0$ for all $n \in A$ and n_0 is an upper bound for A. If $n_0 = 1$ then $A = \{1\}$ and, since A is non-empty, $n_0 \in A$. If $n_0 > 1$ and $n_0 \notin A$ then, letting $n_1 + 1 = n_0$, we see that $n \leq n_1 < n_0$ for all $n \in A$ and $n_1 \in B$. This contradicts the fact that n_0 is the *greatest lower bound* for B. Hence $n_0 \in A$ and n_0 is the *least upper bound* for A. We have proved the following:

A set of positive integers which is bounded above has a least upper bound.

After names[2] such as *one, two, three, four,...* had been given to the smaller numbers came the realization that the set of natural numbers is extremely large and the conclusion that it would be physically impractical to individually name all positive integers. At some stage, it was realized that this was physically impossible[3] and this led to the acceptance, at least implicitly and perhaps reluctantly, of a non-trivial mathematical fact: *the set of natural numbers is infinite.* The ancient Greeks had difficulties with the concept of infinity. They accepted that certain sets, such as $\mathbf{N}$, were *not finite* but did not acknowledge the existence of *infinity*.[4] The ordered display of all natural numbers[5]

$$\mathbf{N} := \{1, 2, 3, 4, \ldots, 10, 11, 12, \ldots, 101, 102, \ldots\} \tag{5.1}$$

is so well known today that it is easy to forget the immense amount of abstract mathematics it contains. The ancient Greeks did not have such a system and used the letters of the alphabet to denote the smaller natural numbers. The role of 0 and the idea of using *position* to relay information[6] are crucial in (5.1). The *same* symbol may be used with *different* meanings, e.g. the number 2 has a different meaning in the numbers 12 and 2673. The analogy with language is perfect, the same letters are used in different places to produce different words. On rewriting (5.1) as (5.2) we connect powers

---

[2]Abstraction is often achieved by changing *adjectives* into *nouns.*

[3] A short story *Funes, The Memorious* by *Jorge Luis Borges* examines the inevitable failure of such a project.

[4]The prefix *in* is used in English to negate and thus *infinite = not finite.*

[5]The symbols $1, 2, 3, \ldots$, the negative numbers and 0 were introduced into Europe by the Arabs in the Middle Ages. They are based on Hindu symbols and mathematics. The Baylonians used an empty space in place of zero. Some primitive tribes, even in the 21th century, still use a counting system consisting of *one, two, few, many.*

[6]Street numbering systems became necessary with the arrival of a popular postal system, the *penny post*, in the middle of the nineteenth century and are another example where numbers, position and information come together. Countries differ in their approach and the information they convey. In the prosaic Anglo-Saxon system, introduced in the middle of the twentieth century, one side of the street is reserved for odd and the other for even numbered houses and houses are numbered consecutively. In the southern suburbs of Rio de Janeiro, house numbered 214 is 214 meters from the city end of the street if the street points towards the center of the city otherwise it is 214 meters from the beach.

with position and display natural numbers concisely:

$$67282 = 6 \times 10^4 + 7 \times 10^3 + 2 \times 10^2 + 8 \times 10 + 2. \qquad (5.2)$$

This uses the base 10, involves the operations of addition and multiplication, and suggests we let $10^1 = 10$ and $10^0 = 1$ (see Theorem 7.20).[7]

We have seen how natural it is to *subtract* a positive integer from a larger one and from there to move to the notion of *addition* of positive integers. Afterwards came the operations of *multiplication* and *division*. In some ways division is more basic even if it is not always possible. For instance six pieces of fruit can easily be divided between two, three or six people but not so easily between five or seven. On the other hand as $n \cdot m = n + n + \cdots + n$ (m times) multiplication may have evolved as a shortcut for addition. The main properties for these operations are very simple. Let $a, b, c \in \mathbf{N}$:

(a) **closure** $a + b, a \cdot b \in \mathbf{N}$,

(b) **commutativity** $a + b = b + a$, $a \cdot b = b \cdot a$,

(c) **associativity** $(a + b) + c = a + (b + c)$, $(a \cdot b) \cdot c = a \cdot (b \cdot c)$,

(d) **distributive law** $a \cdot (b + c) = a \cdot b + a \cdot c$,

(e) **cancellation laws** if $m + a = m + b$ then $a = b$, if $a \cdot c = b \cdot c$ then $a = b$.

---

[7]The almost unconscious physical process of finger counting promoted the use of the base 10. As civilization progressed some cultures considered the base $12 = 2 \times 6 = 3 \times 4$ more convenient when dealing with items, such as money, that required frequent dividing. The Romans used the base 5, but did not have a system similar to (5.2). One can still see the Roman numbering symbols,

$$I, II, III, IV = 4, V = 5, VI = 6, \ldots, XI = 9, X = 10, XI = 10 + 1 = 11, \ldots$$

in use in clocks, chapter headings, etc. The Babylonians divided one complete circular revolution into 360 degrees, presumably as an approximation to the number of days in the year, and used the base 60, the smallest number divisible by the numbers 1 to 6. As a result we have today 60 seconds in a minute, 60 minutes in an hour and buy eggs by the dozen. The Romans produced practically no original mathematics and, during the whole period of the Roman empire, it was the conquered Greeks who made mathematical advances. As educated Romans could read and speak Greek, it was not felt necessary to translate the classical Greek mathematical texts into Latin. Translations were only made when European universities were established in the middle ages.

The relationship between these algebraic operations and order is given in the following list where a, b and $c \in \mathbf{N}$:

(i) if $a < b$ then $a + c < b + c$,

(ii) if $a < b$ then $a \cdot c < b \cdot c$,

(iii) if $a < b$ and $b < c$ then $a < c$.

Property (a) means that the sum and product of positive integers are positive integers, (b) says that we can add and multiply in whatever order we wish, (c) and (d) remove the need for brackets in many cases while (e) is useful in solving equations. Using set theoretic considerations we will show how to extend $\mathbf{N}$ to include new collections, the integers $\mathbf{Z}$, the rational numbers $\mathbf{Q}$ and the real numbers $\mathbf{R}$. All the above rules and relationships extend apart from the cancelation rule (e) for multiplication. Rule (e) is, as we saw in Chapter 1, replaced by $a \cdot c = b \cdot c$ implies either $a = b$ or $c = 0$. We also note that the number 1 is special with regard to multiplication since $n \cdot 1 = n$ for any positive integer n, that is 1 is an identity for multiplication. Later we see that 0 is an identity for addition.

Of course, if we can multiply numbers we can also ask which numbers are the product of positive integers other than 1. Many interesting questions can now be posed and from the methods we have developed in previous chapters we can at least try to answer them in a reasonable fashion. It is also instructive to try and obtain such results using only positive integers, that is without 0, the negative integers, the rational numbers, subtraction or division. The results obtained in this fashion are a good training in the art of thinking and, sooner or later, they lead to non-trivial results. We take a look at one such simple result. A positive integer is called *even* if it is a multiple of 2. This means that the even positive integers are the numbers $2, 4, 6, \ldots$. a number which is not even is called *odd*. Clearly, if n is even then $n = 2m$ for some positive integer m and if n is odd then $n = 1$ or $n = 2r + 1$ for some positive integer r. We now prove a simple theorem.

Theorem 5.1. *A positive integer n is even if and only if n^2 is even.*

Proof.   Let $\mathbf{P}$ denote the statement n *is an even integer* and let $\mathbf{Q}$ denote the statement n^2 *is an even integer*. We aim to show $\mathbf{P} \implies \mathbf{Q}$ and $\mathbf{Q} \implies \mathbf{P}$, that is $\mathbf{P} \iff \mathbf{Q}$.

Suppose $\mathbf{P}$ holds. Then $n = 2m$ for some positive integer m. Hence

$$n^2 = (2m)^2 = 4m^2 = 2(2m^2)$$

and n^2 is even. Hence $\mathbf{P} \implies \mathbf{Q}$.

In place of showing $\mathbf{Q} \implies \mathbf{P}$, we show that $\mathbf{NP} \implies \mathbf{NQ}$. If $\mathbf{NP}$ holds for the positive integer n then $n = 1$ or $n = 2m + 1$ for some positive integer m. If $n = 1$ then $n^2 = 1$ and $\mathbf{NQ}$ holds. If $n = 2m + 1$ then $n^2 = (2m + 1)^2 = 4m^2 + 4m + 1 = 2(2m^2 + 2m) + 1$ and, by our remarks above, n^2 is odd. Hence $\mathbf{NQ}$ holds. This completes the proof. $\qquad\square$

We can rephrase the above using the concept of a divisor. If n and m are positive integers, we say that m divides n and write $m|n$ if $n = r \cdot m$ for some positive integer r. We call m a *divisor* of n. If n and m are positive integers, $n > m$ and $n \neq r \cdot m$ for any $r \in \mathbf{N}$ let $A := \{r \in \mathbf{N} : r \cdot m > n\}$. By the *Well Ordered Principle* A contains a smallest element s and, by our hypothesis, $s > 1$. Hence $(s - 1)m < n < sm$ and $1 \leq n - (s - 1)m < m$. We have proved the following result: *if m and n are natural numbers then either $m < n$, $m|n$ or there exist positive integers r and s, $1 \leq r < m$, such that $n = s \cdot m + r$.* Clearly n is even if and only if $2|n$. Theorem 5.1 can now be restated as follows: $2|n$ if and only if $2|n^2$.

Although addition and multiplication were introduced for computational convenience, frequent use led to the observation that certain numbers could be obtained by multiplying together *smaller* numbers while others could not. This led to distinguishing between different classes of natural numbers and the following definition.

Definition 5.2. Positive integers greater than 1 which have no divisors other than 1 and themselves are called *prime numbers*.

One of the most important areas within mathematics is *number theory* and this is motivated by the desire to understand prime numbers. This has led to deep insights within mathematics and surprisingly to important applications. To see the process by which mathematics evolves we mention that, over two thousand years ago, the ancient Greeks proved that there were an infinite number of primes and moreover that every natural number could be written in a unique way, apart from the order of the terms, as a product of primes (see Exercise 5.9). Thus, the primes are the building blocks for the natural numbers.

Towards the end of the eighteenth century, Legendre[8] conjectured that $\pi(n) \approx n/\log(n)$ when n is large. At the same time and independently of Legendre, Gauss, after inspecting tables of primes, conjectured the following result, known today as the *Prime Number Theorem*. It was finally proved, using complex analysis and the Riemann Zeta function, at the end of the nineteenth century.[9]

Theorem 5.3. *If n is a positive integer and $\pi(n)$ denotes the number of primes less than or equal to n then, for large n, we have approximately*

$$\pi(n) \approx \int_2^n \frac{dx}{\log x}.$$

Legendre's conjecture, while not as profound as that of Gauss, has a more immediate impact and does not require a knowledge of integration theory.

---

[8]Adrien Marie Legendre, 1752-1833, came from a well off background but lost his financial security during the French revolution. He made contributions to many areas of pure and applied mathematics including number theory, geometry, astronomy, integration theory, mechanics and his book on elementary geometry was the standard text on the subject for almost a hundred years. He was very involved in establishing the metric system for weights and measures. In 1824, after refusing to vote for a government candidate, his pension was stopped and he died in poverty.

[9]Johann Carl Frederick Gauss, 1777-1855, from Brunswick in Germany was a student and later professor of astronomy at Gottingen. He is regarded as one of the great mathematicians. He was a gifted astronomer, a talented experimenter, and a skillful observer. He had a tremendous capacity for mental arithmetic and his financial acumen helped his university achieve financial stability. In mathematics he contributed to number theory, potential theory, differential geometry, approximation theory and statistics. Many different results are named after him: the *Gauss-Bonet formula* and *Gaussian curvature* in differential geometry, the *Gaussian or normal distribution* in statistics, the unit of *magnetic flux density*, the *Gaussian or complex plane*, *Gaussian integration* in approximation theory, etc.

To complete this section on the *natural numbers* we note that the equations

$$n + x = m \tag{5.3}$$

and

$$p \times y = q \tag{5.4}$$

where n, m, p and q are in **N** sometimes do and sometimes do not have solutions. Equation (5.3) has a solution if and only if $n < m$, while (5.4) has a solution precisely when p divides q. These equations involve addition and multiplication and we have already informally used functions based on these operations. To obtain inverse functions we require solutions to the above equations.

5.2   The Integers and Rational Numbers

Mathematicians throughout history have had different opinions on the degree of detail and transparency that should be included for the benefit of their readers.[10] Descartes feared that by being clear others might steal his results, Archimedes and Fermat had the not unreasonable idea of withholding proofs to give others the benefit of discovering them, Archimedes circulated false propositions to expose those who claimed his results, while Newton had to be convinced to publish. Moreover, Viète, Descartes, Newton and others down through the ages, firmly believed that the ancient Greeks had methods of analysis that they deliberately kept hidden. These approaches generated mixed reactions. We make these remarks as the material in this section is often avoided and sometimes relegated to appendices and the reader may wish, especially on a first reading, to pass rapidly to

---

[10]The following comment on Euclid, is taken from a 1930's school text by G. W. Spriggs and R.F. Ward: *The proofs of his theorems show the finished product without any indication of the preliminary analysis with its tentative endeavors and frequent failures, and satisfying as it is for the more mature to observe the beautiful enfolding of his subject with logical inevitability, it cannot be too clearly emphasized that intelligent analysis must first supply the elements which may be ultimately ordered in logical form.*

the next section or, indeed, to the next chapter. To generate situations in which (5.3) and (5.4) are always, or practically always, solvable we require the negative integers and rational numbers. These may be introduced in various ways. Many modern textbooks first introduce the negative integers and then the rationals, but historically the positive rationals preceded the negative integers. We found, having experimented with both approaches, the historical route the more natural. Thus we start by using $\mathbf{N}$ to construct $\mathbf{Q}^+$, the positive rational numbers[11] and then use $\mathbf{Q}^+$ to construct $\mathbf{Q}$. Note that (5.4) has a solution whenever $p, q \in \mathbf{Q}^+$. In Chapter 12 we use subsets of $\mathbf{Q}$ to construct the real numbers. We have to be careful. We have already been using properties of the rational numbers. We need to construct them using only the positive integers and the operations of addition and multiplication of positive integers. In particular we may not use 0, subtraction, or division. We may use, informally, any knowledge and experience that we have in order to see how to proceed but must not become involved in any circular argument.

We do know that positive rational numbers are written as one positive integer over another, we know how to add, multiply and order these numbers and that we may, for instance, identify the following:

$$\frac{1}{2} = \frac{2}{4} = \frac{3}{6} = \ldots = \frac{11}{22} = \ldots = \frac{57}{114} = \ldots = \frac{211}{422} = \ldots . \qquad (5.5)$$

We cannot use, but will be guided by what we see in (5.5). To begin, we note that a rational number defines a relationship between pairs of integers and a particular case of this relationship is outlined in (5.5). It must be abstractly introduced using only the integers. In the following definition the expression '*is equivalent to*' is replaced by the symbol $\sim$.

Definition 5.4. If (n, m) and (p, q) are pairs of positive integers let $(n, m) \sim (p, q)$ if $nq = mp$. Let $\left[\frac{n}{m}\right] = \{(p, q) \in \mathbf{N} \times \mathbf{N} : (n, m) \sim (p, q)\}$.

---

[11]Rational is derived from *ratio*. As $\mathbf{R}$ has already been reserved for the real numbers, we denote the rational numbers by $\mathbf{Q}$ since rationals are *quotients*, p/q, of integers with $q \neq 0$. Rational numbers are also called *fractions*, a word derived from the Latin word *fractum*, to break, e.g. 1/3 is obtained by breaking 1 into 3 equal parts.

Note that $\left[\frac{n}{m}\right]$ is a *subset* of $\mathbf{N} \times \mathbf{N}$. When we have shown that the collection of all such subsets of $\mathbf{N} \times \mathbf{N}$ can be identified with the positive rationals we will revert to the standard notation for rationals and write $\frac{n}{m}$ in place of $\left[\frac{n}{m}\right]$ and n in place of $\left[\frac{n}{1}\right]$.[12]

The relationship $\sim$ is of a special kind, an *equivalence relationship*. It occurs regularly in other parts of mathematics, for instance in algebra and probability theory, and we now briefly consider the general situation.

Definition 5.5. A relationship between elements of a set Ω is called an equivalence relationship if the following axioms hold:

$$x \sim x \quad \text{(reflexive)}, \tag{5.6}$$

$$x \sim y \Longleftrightarrow y \sim x \quad \text{(symmetric)}, \tag{5.7}$$

$$x \sim y \text{ and } y \sim z \Longrightarrow x \sim z \quad \text{(transitive)}, \tag{5.8}$$

where we have written $x \sim y$ if x and y are equivalent.

We call $[x] := \{y \in \Omega : x \sim y\} = \{y \in \Omega : y \sim x\}$ the *equivalence class* containing x.

Lemma 5.6. *If $\sim$ is an equivalence relationship on Ω, then for x, y in Ω*

(a) $[x] = [y] \Longleftrightarrow x \sim y$,

(b) *either $[x] = [y]$ or $[x] \cap [y] = \emptyset$ (that is two equivalence classes either coincide or are totally disjoint).*

Proof. (a) Suppose $[x] = [y]$. By reflexivity $x \in [x] = [y]$ and hence $x \in [y]$ and $x \sim y$.

Conversely, suppose $x \sim y$. If $z \in [x]$, then $z \sim x$, and since $x \sim y$, transitivity implies $z \sim y$ and hence $z \in [y]$. Since z was an arbitrary element of $[x]$, this shows $[x] \subset [y]$. The same argument shows that $[y] \subset [x]$. Hence $[x] = [y]$.

---

[12]Clearly $(n, m) \sim (p, q)$ if and only if $\frac{n}{m} = \frac{p}{q}$. We may use this information in our investigation but not yet in our presentation. We see this principle in operation in the courts, lawyers usually know facts that they cannot prove and hence cannot introduce as evidence, but of course what they know does influence their approach and helps them prepare what they do present.

(b) It suffices to show that $[x] \cap [y] \neq \emptyset$ implies $[x] = [y]$. If $z \in [x] \cap [y]$, then $z \sim x$ and $z \sim y$. By (5.7) and (5.8) this implies $x \sim y$. Hence, by (a), $[x] = [y]$, and this completes the proof. $\qquad\square$

A *partition* of a set Ω is a collection of non-overlapping subsets of Ω whose union covers the whole space. Formally we have the following definition.

Definition 5.7. A collection of subsets of Ω, $(A_\alpha)_{\alpha \in \Gamma}$, is called a partition of Ω if

$$A_\alpha \cap A_\beta = \emptyset \text{ if } \alpha \neq \beta, \tag{5.9}$$

$$\bigcup_{\alpha \in \Gamma} A_\alpha = \Omega. \tag{5.10}$$

Condition (5.9) says that the sets in the partition are *pairwise disjoint* while (5.10) says that they cover the whole space. The set Γ is called the indexing set for the collection of subsets, if the number of sets is finite we usually write $(A_n)_{n=1}^k$. Lemma 5.6 is significant because it establishes a correspondence between equivalence relationships and partitions. Since $x \in [x]$, each point in Ω is in some equivalence class and, as equivalence classes are disjoint by Lemma 5.6(b), we obtain a partition of Ω. Conversely, if we are given a partition of Ω, $(A_\alpha)_{\alpha \in \Gamma}$, we obtain an equivalence relationship by letting $x \sim y$ if x and y belong to the same A_α. Thus we can pass back and forth, as required, between *partitions* and *equivalence relationships*. A map is a common example of a partition: a map of Ω divides Ω into counties, states, or countries, etc., and points which are not clearly in one subdivision are usually a source of dispute.

We now return to the specific relationship $\sim$ given in Definition 5.4. We have $(n, m) \sim (n, m)$ and clearly $(n, m) \sim (p, q)$ implies $(p, q) \sim (n, m)$. If $(n, m) \sim (p, q)$ and $(p, q) \sim (r, s)$ then $nq = mp$ and $ps = qr$. Hence $nqps = pmrq$ and, by the multiplicative cancellation law for positive integers, $ns = mr$. This implies that $(n, m) \sim (r, s)$ and shows that $\sim$ is an equivalence relationship. In conformity with modern notation we let

$$\mathbf{Q}^+ := \{[\frac{n}{m}] : (n, m) \in \mathbf{N} \times \mathbf{N}\}$$

and guided by hindsight, we define addition, multiplication and an order on $\mathbf{Q}^+$ as follows:

$$[\frac{n}{m}] + [\frac{p}{q}] = [\frac{nq + mp}{mq}],$$

$$[\frac{n}{m}] \cdot [\frac{p}{q}] = [\frac{np}{mq}]$$

and

$$[\frac{n}{m}] < [\frac{p}{q}] \iff nq < mp.$$

We now come to a subtle point that should not be glossed over. Are there any ambiguities in the above definitions? In mathematical terminology we ask if the above are *well defined*. A first reaction is to say, of course they are, we have been very clear in the above statements, but this is not sufficient. If we wish to show, for instance, that addition is well defined then we need to prove the following: if $[\frac{n}{m}] = [\frac{n_1}{m_1}]$ and $[\frac{p}{q}] = [\frac{p_1}{q_1}]$ then,

$$[\frac{nq + mp}{mq}] = [\frac{n_1 q_1 + m_1 p_1}{m_1 q_1}] \tag{5.11}$$

or, equivalently, that $(nq + mp, mq) \sim (n_1 q_1 + m_1 p_1, m_1 q_1)$. Using the substitutions $nm_1 = mn_1$ and $pq_1 = p_1 q$, we obtain (5.11) from

$$(nq + mp)m_1 q_1 = n_1 mqq_1 + mm_1 p_1 q = (n_1 q_1 + m_1 p_1)mq.$$

It is a tedious but easy exercise to verify that multiplication and order are well defined and obey the laws of closure, commutativity, cancellation for both addition and multiplication, etc., previously discussed for $\mathbf{N}$. For example, by the cancellation law for multiplication in $\mathbf{N}$, we have for positive integers p, q, r, s, n and m

$$[\frac{p}{q}] \cdot [\frac{r}{s}] = [\frac{n}{m}] \cdot [\frac{r}{s}] \iff [\frac{pr}{qs}] = [\frac{nr}{ms}]$$

$$\iff prms = qsnr \iff pm = qn$$

$$\iff [\frac{p}{q}] = [\frac{n}{m}]$$

which proves the cancellation law for multiplication in $\mathbf{Q}^+$.

Rewriting Equation (5.4) as $a \cdot x = b$ with $a, b \in \mathbf{Q}^+$, we obtain the following: if $a = [\frac{n}{m}]$ and $b = [\frac{p}{q}]$ where $n, m, p, q \in \mathbf{N}$ then the cancellation law shows that $x = [\frac{pm}{qn}]$ is the unique solution of $a \cdot x = b$. If $\widehat{1} := [\frac{1}{1}]$ then

$a \cdot \widehat{1} = a$ for all $a \in \mathbf{Q}^+$ and we let a^{-1} denote the unique solution of the equation $a \cdot x = \widehat{1}$. With this more familiar notation we see that $x = a^{-1}b$ is the unique solution to the equation $a \cdot x = b$ in $\mathbf{Q}^+$.

Using the analogous property for $\mathbf{N}$, one easily proves a cancellation law for addition in $\mathbf{Q}^+$. In this case we only obtain a solution to the equation $a + x = b$ when $b > a$. If $a = \left[\frac{n}{m}\right], b = \left[\frac{p}{q}\right] \in \mathbf{Q}^+$ and $a < b$ then there exists $s \in \mathbf{N}$ such that $nq + s = pm$. If $c = \left[\frac{s}{qm}\right]$ then

$$\left[\frac{n}{m}\right] + \left[\frac{s}{qm}\right] = \left[\frac{nqm + sm}{mqm}\right] = \left[\frac{nq + s}{mq}\right] = \left[\frac{pm}{mq}\right] = \left[\frac{p}{q}\right]$$

and $a + c = b$. By the cancellation law for addition, c is the unique solution of the equation $a + x = b$.

Our final task is to place $\mathbf{N}$ inside $\mathbf{Q}^+$ and to verify that the operations on $\mathbf{Q}^+$ extend those already defined on $\mathbf{N}$. We embed $\mathbf{N}$ in $\mathbf{Q}^+$ by means of the function $\phi : \mathbf{N} \longrightarrow \mathbf{Q}^+$, $\phi(n) = \left[\frac{n}{1}\right]$.

If $\phi(n) = \phi(m)$ then $\left[\frac{n}{1}\right] = \left[\frac{m}{1}\right]$ and $n = m$. Hence ϕ is *injective*. Since

$$\phi(n) + \phi(m) = \left[\frac{n}{1}\right] + \left[\frac{m}{1}\right] = \left[\frac{n+m}{1}\right] = \phi(n+m),$$

$$\phi(n) \cdot \phi(m) = \left[\frac{n}{1}\right] \cdot \left[\frac{m}{1}\right] = \left[\frac{nm}{1}\right] = \phi(nm),$$

and

$$\phi(n) = \left[\frac{n}{1}\right] < \left[\frac{m}{1}\right] = \phi(m) \Longleftrightarrow n < m,$$

we may, as far as the operations of addition, multiplication and order are concerned, identify $\mathbf{N}$ with its image under ϕ in $\mathbf{Q}^+$.

We have lifted familiar properties of $\mathbf{N}$ to $\mathbf{Q}^+$ and completed our construction of the *positive rational numbers*.

To define $\mathbf{Q}$ we follow a similar process but use an equivalence relationship on $\mathbf{Q}^+ \times \mathbf{Q}^+$ and lift properties from $\mathbf{Q}^+$ to $\mathbf{Q}$. We again use the symbol $\sim$ but, now, let

$$(p, q) \sim (r, s) \Longleftrightarrow p + s = q + r.$$

This is easily seen to be an equivalence relationship. Let

$$[p, q] := \{(r, s) \in \mathbf{Q}^+ \times \mathbf{Q}^+ : (r, s) \sim (p, q)\}$$

and let $\mathbf{Q} := \{[p,q]; (p,q) \in \mathbf{Q}^+ \times \mathbf{Q}^+\}$. On $\mathbf{Q}$ we define addition, multiplication, and an order as follows:[13]

$$[p,q] + [r,s] : = [p+r, q+s],$$

$$[p,q] \cdot [r,s] : = [pr + qs, ps + qr],$$

$$[p,q] > [r,s] : \Longleftrightarrow p + s > q + r.$$

It is necessary, but tedious, to show that these are *well defined* operations and satisfy the standard properties of closure, commutativity, associativity, etc. Since

$$[p,q] + [1,1] = [p+1, q+1] = [p,q]$$

$[1,1]$ is the identity for *addition* and, as $[p,q] + [q,p] = [p+q, p+q] = [1,1]$, $[q,p]$ is an additive inverse for $[p,q]$. This implies a cancellation law for addition.

If $[p,q] > [1,1]$, that is if $p > q$, we say that $[p,q]$ is *positive*, if $[1,1] > [p,q]$ we say that $[p,q]$ is *negative*, and call $[p,p] = [1,1]$, the identity for addition, the *zero* element in $\mathbf{Q}$.

Since $[p,q] \cdot [1,1] = [p+q, p+q] = [1,1] = [r,s] \cdot [1,1]$ for any $[p,q], [r,s] \in \mathbf{Q}$ we do not have an unrestricted cancellation law for multiplication. If $r, s \in \mathbf{Q}^+$ and $r > s$ then there exists $t \in \mathbf{Q}^+$ such that $s + t = r$. If $[p,q], [p_1, q_1] \in \mathbf{Q}$ and $[p,q] \cdot [r,s] = [p_1, q_1] \cdot [r,s]$ then

$$[pr + qs, ps + qr] = [p_1 r + q_1 s, p_1 s + q_1 r].$$

Hence

$$(pr + qs) + (p_1 s + q_1 r) = (ps + qr) + (p_1 r + q_1 s)$$

and

$$(p + q_1)r + (p_1 + q)s = (p_1 + q)r + (p + q_1)s.$$

---

[13]The analysis becomes transparent by noting that we will eventually identify $[p,q]$ with $p - q$. Also note that addition and multiplication are based on the identities $(n - m) + (r - s) = (n + r) - (m + s)$ and $(n - m) \cdot (r - s) = (nr + ms) - (ns + mr)$. Note that this identifies $[1,1]$ with $1 - 1 = 0$, the identity for addition, and $[2,1]$ with $2 - 1 = 1$, the identity for multiplication.

Replacing r by $s+t$ and using the multiplicative cancellation law in $\mathbf{Q}^+$ we obtain $(p+q_1)t = (p_1+q)t$ and a further application of the same law implies $p+q_1 = p_1+q$, that is $[p,q] = [p_1,q_1]$. A similar argument implies the same result when $s > r$ and we have proved a multiplicative cancellation law by *non-zero elements* in $\mathbf{Q}$.

Now suppose $x, y \in \mathbf{Q}$ and $x \cdot y = 0$. If x is non-zero then $x \cdot y = 0 = x \cdot 0$ and, by cancellation, $y = 0$ (we have used the real number version of this rule extensively in Chapter 1). Note that $[2, 1]$ is a multiplicative identity for $\mathbf{Q}$. If $a = [r, s] \in \mathbf{Q}$ is non-zero, that is if $r \neq s$, we let $a^{-1} = [(r-s)^{-1}+1, 1]$ if $r > s$ and let $a^{-1} = [1, (s-r)^{-1} + 1]$ if $r < s$. It is easily verified that a^{-1} satisfies the equation $a \cdot x = [2, 1]$ and, by the cancellation law for multiplication, it is the unique solution in $\mathbf{Q}$. We call a^{-1} the multiplicative inverse of a.

In place of Equations (5.3) and (5.4) we now see that the single more general equation

$$ax + b = c \tag{5.12}$$

where $a, b, c \in \mathbf{Q}$ and $a \neq 0$ has the unique solution in $\mathbf{Q}$:

$$x = (c - b) \cdot a^{-1} = a^{-1} \cdot (c - b).$$

Finally, to show that $\mathbf{Q}$ extends $\mathbf{Q}^+$, we embed $\mathbf{Q}^+$ in $\mathbf{Q}$ and note that operations of addition, multiplication and order are preserved. We let

$$\varphi : \mathbf{Q}^+ \longrightarrow \mathbf{Q} \quad \varphi(x) = [x + 1, 1].$$

If $\varphi(x) = \varphi(y)$ then $[x+1, 1] = [y+1, 1]$ and $x+2 = y+2$, that is $x = y$ and φ is injective. Again it is routine but tedious to verify, for $x, y \in \mathbf{Q}^+$, that $\varphi(x + y) = \varphi(x) + \varphi(y)$, $\varphi(x \cdot y) = \varphi(x) \cdot \varphi(y)$ and $\varphi(x) < \varphi(y) \Longleftrightarrow x < y$. Note that we have identified $\mathbf{Q}^+$ with the positive elements in $\mathbf{Q}$.

This completes our construction of the rational numbers and we may now revert to standard notation. We let $-n$ denote the additive inverse for n, 1 and 0 will denote the identities for multiplication and addition respectively, and we let $\mathbf{Z} := \{0, \pm 1, \pm 2, \pm 3, \ldots\}$ denote the set of all integers.[14] We also write $p - q$ in place of $[p, q]$.

---

[14]The notation comes from the German word for number *Zahlen*.

We introduced equivalence classes because each rational number has many different representatives. In certain circumstances, some representatives as, for instance, those given in the following theorem, may be more useful than others.

Theorem 5.8. *If r is a positive rational number then $r = \frac{p}{q}$ where $p, q \in \mathbf{Z}$, $q \neq 0$, and at least one of p and q is odd.*

Proof. Let $r = \frac{p}{q}$ where p, q are positive integers. If one of these is odd the proof is complete. Otherwise both are even. Let 2^s denote the highest power of 2 that divides p to give an integer.[15] Hence $p = 2^s \cdot t$ where t is an odd positive integer. Similarly, $q = 2^u \cdot v$ where u is a positive integer and v is an odd positive integer. We then have:

$$r = \frac{t}{v} \text{ if } s = u, \quad r = \frac{2^{s-u}t}{v} \text{ if } s > u \quad \text{and} \quad r = \frac{t}{2^{u-s}v} \text{ if } s < u.$$

This completes the proof. $\square$

5.3 Arithmetic and Geometry

Counting and measuring gave rise to *Arithmetic* and *Geometry* respectively. Both are derived from classical Greek words: Arithmetic from *arithmos* (number) and *techne* (science), while geometry comes from *ge* (the earth) and *metron* to measure. Moreover, mathematics is derived from the ancient Greek word for *knowledge*. From arithmetic we obtained the number systems, $\mathbf{N}$, $\mathbf{Z}$ and $\mathbf{Q}$ and from geometry the real numbers $\mathbf{R}$. We suppose that *the real numbers are in one-to-one correspondence with the points on the number line.* The ordering of $\mathbf{Q}$ introduced in the previous section allows us to display $\mathbf{Q}$ and $\mathbf{R}$ on the same number line. We may now compare the different sets of numbers we have introduced. The ancient Greeks' intrinsic belief in the unity of mathematics led them to suppose that all real numbers were rational or, equivalently, that the number line (see Figure 5.1)

---

[15] Let $A = \{r \in \mathbf{N} : 2^r \text{ divides } p\}$. The set A is non-empty since p is even and it is bounded above since $2^n > p$ for all $n > p$. Hence A has a largest element or a least upper bound s which belongs to A. Then 2^s divides p but 2^{s+1} does not.

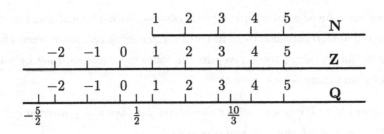

Fig. 5.1

consisted entirely of rational numbers. As sets we have

$$\mathbf{N} \subset \mathbf{Z} \subset \mathbf{Q} \subset \mathbf{R}.$$

The only outstanding question is whether or not $\mathbf{Q} = \mathbf{R}$? In Figure 5.2

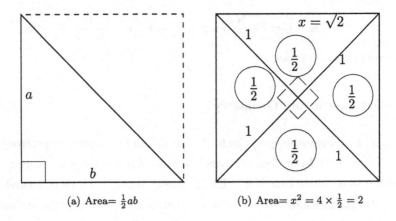

(a) Area= $\frac{1}{2}ab$ (b) Area= $x^2 = 4 \times \frac{1}{2} = 2$

Fig. 5.2

we consider two lines of length 2 which bisect one another at right angles. Now, using only the elementary fact that the area of a right-angled triangle is one half the base times the perpendicular height, we can construct a square with side length x and area 2. This is a geometric proof of the fact that the equation

$$x^2 = 2 \tag{5.13}$$

has a real number as a solution. We write the solution as $\sqrt{2}$. Is $\sqrt{2}$ a rational number? It is generally accepted that the answer was given in the 6th

century BC by the Pythagorean brotherhood who were influential in the development of an axiomatic approach to mathematics. Hippocrates of Chios was a Pythagorean while Plato of Athens, 427-347 BC, who played a pivotal role in the intellectual development of ancient Greece, was strongly influenced by the Pythagoreans.[16] Plato was a friend and follower of Socrates, 470-399 BC, who was the first to use rules and logic to discuss important matters and it is through the writings of Plato that we know the educational ideas of Socrates. Plato founded his academy of higher learning around 387 BC and it survived and flourished for almost 1000 years, until 529 AD. Aristotle of Stagira, 384-322 BC, the teacher of Alexander the Great and the supreme authority on logic and science until the modern era, was a student at Plato's academy for twenty years. Plato's educational approach was based on a thorough study of mathematics and the Socratic method of question and answer. Aristotle, Plato and Socrates were not mathematicians but all three appreciated and promoted mathematics.

Theorem 5.9. $\sqrt{2}$ *is not rational.*

Proof. (By contradiction). Suppose $\sqrt{2}$ is rational. Then $\sqrt{2} = \frac{p}{q}$ and we may suppose by Theorem 5.8 that either p or q is odd. Then $2 = \frac{p^2}{q^2}$ and $p^2 = 2q^2$. Hence p^2 is even and, by Theorem 5.1, p is even. Let $p = 2s$ where $s \in \mathbf{N}$. Then

$$p^2 = (2s)^2 = 4s^2 = 2q^2$$

and $2s^2 = q^2$. Hence q^2 is even and, by Theorem 5.1, q is even. We have shown that both p and q are even. This contradicts the fact that one of them is odd. We have arrived at a contradiction and thus our only assumption–that $\sqrt{2}$ is rational– is false. This completes the proof. $\square$

[16]Pythagoras of Samos, c.576-500 BC, was well educated and well travelled. He studied mathematics in Samos before spending ten years studying religion and geometry in Egypt and afterwards visited Babylon, some say as a prisoner of war, and India. His distaste for the political arrangements in place in Samos on his return prompted him to move permanently to Croton in Italy. There he founded a society devoted to mathematics, music and mysticism. This secretive society, which survived for a couple of hundred years, believed that reality was mathematical in nature and that mathematical secrets would uncover the meaning of life.

Corollary 5.10.

$$\mathbf{Q} \neq \mathbf{R}.$$

The existence of two different kinds of numbers was, by all accounts, devastating for the ancient Greeks. Even though it was not apparent at the time, the key to reconciling the two number systems involved the *infinite*. This was only recognized two thousand years later and, in the meantime, certain difficulties required attention. The Greeks had previously considered *ratios* of natural numbers, that is the relationship of one number to another, and wrote $n : m = p : q$ for natural numbers $\{n, m, p, q\}$ if $nq = mp$. They assumed that this included the ratio of any two real numbers. Eudoxus[17] solved a major problem by showing how to compare different ratios of real numbers using only natural numbers, without assuming that these numbers were rational. He considered positive real numbers $\{a, b, c, d\}$ and said that[18]

$$a : b = c : d$$

if and only if the following three conditions hold:

$$\text{for } n, m \in \mathbf{N}, n \cdot a = m \cdot b \text{ if and only if } n \cdot c = m \cdot d, \qquad (5.14)$$

$$\text{for } n, m \in \mathbf{N}, n \cdot a > m \cdot b \text{ if and only if } n \cdot c > m \cdot d, \qquad (5.15)$$

$$\text{for } n, m \in \mathbf{N}, n \cdot a < m \cdot b \text{ if and only if } n \cdot c < m \cdot d. \qquad (5.16)$$

This is called the *Postulate of Eudoxus*. It allowed Eudoxus and Archimedes to calculate rigorously, that is with proof, the values of certain areas and volumes, avoiding the use of infinity, and to present their results as ratios.

---

[17]Eudoxus of Cnidus, c.408-355 BC, was the inspiration for Book V of *Euclid's Elements* and is regarded, after Archimedes, as the greatest of the classical Greek mathematicians. His method of comparing ratios was the nearest the ancient Greeks came to defining real numbers. He attended lectures on philosophy by Plato in Athens. In Italy, he studied mathematics under Archytas, a follower of Pythagoras. In Sicily, he studied medicine, and, in Egypt, he studied astronomy. He was an astronomer, a physician, a legislator and he founded a school in Cyzicus in Asia Minor. To him we owe the method of exhaustion which eventually led to modern *integration theory*.

[18]In ancient Greece all numbers were positive. At the moment we call any number that arises from a geometrical or mechanical construction a real number.

It is remarkable that this approach is so close to what we do nowadays. We have $a : b = c : d$ if and only if $(a, b) \sim (c, d)$, where $\sim$ is the equivalence given in Definition 5.4 and, in treating all pairs with the same ratio as equal, the ancient Greeks were considering equivalence classes. Moreover, (5.14), (5.15) and (5.16) state that any positive real number is fully determined by the sets of rational numbers to its left and to its right. This is the basic idea behind different constructions of the real numbers (see Chapter 12).

One other consequence of this duality of numbers is almost forgotten[19] nowadays. Multiplication of real numbers was considered, until the 17th century, to be somehow different to multiplication of natural numbers and required a geometrical interpretation.[20] Before the work of Descartes the expression

$$ax^2 + bx + c$$

required some qualification, for instance that x and a are *lengths*, b is an *area* and c is a *volume*. In this way the expression has a geometrical meaning and defines a real number. This was very cumbersome. To Descartes, the origin

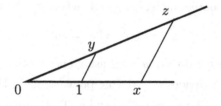

Fig. 5.3

of real numbers, whether as lengths, areas or volumes, was immaterial and he introduced[21] a rigorous way of multiplying *any* two real numbers. Given positive real numbers x and y, consider Figure 5.3 in which we have marked off lengths 1, x and y on identically scaled non-perpendicular lines. The point z is found by drawing a line through x parallel to the line segment

---

[19]One reminder is our use of the terminology x-squared for x^2 and x-cubed for x^3.

[20]How does one define $\sqrt{2} \times \sqrt{3}$? See Theorem 7.20(b).

[21]In view of Figure 5.3 it is not surprising that this innovation is due to one of the founders of coordinate geometry.

$[1, y]$. Using similar triangles we see that

$$z : y = x : 1$$

and hence $z = xy$.

We now consider a result similar to that given in Theorem 5.9 and show that $\sqrt{3}$ is irrational. Our aim, in this case, is to give an example of how one proof may be modified in order to prove a similar result. Our initial strategy is to imitate the proof of Theorem 5.9 until we are forced to do otherwise. That is we write down the proof of Theorem 5.9 with 3 in place of 2 and see if we accept the resulting new statements that appear.

Theorem 5.11. $\sqrt{3}$ *is not rational.*

Proof.   (By contradiction). Suppose $\sqrt{3}$ is rational. Then $\sqrt{3} = \frac{p}{q}$ where p and q are positive integers and either p or q is odd. Then $3 = \frac{p^2}{q^2}$ and hence $p^2 = 3q^2$. The next step in the proof for $\sqrt{2}$ is to conclude that p^2 is even. We cannot draw this conclusion here. What we can say is that 3 divides p^2 and, imitating the proof of Theorem 5.9, we should then conclude that 3 divides p. Continuing we let $p = 3s$ where $s \in \mathbf{N}$ and obtain

$$p^2 = (3s)^2 = 9s^2 = 3q^2$$

and $3s^2 = q^2$. Hence 3 divides q^2 and using once more the same argument we conclude that 3 divides q. At this point, modulo the steps we have assumed, we see that 3 divides both p and q. This would be a contradiction if we can call on a modified version of Theorem 5.8. We need the following results:

(a) *If p is a positive integer and 3 divides p^2 then 3 divides p,*
(b) *If $r \in \mathbf{Q}^+$ then $r = \frac{p}{q}$ where $p, q \in \mathbf{N}$ and 3 does not divide both p and q.*

To prove both of these it is necessary to modify the corresponding earlier results, Theorems 5.1 and 5.8.

Proof of (a). It suffices to show the following: if 3 does not divide p then 3 does not divide p^2. If 3 does not divide p then $p = 3u + v$ where $u \in \mathbf{Z}$ and

$v = 1$ or 2. Hence $p^2 = 9u^2 + 6uv + v^2$ and $v^2 = 1$ or 4. Since 3 divides $9u^2 + 6uv$ this shows that 3 does not divide p^2 and proves (a).

Proof of (b). If 3 does not divide both p and q the proof is complete. Otherwise 3 divides both p and q. Let 3^s denote the highest power of 3 that divides p to give an integer.[22] Hence $p = 3^s t$ where 3 does not divide t. Similarly, $q = 3^u \cdot v$ where u is a positive integer and 3 does not divide v. We then have: $r = \frac{t}{v}$ if $s = u$, $r = \frac{3^{s-u}t}{v}$ if $s > u$ and $r = \frac{t}{3^{u-s}v}$ if $s < u$. This proves (b).

The final step in the proof consists in a careful writing up of the result using (a) and (b) above. We leave this as an exercise. A completely different proof is outlined in Exercise 5.11 □

The above proof suggests further possible results, for instance from (a) above we might consider the following statement **P**:

If n is a positive integer and 4 divides n^2 then 4 divides n.

The obvious approach of imitating the proof of (b) above is not satisfactory. Indeed, this is not a correct result and one way to show this is by giving an example. If we let $n = 2$ then 4 divides 2^2 but 4 does not divide 2. We call this example a *counter-example* to statement **P**. Counter-examples are an effective technique in developing an understanding of mathematical processes. In many cases, they are simple to construct. If you have any doubts about a result you wish to prove or use, try a few examples, as we did in Chapter 1. These may suggest how to construct a counter-example or may lead to a proof.

Our final result, known as the Postulate of Archimedes, says that the set of multiples of a positive rational number is not bounded above. We use it in Chapter 12 in constructing the real numbers.

Theorem 5.12. *If r and s are positive rational numbers then there exists a positive integer n such that $nr > s$.*

---

[22] Let $A = \{r \in \mathbf{N} : 3^r$ divides $p\}$. The set A is non-empty since 3 divides p and it is bounded above since $n \notin A$ for $n > p$. Hence A has a largest element, s say. Then 3^s divides p and 3^{s+1} does not.

Proof. Let $r = \frac{p}{q}$ and $s = \frac{n}{m}$ where p, q, m and n are positive integers. The set $A := \{u \in \mathbf{N} : upm \leq nq\} = \{u \in \mathbf{N} : ur \leq s\}$ is bounded above by nq and hence has a least upper bound n_0. Then $(n_0 + 1)pm > nq$ and hence $(n_0 + 1) \cdot \frac{p}{q} = (n_0 + 1)r > s = \frac{n}{m}$. $\qquad\qquad\Box$

We have constructed the integers and the rational numbers using the natural numbers. Dedekind[23] showed, in 1888, how to construct the natural numbers using set theory. A year later Peano[24] gave an axiomatic presentation of Dedekind's construction using only *formal logic* and what we call today the *Peano Axioms* for $\mathbf{N}$.

In Chapter 12 we construct the real numbers using only properties of rational numbers and Theorem 5.12. Until then we rely on an intuitive notion of real number and assume that the real numbers obey the associative, commutative, etc. laws and the order relations that we have already verified for $\mathbf{Q}$ and, most important, we suppose that the real numbers satisfy an *Upper Bound Principle* (see Theorems 6.7 and 12.8).

---

[23]Richard Dedekind (1831-1916) from Brunswick (Germany) was the last research student of Gauss at Göttingen. He also studied at Berlin and, held teaching positions in Göttingen, Zürich and Brunswick. He introduced what we now call Dedekind cuts and provided a logical foundation for the real numbers. He made fundamental contributions to *set theory*, *number theory* and *abstract algebra*. Dedekind's style of presentation and his clarity of expression have been, and still are, a source of inspiration to many mathematicians.

[24]Giuseppe Peano, 1858-1932, from Cuneo in Italy is regarded as the greatest Italian mathematician of his generation and, in addition to his work on foundational questions, he is known for his many examples and counterexamples including one of a space filling curve. He was interesting in teaching and, motivated partly by the many errors he found in contemporary textbooks, he set out to see if he could, by reducing mathematics to the basic ingredients of axioms and logic, make the subject absolutely transparent to his students. This led to the important, but complicated, notion of a symbolic language devoid of content and the concept of *formal proof* that we discuss in Chapter 12. However, the courses that he himself gave to engineering students following this fundamentalist teaching philosophy were a disaster and we too have to be careful that we do not end up in a similar situation.

5.4   Exercises

(5.1) Show that the number of primes is infinite.

(5.2) Show that $a - b$ divides $a^n - b^n$ for any positive integer n.

(5.3) If $n = a_m a_{m-1} \cdots a_0$ is the usual expansion of the positive integer n (that is to the base 10) show that $9|n$ if and only if 9 divides $a_1 + a_2 + \cdots + a_m$.

(5.4) Prove that $\sqrt{2} + \sqrt{4}$ and $\sqrt{2} + \sqrt{4} + \sqrt{8}$ are irrational.

(5.5) For a, b in $\mathbf{N}$ let $A := \{an + bm : n, m \in \mathbf{Z}\}$ and let $A^+ = A \cap \mathbf{N}$. Show that A^+ is non-empty. If c is the smallest element in A^+ show that A^+ consists of all positive multiples of c and that c divides a and b. If $d \in \mathbf{N}$ divides a and b show that $d \mid c$. The integer c is called the *greatest common divisor* of a and b and is written (a, b). If $a = (a, b)r$ and $b = (a, b)s$ show that $(r, s) = 1$. If p is prime and $p|ab$ show that either $p|a$ or $p|b$.

(5.6) Is $\sqrt{2} + \sqrt{n}$ rational for any positive integer n? Must the sum or product of irrational numbers be irrational? Justify your answers.

(5.7) Let $x \in \mathbf{R}$. If x^2 is irrational show that x is irrational.

(5.8) Let $P(x) = x^2 + x + 41$. Compute $P(1), P(2), \ldots, P(39)$. Comment.

(5.9) Show that every positive integer $n, n \geq 2$, can be written in a unique way as a product of primes[25], that is if $(p_m)_{m=1}^{\infty}$ are the primes written in increasing order with $p_1 = 2$ then there exists for each $n \in \mathbf{N}$ a unique set of non-negative integers $(n_i)_{i=1}^{\infty}$, with all except a finite number zero, such that $n = p_1^{n_1} p_2^{n_2} \cdots p_i^{n_i} \cdots$ and where $p_i^0 = 1$ for all i. This is known as the *Fundamental Theorem of Arithmetic*.

(5.10) Use Exercise 5.5 to show that any non-zero rational number can be

---

[25]Cryptography deals with the design of systems for the safe transfer of information. The most successful cryptosystems, including the public-key RSA system which is the basis for security on the internet, are based on the fact that it is much more difficult to factor positive integers than to find and multiply primes. The ancient Greeks, Fermat, Euler, Dirichlet and Gauss made important contributions to number theory and it is today an active area of research. While many major results in number theory can be presented using the positive integers it is of interest to note that irrational, complex and imaginary numbers often play a role in the proofs.

written as n/m where n and m are non-zero integers with $(n, m) = 1$. If $(n, m) = 1$ and $n/m = r/s$ for positive integers r and s, $1 \leq s \leq m$, show that $s = m$.

(5.11) Let m denote a positive integer which is not the square of any positive integer. Let n denote the positive integer such that $n < \sqrt{m} < n+1$. If $\sqrt{m} = \frac{r}{s}$, r and s positive integers, show that $1 \leq r - ns < s$. Use the identity,

$$(ms - nr)^2 - m(r - ns)^2 = (n^2 - m)(r^2 - ms^2)$$

to show that $\sqrt{m} = (ms - nr)(r - ns)^{-1}$. Use this result and the previous exercise to show that $\sqrt{m}$ is not rational.

(5.12) If $\sim$ denotes the equivalence relationship on $\mathbf{Q}^+$ used to define $\mathbf{Q}$ show and interpret, in the usual notation, $[1, 2]^2 = [2, 1]$. Show that $x^2 \geq 0$ for any $x \in \mathbf{Q}$.

(5.13) Is $n(n^2 - 1)(n^2 + 1)$ always a multiple of 10? Justify your answer.

(5.14) If $x \in \mathbf{N}$ and $\sqrt{x} \in \mathbf{Q}^+$ show that $\sqrt{x} \in \mathbf{N}$.

(5.15) If a, b, c and d are positive rational numbers and $\frac{a}{b} < \frac{c}{d}$ show, using the equivalence class definition of rational number, that

$$\frac{a}{b} < \frac{a+c}{b+d} < \frac{c}{d}.$$

(5.16) (The Binomial Theorem) For any positive integer n let $n! = 1 \cdot 2 \cdots n$ and let $0! = 1$. If x and y are real numbers and $n \in \mathbf{N}$ show that

$$(x + y)^n = \sum_{j=0}^{n} \frac{n!}{j!(n-j)!} x^{n-j} y^j.$$

(5.17) Use equivalence classes to prove the cancelation law for addition in $\mathbf{Q}^+$.

(5.18) Let $a \in \mathbf{A}$ and let $\phi : \mathbf{A} \longrightarrow \mathbf{Q}, \phi(x) = x + a$ and $\varphi : \mathbf{A} \longrightarrow \mathbf{Q}, \varphi(x) = x \cdot a$ where $\mathbf{A}$ is one of the sets $\mathbf{N}, \mathbf{Z}, \mathbf{Q}^+$ and $\mathbf{Q}$. Find for which a and $\mathbf{A}$ the mappings ϕ and φ are injective, surjective, bijective? When ϕ or φ is bijective find its inverse.

(5.19) If $r \in \mathbf{Q}$ and $s \in \mathbf{Q}^+$ and $nr \leq s$ for all $n \in \mathbf{N}$ show that $r \leq 0$.

References [1; 4; 7; 8; 9; 11; 18; 20; 21; 22; 27; 29; 30; 31; 33; 37]

Chapter 6

SETS OF REAL NUMBERS

If a property M does not apply to all values of a
variable quantity x, but to all those which are
smaller than a certain u: so there is always
a quantity U which is the largest of those of
which it can be asserted that all smaller
x possess the property M.

Bernard Bolzano, 1817

Summary

We consider properties of sets relating to size: small (finite), medium
(countable), large (uncountable); and properties of dispersion such as
boundedness or being an interval.

6.1   Countable Sets

The infinitely large was recognized but, following Zeno's criticisms, not
accepted by the ancient Greeks as a legitimate mathematical concept. This
had to wait until the investigations of Cantor and Dedekind in the second
half of the 19th century. Instead infinite sets were said to be larger than any
finite set while Archimedes said that it was permissible to use the infinite
to find a result if its validity could be justified later using proofs of an

acceptable standard, including the postulate of Eudoxus.

Archimedes[1] mentioned in a letter to Eratosthenes, 275-195 BC, a technique, now known as *The Method*, which he had used frequently and effectively. It was apparently lost forever. Remarkably it came to light in 1906 when the significance of a 10th century recycled parchment or *palimpsest*, which had been commuting between Greek Orthodox libraries in Jerusalem and Constantinople (now Istanbul) for centuries, was recognized and partially deciphered by the Danish classical scholar Johan Ludvig Heiberg, 1854-1928. The parchment disappeared soon afterwards and then reappeared in very poor condition at an auction in 1998. It was purchased anonymously and deposited in *The Walters Art Museum* in Baltimore (USA) for study, interpretation and perhaps restoration. Using photographs taken by Heiberg in 1906, multi-spectral imaging, x-ray florescence imaging and electro-magnetic radiation, scientists have been able in the last few years to read 80% of the writings due to Archimedes in the parchment. The restored work showed that Archimedes used mechanics, particularly the law of the lever, and infinitesimals to find results that were then subjected to a geometric proof.

Archimedes while finding the area within a parabola showed that

$$1 + \frac{1}{4} + \frac{1}{4^2} + \cdots + \frac{1}{4^n} + \frac{1}{3} \cdot \frac{1}{4^n} = \frac{4}{3} \tag{6.1}$$

for any positive integer n. Today, we recognize that (6.1) essentially sums

---

[1] Archimedes, c.287-212 BC, was from Syracuse in Sicily. He studied in Alexandria for a considerable period with the successors of Euclid and made friends there with whom he maintained a lifelong scientific correspondence. On returning to Syracuse he devoted the rest of his long life to mathematical research. He was an excellent engineer and used practical intuitions to sharpen his mathematics and to help in civil and military matters. The stories about him are legendary. According to Plutarch, Archimedes once told King Hieron that any weight could be moved by a given force. Hieron, amazed, asked for an illustration of a small force moving a great weight. Archimedes loaded a ship, which all the Syracusans had been unable to launch, with many passengers and full freight and while far off and with no great endeavor but only holding the end of a compound pulley drew the ship as smoothly and safely as if it was moving through the sea. Hieron declared afterwards: *from this day forth Archimedes was to be believed in everything that he might say.*

an *infinite series*:

$$1 + \frac{1}{4} + \frac{1}{4^2} + \frac{1}{4^3} + \cdots + \frac{1}{4^n} + \cdots = \sum_{n=0}^{\infty} \frac{1}{4^n} = \frac{4}{3}.$$

Two thousand years later a similar role was played by the infinitely small in the development of the differential and integral calculus. The skillful and intuitive use of infinitely small numbers which are not zero but yet smaller than any positive numbers, the *infinitesimals*, led to important results and also to unease regarding their apparent lack of mathematical precision. The development of analysis in the 19th century was a response to this unease and to the criticism of the philosopher Bishop George Berkeley.[2]

In this chapter we take our first steps towards understanding the infinite.[3] We discuss properties of *infinite sets*, that is we consider the *infinitely large*. Unexpectedly, functions provide us with the language to clearly and precisely articulate the different concepts that we need. Clearly, the smallest sets are the finite sets. A set A is finite if there exists a positive integer n such that the points in A are $x_1, x_2, \ldots, x_n$, that is

$$A = \{x_1, x_2, \ldots, x_n\}.$$

From this we see that A is finite if and only if, for some positive integer n, there exists a surjective mapping $f : \{1, 2, \ldots, n\} \longrightarrow A$. We have been using the weakest possible definition of infinite set: *a set is infinite if it is not finite* and now introduce a more useful definition due to Richard Dedekind, although it could be said that this definition had, at least informally, been

---

[2] A satisfactory mathematical theory of *infinitesimals* was developed in the middle of the 20th century by Abraham Robinson, see Chapter 12.

[3] Galileo Galilei, 1564-1642, was the first to write on infinite sets in *Dialogs Concerning the New Sciences*. In 1636, when writing about positive integers and positive square roots, he said: *if I ask how many squares there are, one might reply truly that there are as many as the corresponding numbers of roots, since every square has its own root and every root has its own square, while no square has more than one root and no root has more than one square.* He concluded that the terms *less*, *greater* and *equal* did not apply to the infinite. Cantor, almost two and a half centuries later, applied a similar analysis but drew different conclusions and proved that an infinite set could have the same size as a proper subset of itself. The set A is smaller than the set B if there exists an injective function $f : A \longrightarrow B$. This is one of the first definitions in Cantor's theory of cardinal and ordinal numbers.

considered by Bolzano and Cantor. A subset B of a set A is *proper* if it is non-empty and $B \neq A$.

Definition 6.1. A set A is infinite if there exists a bijective function $\phi :$ $B \longrightarrow A$, where B is a proper subset of A.

If $B := \{n^2 : n \in \mathbf{N}\}$, then B is a proper subset of $\mathbf{N}$ and $\phi : B \longrightarrow \mathbf{N}, \phi(n^2) = n$ is bijective and, thus, the set of natural numbers is infinite. The importance of the next two definitions, of a *countable set* and a *sequence*, stems from the fact that $\mathbf{N}$ is both *infinite* and *countable*, and hence the most manageable of all sets which are not finite.

Definition 6.2. A set A is countable if there exists a surjective function from $\mathbf{N}$ to A.

Because we do not require an injective function, finite sets are countable. If $A = \{x_1, x_2, \ldots, x_n\}$ and $f : \mathbf{N} \longrightarrow A$ is given by $f(i) = x_i$ for $i \leq n$ and $f(i) = x_n$ for $i > n$, then f is surjective and A is countable. Thus A is countable if there exists a procedure that counts all elements in A. It may require an *infinite* number of steps to count the whole set but any *fixed* element in A will be counted after a *finite* number of steps. Another interpretation uses the familiar concept of a *queue*. A set is countable if and only if *all* its elements can be arranged to form a queue. Everyone in a queue, no matter where in the queue, is only behind a finite number of people even if the queue is infinite. In some queues, numbers are assigned to each person in the queue. This may be regarded, for infinite queues, as the function in Theorem 6.3(c). The integers $\mathbf{Z}$ form a countable set since

$$\mathbf{Z} = 0, 1, -1, 2, -2, \ldots.$$

Our next result clarifies the relationship between infinite and countable sets.

Theorem 6.3. *Consider the following properties of the set A:*

  (a) there exists a surjective function from $\mathbf{N}$ onto A,
  (b) there exists an injective function from A into $\mathbf{N}$,

(c) there exists a bijective function from **N** *onto A,*

(d) there exists a bijective function from A onto **N**.

Then (a) *and* (b) *are equivalent and hold if and only if A is countable. Conditions* (c) *and* (d) *are equivalent and hold if and only if A is countable and infinite.*

Proof. (a) implies (b). Let $f : \mathbf{N} \longrightarrow A$ be surjective. For each n let $a_n = f(n)$. Let $g(a_1) = 1$. If $f(2) = f(1)$ then $a_2 = a_1$ and $g(a_2)$ has already been defined, otherwise $f(2) \neq f(1)$ and we let $g(a_2) = 2$. If $a_3 = f(3) \notin \{a_1, a_2\}$ let $g(a_3) = 3$ and so on. In this way we define, since f is surjective, an injective mapping $g : A \longrightarrow \mathbf{N}$. Hence (b) holds.

Suppose (b) is satisfied. Let $g : A \longrightarrow \mathbf{N}$ denote an injective function. Then $h : A \longrightarrow im(g)$, where $h(x) = g(x)$ for all $x \in A$, is bijective. Choose an element from A and denote it by a. Let $f : \mathbf{N} \longrightarrow A$ be defined as follows: if $n \in im(g)$ let $f(n) = h^{-1}(n)$ and if $n \notin im(g)$ let $f(n) = a$. Then $im(f) = im(h^{-1}) = A$ and $f : \mathbf{N} \longrightarrow A$ is surjective. Hence, (a) holds and (a) and (b) are equivalent. By Definition 6.2, (a) holds if and only if A is countable.

Clearly, (c) and (d) are equivalent and imply (a) and (b). If A is countable and infinite, then $A = \{a_n\}_{n=1}^{\infty}$ where $a_n \neq a_m$ whenever $n \neq m$. If $f : \mathbf{N} \longrightarrow A$ is given by $f(n) = a_n$ then f is bijective and (c) holds. Conversely, if $f : \mathbf{N} \longrightarrow A$ is bijective then A is countable by (a). Moreover, the set $B := \{f(n)\}_{n=2}^{\infty}$ is a proper subset of A and the mapping $\varphi : B \longrightarrow A, \varphi(f(n)) = f(n-1)$ for all $n \geq 2$ is surjective. Hence, A is infinite. This completes the proof. $\square$

Example 6.4. Let $f : \mathbf{N}^2 := \mathbf{N} \times \mathbf{N} \longrightarrow \mathbf{N}$ be given by $f(n, m) = 2^n 3^m$. Suppose $f(n, m) = f(p, q)$. If $n > p$ then $2^{n-p} 3^m = 3^q$ and as $2^{n-p} 3^m$ is even and 3^q is odd this is impossible. Hence $n \leq p$. The same argument shows that $p \leq n$ and thus $p = n$. This implies $3^m = 3^q$. If $m > q$ then $3^m > 3^q$ and, if $m < q$, then $3^m < 3^q$. Both of these are untrue. Hence $m = q$ and f is injective. By Theorem 6.3, $\mathbf{N}^2$ is countable and there exists a surjective function $g : \mathbf{N} \longrightarrow \mathbf{N}^2$. Let $h : \mathbf{N}^2 \longrightarrow \mathbf{Q}^+$ (the positive rational numbers) where $h(n, m) = n/m$. By Theorem 4.9(b), the composition of surjective functions is surjective and hence $h \circ g : \mathbf{N} \longrightarrow \mathbf{Q}^+$ is surjective. Theorem 6.3 shows that $\mathbf{Q}^+$ is countable and from this it is easily seen that $\mathbf{Q}$ is countable (see Exercise 6.10).

The functions considered in the proof of Theorem 6.3 are rather special. We will see that the more general notion in Definition 6.5 of a possibly non-surjective mapping with domain $\mathbf{N}$ is very adaptable.

Definition 6.5. A function with domain $\mathbf{N}$ is called a sequence.

Clearly the image of a sequence is a countable set. It is customary to let n denote the variable for functions with domain $\mathbf{N}$ and when we take a fixed point in $\mathbf{N}$ we often denote it by n_0 or n_1. It is traditional to write x_n in place of $f(n)$ when dealing with $f : \mathbf{N} \longrightarrow A$ and we may thus write, without ambiguity, the sequence as either $(x_n)_{n=1}^{\infty}$ or as $x_1, x_2, \ldots$.

6.2   Uncountable Sets

In the previous section we looked at countable sets and saw that the set of natural numbers $\mathbf{N}$ is countable and infinite. If a set is not countable we say it is *uncountable*. The main result in this section is that $\mathbf{R}$ is uncountable. This result was first proved in 1874 by Georg Cantor and ushered in a new era in mathematics. Cantor[4] went on to show how one could distinguish between the sizes of different infinite sets. Later he used equivalence classes of Cauchy sequences to construct the real numbers. The

---

[4]Georg Ferdinand Ludwig Phillipp Cantor was born in St. Petersburg in 1845, but moved to Germany in 1856 and lived there until his death in 1918. He studied in Zurich and Berlin, and was a professor of mathematics at Halle (Germany) from 1869 until he retired in 1913. Cantor is regarded as the founder of *set theory*, and his discovery and development of transfinite and ordinal numbers had profound implications for mathematics, logic and philosophy. His two main articles have been translated and published in book form, *Contributions to the Founding of the Theory of Transfinite Numbers*, by Dover Publications. This book contains an excellent introduction by Philip E.B. Jourdain and includes both an overview of Cantor's contributions and a survey of the conceptual continuity of mathematics during the 19th century.

distinction between the countable and the uncountable proved crucial for the development of measure theory by Lebesgue at the beginning of the 20th century and this in turn led to an axiomatic basis for probability theory. However, the theory put forth by Cantor were initially met with scepticism and, in some cases, outright hostility as it was so different to the accepted wisdom. It was generally agreed, at the time, that *infinity* was just a figure of speech and there was much philosophical speculation on the *potential* and the *actual* infinite. Cantor, on the other hand made it concrete, showed how to distinguish between different infinities, and even developed an arithmetic for infinite numbers. The work of Cantor had its proponents and eventually his results were accepted. Since then they have led to many extremely important developments in mathematical logic and in the foundations of mathematics. It is of interest to note that Cantor was researching Fourier series problems when he became interested in set theory.

Our next result shows that the real numbers are not countable. This means that, no matter how we attempt to count the real numbers, we will always miss some of them, and it is impossible to form a queue containing *all the real numbers.* Since the rational numbers are countable, we may also say, and this can be made mathematically precise, that most real numbers are not rational.

We recall that any real number a can be written in a unique way as $b + c$ where b is an integer and $0 \le c < 1$. The number b is called the integer part of a and c has a *decimal expansion* $c = .c_1 c_2 c_3 c_4 \cdots$ where each c_i is an integer, $0 \le c_i \le 9$. The decimal expansion is not necessarily unique.[5]

Theorem 6.6. *The real numbers* $\mathbf{R}$ *are uncountable.*

Proof. If $\mathbf{R}$ is countable then $\mathbf{R} = (x_n)_{n=1}^{\infty}$ and each real number occurs somewhere in this sequence. For each n let $x_n = y_n + z_n$ where y_n is the integer part of x_n and $z_n = .z_n^1 z_n^2 z_n^3 \cdots$. If z_n has two expansions, we use

---

[5] A real number has more than one decimal expansion if and only if it has an expansion in which all entries are 9 after a certain point. The proof uses geometric series that we discuss in Chapter 7.

the one containing all 9's after a certain point. The method of finding a real number which does not belong to the sequence $(x_n)_{n=1}^{\infty}$ is called a *diagonal process*, a name suggested by the following display:

$$z_1 = .\underline{z_1^1} z_1^2 z_1^3 z_1^4 \cdots$$

$$z_2 = .z_2^1 \underline{z_2^2} z_2^3 z_2^4 \cdots$$

$$z_3 = .z_3^1 z_3^2 \underline{z_3^3} z_3^4 \cdots$$

$$. = .\cdots\cdots\cdots$$

$$z_n = .z_n^1 z_n^2 z_n^3 z_n^4 \cdots \underline{z_n^n} \cdots$$

$$. = .\cdots\cdots\cdots\cdots\cdots\cdots .$$

Let $w_n = 6$ if $z_n^n \leq 5$ and let $w_n = 2$ if $z_n^n > 5$. We let $w = .w_1 w_2 w_3 \cdots$. By our hypothesis, there exists a positive integer n_0 such that $w = x_{n_0} = z_{n_0}$. The decimal expansion of w does not contain either a nine or a zero. Hence w and x_{n_0} have unique decimal expansions. This means, in particular, that $w_{n_0} = z_{n_0}^{n_0}$. However, by our construction, $w_n \neq z_n^n$ for all n. We have arrived at a contradiction, and thus our original assumption that **R** is countable is false. Hence **R** is uncountable. This completes the proof. $\Box$

6.3   Sets Which Are Bounded Above

Bernard Bolzano was probably the first modern mathematician to make significant contributions to the foundations of real analysis. He appreciated the need to provide proofs of quality, similar to those given two thousand years earlier by the classical Greek mathematicians and, in this, he anticipated Dedekind, Cantor and Weierstrass. In 1817, he gave an analytic, as opposed to the usual geometric, proof of the Intermediate Value Theorem and while doing so investigated convergence of series, introduced *upper bounds* and stated the *Upper Bound Principle*. Bolzano published his results in pamphlet form in Prague, far from the center of mathematical activity in Paris. His results were not widely known during his own lifetime and the extent of his influence is unclear. To quote Grattan-Guinness, *his ideas were so far ahead of their times that nobody could properly understand their purpose or develop them further.*[6]

A set of real numbers, A, is bounded above if there exists a real number M such that $x \leq M$ for all x in A. We call M an *upper bound* for A. An upper bound which is less than all other upper bounds is called a *least upper bound*. Hence U is a least upper bound for the set A if and only if it satisfies the following two conditions:

$$x \leq U \text{ for all } x \in A,$$

$$\text{if } x \leq M \text{ for all } x \in A \text{ then } U \leq M.$$

We denote the least upper bound of the set A by $lub(A)$. If m_1 and m_2 are two least upper bounds for the set A, then $m_1 \leq m_2$ and $m_2 \leq m_1$ since least upper bounds are less than or equal to any upper bound. Hence $m_1 = m_1$ and least upper bounds are unique.

If $x \geq m$ for all x in A, then m is called a *lower bound* for A and A is said to be bounded below. A *lower bound* which is greater than all other lower bounds for A is called the *greatest lower bound* of the set A and is denoted by $glb(A)$. If a set A does not have an upper bound (respectively a lower bound) we let $lub(A) = +\infty$ (respectively $glb(A) = -\infty$). A set A has a lower bound and an upper bound if and only if $A \subset [a, b]$ for real numbers a and b. When this happens, we say that A is a *bounded* subset of **R**. We will prove the following theorem in Chapter 12 when we construct

---

[6]Bernard Placidus Johann Nepomuk Bolzano, 1781-1848, from Prague was a teacher, philosopher, logician, political activist, writer, and a Roman Catholic priest. He was a student and later a professor at Charles University in Prague. His strong, publicly-aired views supporting human rights, socialism, pacifism, equality for Czech speaking Bohemians and his free thinking intellectual approach were a thorn in the side of the political, religious, and academic establishments and he was, at various times, suspended from his professorship, put under house arrest, tried for heresy, and banned from publishing. On the other hand, he must have enjoyed scholarly support as he was elected Dean of the Faculty of Philosophy at Charles University and president of the Royal Bohemian Society of Sciences. His mathematical interests were broad and deep. Apart from his original ideas on the foundations of mathematics, he wrote substantial volumes on topics such as the *Theory of Science*, the *Theory of Scientific Discovery*, the *Philosophy of Logic* and the *Methodology of Writing Textbooks*. Moreover, he left many unpublished manuscripts, one of which contained the first example of a continuous nowhere-differentiable function (see Exercise 10.12).

the real numbers.

Theorem 6.7. (*Upper Bound Principle*). *A set of real numbers which is bounded above has a least upper bound.*

Example 6.8. Let $a \in \mathbf{R}^+$. If the set $A := \{na\}_{n=1}^{\infty}$ is bounded above it has a least upper bound m. Hence $na \leq m$ for all positive integers n. This implies $(n+1)a \leq m$ and hence $na \leq m - a < m$. This contradicts the fact that m is the least upper bound for A. Our only assumption was that A had an upper bound. This is false and we conclude that the set $\{na\}_{n=1}^{\infty}$ is not bounded above. If $b \in \mathbf{R}$ then, since b is not an upper bound for A, we can find a positive integer n such that $na > b$. Thus the Upper Bound Principle implies the Postulate of Archimedes (Theorem 5.12) for real numbers (see also Exercise 12.3).

Example 6.9. The set $A := \{n^2\}_{n=1}^{\infty}$ is not bounded above.

First Proof. Suppose A is bounded above by m. Then $n^2 \leq m$ for all n. Hence, $n \leq \sqrt{m}$ for every positive integer n and $\mathbf{N}$ is bounded above. This contradicts the result in Example 6.8. Hence, our assumption is false and A is not bounded above.

Second Proof. We again prove the result by contradiction. If A is bounded above, it has a least upper bound m. Then, $n^2 \leq m$ for every integer n. Hence, $n^2 + 2n + 1 = (n+1)^2 \leq m$ and $n^2 \leq n^2 + 2n \leq m - 1$ and $m - 1$ is an upper bound that is smaller than the least upper bound. This contradiction shows that A is not bounded above.

Example 6.10. This deceptively simple example is frequently used. Let a denote a real number and let $A = \{x \in \mathbf{R} : x < a\}$. Clearly a is an upper bound for A. We claim that $a = lub(A)$. Suppose otherwise. Then $b := lub(A) < a$. This is impossible since $b < \frac{a+b}{2} < a$ and $\frac{a+b}{2} \in A$. We can derive a further useful consequence of this result: if $x < b$ whenever $x < a$ then $a \leq b$. This holds since b is an upper bound for $\{x : x < a\}$ and, hence, is greater than or equal to the least upper bound a.

In the proofs of the following two theorems we suppose, for convenience, that the least upper bounds and the greatest lower bounds are all finite.

Theorem 6.11. *If A and B are non-empty subsets of $\mathbf{R}$ and $A + B = \{x + y : x \in A, y \in B\}$, then*

$$lub(A + B) = lub(A) + lub(B)$$

and

$$glb(A + B) = glb(A) + glb(B).$$

Proof. If $x \in A$, then $x \leq m_A := lub(A)$ and if $y \in B$, then $y \leq m_B := lub(B)$. Hence, $x + y \leq m_A + m_B$ and $m_A + m_B$ is an upper bound for $A + B$. Then $A + B$ is bounded above and

$$m := lub(A + B) \leq lub(A) + lub(B). \tag{6.2}$$

If $2c := m_A + m_B - m > 0$, choose $x \in A$ such that $x > m_A - c$ and $y \in B$ such that $y > m_B - c$. Then $x + y \in A + B$ and $x + y > m_A + m_B - 2c$. Hence $m = lub(A + B) > m_A + m_B - 2c = m$. This is impossible and hence $c \leq 0$. We then have

$$lub(A) + lub(B) \leq lub(A + B). \tag{6.3}$$

Combining (6.2) and (6.3) completes the proof for least upper bounds.

The same direct approach, or Exercise 6.5 and the least upper bound result, completes the proof. $\qquad\square$

Theorem 6.12. *If A and B are subsets of the strictly positive real numbers $\mathbf{R}^+$ and $A \cdot B := \{x \cdot y : x \in A, y \in B\}$, then*

$$lub(A \cdot B) = lub(A) \cdot lub(B).$$

Proof. Let $m_A = lub(A)$ and $m_B = lub(B)$. Then $0 \leq x \leq m_A$ for all $x \in A$ and $0 \leq y \leq m_B$ for all $y \in B$. Hence, $x \cdot y \leq m_A \cdot m_B$ and

$$lub(A \cdot B) \leq lub(A) \cdot lub(B). \tag{6.4}$$

Since A and B contain strictly positive numbers, we have $m_A > 0$ and $m_B > 0$. Choose c such that $0 < c < m_A \cdot m_B$. Then $\frac{c}{m_A} < m_B$ and there exists $y \in B$ such that $0 < \frac{c}{m_A} < y \leq m_B$. Hence $\frac{c}{y} < m_A$ and there exists $x \in A$ such that $\frac{c}{y} < x \leq m_A$. Hence, $c < x \cdot y \leq m_A \cdot m_B$ and $lub(A \cdot B) \geq c$. Since c was any number less than $m_A \cdot m_B$, Example 6.10 implies that

$$lub(A) \cdot lub(B) \leq lub(A \cdot B). \tag{6.5}$$

Combining (6.4) and (6.5) completes the proof. $\square$

6.4   Intervals

We define intervals abstractly and show that this definition leads to the same concrete collections of intervals already given in Chapter 4.

Definition 6.13. A non-empty set $A \subset \mathbf{R}$ is an interval if x, y in A and $x < z < y$ imply $z \in A$.

If A is an interval which is bounded above and below, we let b denote the least upper bound of A and let a denote the greatest lower bound of A. Clearly $a < b$. If $a < z < b$ then, by the definition of upper and lower bounds, we can find $x, y \in A$ such that $x < z < y$. This implies $z \in A$. Using once more the definition of upper and lower bounds, we see that

$$\{x \in \mathbf{R}, a < x < b\} \subset A \subset \{x \in \mathbf{R} : a \leq x \leq b\}$$

and a case by case examination shows that all possible inclusions are admissible and listed in Chapter 4.[7] It is easily seen that $\mathbf{R}$ is the only interval which is neither bounded above nor below. If A is bounded below but not bounded above, then $A = \{x \in \mathbf{R} : a < x < \infty\} =: (a, +\infty)$ or $A = \{x \in \mathbf{R} : a \leq x < \infty\} =: [a, +\infty)$ and, if A is bounded above but not bounded below, then $A = \{x \in \mathbf{R} : -\infty < x < b\} := (-\infty, b)$ or $A = \{x \in \mathbf{R} : -\infty < x \leq b\} := (-\infty, b]$.

Example 6.14. If the non-empty set of real numbers, A, is bounded above we let U_A denote the set of all upper bounds for A. If $a \in A$ and $u \in U_A$, then $a \leq u$ and a is a lower bound for U_A. Let L_A denote the set of all

---

[7]It is sometimes convenient to write $[a, b]$ where $a > b$. In such cases we interpret $[a, b]$ as $[b, a]$ etc.

lower bounds for U_A. We have found two non-empty subsets of $\mathbf{R}$, U_A and L_A, such that $l \leq u$ for all $u \in U_A$ and all $l \in L_A$. Since U_A is bounded below, it has a greatest lower bound m and as L_A consists of all lower bounds for U_A, $m \in L_A$ and, as m is greater than every other lower bound, m is also the least upper bound for L_A. A closer examination shows that $m = lub(A)$, $U_A = \{x \in \mathbf{R} : x \geq m\}$ and $L_A = \{x \in \mathbf{R} : x \leq m\}$. Note that we are very close to having a partition of the real numbers.

6.5 Exercises

(6.1) Show that there are no surjective functions from $\mathbf{N}$ to $\mathbf{R}$ and no injective functions from $\mathbf{R}$ to $\mathbf{N}$.

(6.2) Show that the sequence $\{(n + 1)^{10}; n \in \mathbf{N}\}$ is not bounded above. Show that the sequence $(2^n n^{-3})_{n=1}^{\infty}$ is bounded below and not bounded above.

(6.3) Show that the set of real numbers between 0 and 1 whose decimal expansions consist only of fives and sixes is not countable.

(6.4) Give an example of a set A which is bounded above and such that $lub(A) \in A$. Give an example of a set B which is bounded above and such that $lub(B) \notin B$. Justify your examples. If A is a set of real numbers and $a \in A$ is an upper bound for A show that $a = lub(A)$.

(6.5) If $A \subset \mathbf{R}$ let $-A := \{-x; x \in A\}$. Show that A is bounded above if and only if $-A$ is bounded below and, moreover, that $glb(-A) = -lub(A)$. Hence show that a subset of $\mathbf{R}$ which is bounded below has a greatest lower bound.

(6.6) If A and B are subsets of $\mathbf{R}$, $A \subset B$, show that $lub(A) \leq lub(B)$ and $glb(A) \geq glb(B)$.

(6.7) If A and B are subsets of $\mathbf{R}$ and for all $x \in A$ there exists $y \in B$ such that $x \leq y$ show that $lub(A) \leq lub(B)$. Use this result to give a further proof of Exercise 6.6 and to show that $lub\big(\{n^a\}_{n=1}^{\infty}\big) = +\infty$ for any positive real number a.

(6.8) Let A and B denote sets of real numbers such that $x \leq y$ whenever

$x \in A$ and $y \in B$. Show that $lub(A) \leq glb(B)$.

(6.9) Show that there exists a rational number between any two different real numbers.

(6.10) Show (a) that a subset of a countable set is countable and (b) a countable union of countable sets is countable.

(6.11) If A and B are sets, let $A \times B = \{(x,y) : x \in A, y \in B\}$. If A and B are countable, show that $A \times B$ is countable. Show that the product of any finite number of countable sets is countable.

(6.12) If $A = \bigcup_{n=1}^{\infty} A_n$, $A_n \subset \mathbf{R}$ and $x_n = lub(A_n)$ for all n, show that $lub(A) = lub(\{x_n\}_{n=1}^{\infty})$.

(6.13) Let $A \subset \mathbf{R}$ have the following property: if $x \in A$, then there exists $\delta_x > 0$ such that $(x - \delta_x, x + \delta_x) \subset A$. Show that A is a countable union of open intervals.

References [9; 16; 20; 27; 30]

Chapter 7

SERIES WITH POSITIVE ENTRIES

But it is of course easier, when we have
previously acquired, by the method,
some knowledge of the questions, to
supply the proof than it is without
any previous knowledge.

Archimedes

Summary

We discuss convergence of increasing and decreasing sequences and of series[1] with positive entries. We consider absolute convergence of series, the geometric and exponential series, and the rearrangement and multiplication of series.

7.1 Increasing and Decreasing Sequences

Having defined infinite sets and sequences in the previous chapter, we proceed by formally defining limits of increasing and decreasing sequences

---

[1]Historically there has always been more interest in series than sequences. However, the rigorous treatment of infinite series required the use of *limits*, which naturally use sequences. Thus, logically, sequences precede series and modern presentations follow this approach. This sometimes happens in mathematics: the logical and historical order are reversed.

of real numbers. Mathematicians have been interested in summing infinite series for over two thousand years. It was, for many, a natural extension of the process of addition and often led to results, whose truth could be independently verified. Gregory and Leibniz used the calculus to sum series and proved independently in 1671 and 1673, respectively, that $\sum_{n=1}^{\infty}(-1)^{n+1}/(2n+1) = \pi/4$. Euler has been described as the first master of infinite series. His landmark book on infinite series, which was both a research monograph and a textbook, is as readable today as when it was first written. It was written in Latin and, despite being included among the great printed books in the history of civilization by the British Museum in 1963, it was only translated into English in 1998. The notation $f(x)$, the use of e for $\exp(1)$, and much more first saw the light of day in this book. To hint at some of the uses to which Euler put infinite series we mention that he used the divergence of the harmonic series $\sum_{n=1}^{\infty}(1/n)$ to show that the number of primes was infinite and he showed, when $s > 1$, that

$$\zeta(s) := \sum_{n=1}^{\infty} \frac{1}{n^s} = \prod_{i=1}^{\infty} \frac{1}{1 - p_i^{-s}}$$

where $(p_i)_{i=1}^{\infty}$ is the set of all primes (note $p_1 = 2$) and $s > 1$. Using complex analysis Riemann extended this function in a 'regular' fashion to the complex plane with the point 1 removed and today it is known as the Riemann Zeta function. Many questions about prime numbers can be formulated in terms of the zeros of this function and the conjecture that

$$\zeta(s + t\sqrt{-1}) = 0 \implies s = 1/2$$

is currently the most famous unsolved problem within mathematics. It is known as *The Riemann Hypothesis*.

Until the 19th century, there was no rigorous, or even systematic, approach to the summation of series and it is difficult to attribute, with confidence, priorities of discovery. Gauss is credited with providing, in his 1812 study of hypergeometric series, the first adequate study of infinite series. The 1817 investigations of Bolzano and the 1821 approach of Cauchy to defining limits were a big improvement and clarification. The final step was taken in 1859 by K. Weierstrass who introduced the $\epsilon - \delta$ method.

This replaced the intuitive approach using *infinitesimals* with rigorous arguments. Of the tests for convergence that we use, the condensation tests is often assigned to Cauchy, the comparison test to Abel, the integral test to Maclaurin, the alternating series test to Leibniz, while the ratio test was, apparently, discovered independently by a number of people in the 18th century. Weierstrass was reluctant to publish and his contributions are mainly found in versions of his lectures notes prepared by his students. We consider the $\epsilon - \delta$ approach too severe for an introductory one variable calculus course and choose instead to base our analysis on monotonic sequences. In several areas of mathematics, e.g. several variables calculus, metric space theory, complex analysis, functional analysis and measure theory, the approach of Weierstrass[2] is widely used.

In this chapter we use *least upper bounds* to define *convergent increasing sequences*. We have found this approach, based on the *ordering of the real numbers*, intuitive and a good foundation for the general theory considered in Chapter 8. We are able, in this chapter, to analyze series with positive entries and absolutely convergent series and to address two questions that are only of interest for absolutely convergent series; the rearrangement of series and the multiplication of series. We define two very important series,

---

[2]Karl Theodor Weierstrass, 1815-1897, was born in Otenfelde in Westfhalia, studied at Bonn and Münster, taught at high school for fifteen years and eventually, in his forties and after having published some important mathematical articles, received a university appointment in Berlin. He might have achieved his university post earlier had he mainly studied mathematics at university but he was obliged to study the *practical subjects* of finance and administration. As a form of rebelion, he indulged in drinking and sword fighting and left Bonn after four years without a qualification. Weierstrass is noted for the degree of rigor which he injected into analysis and for having settled, often surprisingly and always impeccably, questions that no one else had even considered. He gave an example of a *continuous nowhere-differentiable function* which caused consternation and outrage among his contemporaries. Ironically, the modern theory of *Financial Mathematics* uses a form of integration, the *Itô integral*, which shows in a precise mathematical way that *almost all* continuous functions are nowhere-differentiable. Mathematics has always had, and probably always will have, controversies. Today we have controversy regarding the use of computers to prove results: how acceptable is a proof involving a computer program which takes so long that no one individual can check it completely? Cantor was a student of Weierstrass.

the geometric and exponential series. The exponential series gives us the exponential and log functions and allows us to define a^b for $a > 0$ and all $b \in \mathbf{R}$ and the geometric series is the basis for the ratio test in Chapter 8.

We give two equivalent formulations in the following definition, the first is formal and refers to the initial definition of sequence from the previous chapter while the second uses more adaptable notation (see also Definition 4.13).

Definition 7.1. A sequence $f : \mathbf{N} \longrightarrow \mathbf{R}$ is increasing (respectively decreasing) if it preserves (respectively reverses) order, that is if $f(n) \leq f(n+1)$ (respectively $f(n) \geq f(n+1)$) for all $n \in \mathbf{N}$. Equivalently, a sequence of real numbers $(a_n)_{n=1}^{\infty}$ is increasing (respectively decreasing) if $a_n \leq a_{n+1}$ (respectively $a_n \geq a_{n+1}$) for all n. A sequence which is either increasing or decreasing is said to be *monotonic*.

Example 7.2. Consider the sequence $(n + \frac{1}{n})_{n=1}^{\infty}$. Is it increasing or decreasing? The answer is not obvious since the sequence is the sum of an increasing sequence $(n)_{n=1}^{\infty}$ and a decreasing sequence $(\frac{1}{n})_{n=1}^{\infty}$. The sequence is increasing if and only if

$$(n+1) + \frac{1}{n+1} \geq n + \frac{1}{n}$$

that is if and only if

$$(n+1) - n = 1 \geq \frac{1}{n} - \frac{1}{n+1} = \frac{n+1-n}{n(n+1)} = \frac{1}{n(n+1)}$$

for all n. Since $n(n+1) = n^2 + n \geq 1$ we have $\frac{1}{n(n+1)} \leq 1$. This shows the sequence is increasing.

In some cases we may, by *interpolation*,[3] relate increasing and decreasing sequences to the graphs of functions. This will prove quite useful when we

---

[3] Graph sketching and interpolation are dual operations. The purpose of graph sketching is to display a sufficiently representative set of values of a given function to aid further analysis. Interpolation seeks to place data in a fertile environment. There are, generally, many ways to do this, some obvious as in Figure 7.1, others depend on the ends being sought. Interpolation theory was popular in the 18th century, a time when functions and formulae were synonymous, and interpolation meant finding a mathematical expression for the extended function. This led to the discovery of many new still useful functions. We describe one such function, the Gamma function of Euler, in Chapter 14. Interpolation is also the key to the *integral test* for series convergence.

have the differential and integral calculus at our disposal. For example, if we consider the function $f : \mathbf{R} \longrightarrow \mathbf{R}, f(x) = x^2$, then $f(n) = n^2$ and, as the graph of f is increasing for $x > 0$, it follows that the sequence $(n^2)_{n=1}^{\infty}$ is increasing and the sequence $(1/n^2)_{n=1}^{\infty}$ is decreasing (see Figure 7.1). The sequence in Example 7.2 is interpolated by the function $f(x) = x + (1/x)$ (see Figure 2.13). We list in the following theorem a number of simple but

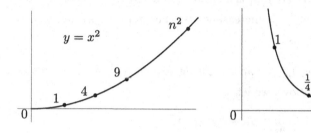

Fig. 7.1

useful results. The technique in (c) is frequently useful.

Theorem 7.3. *Let* $(a_n)_{n=1}^{\infty}$ *and* $(b_n)_{n=1}^{\infty}$ *be two increasing sequences of real numbers and let c be a real number. Then*

(a) $(a_n + b_n)_{n=1}^{\infty}$ *is an increasing sequence,*

(b) $(ca_n)_{n=1}^{\infty}$ *is increasing if* $c > 0$ *and decreasing if* $c < 0$,

(c) *if* $a_n > 0$ *and* $b_n > 0$ *for all* n, *then* $(a_n b_n)_{n=1}^{\infty}$ *is an increasing sequence,*

(d) *if* $a_n > 0$, *then* $(\frac{1}{a_n})_{n=1}^{\infty}$ *is a decreasing sequence.*

Proof.   (a) $a_n \leq a_{n+1}$ and $b_n \leq b_{n+1}$ imply $a_n + b_n \leq a_{n+1} + b_{n+1}$.

(b) If $a_n \leq a_{n+1}$, then $a_{n+1} - a_n \geq 0$. Hence, if $c > 0$, then $c(a_{n+1} - a_n) \geq 0$, $ca_{n+1} \geq ca_n$ and the sequence $(ca_n)_{n=1}^{\infty}$ is increasing. If $c < 0$, then $c(a_{n+1} - a_n) \leq 0$, $ca_{n+1} \leq ca_n$ and the sequence $(ca_n)_{n=1}^{\infty}$ is decreasing.

(c) It suffices to use

$$a_{n+1}b_{n+1} - a_n b_n = a_{n+1}(b_{n+1} - b_n) + b_n(a_{n+1} - a_n).$$

(d) The result follows since $a_{n+1} \geq a_n \iff \frac{1}{a_n} \geq \frac{1}{a_{n+1}}$.   □

Example 7.4. The sequence

$$\left(\frac{1}{n^4 2^n + n^3 + (n+1)^{15}}\right)_{n=1}^{\infty} \tag{7.1}$$

is decreasing. Since $n + 1 \geq n$, the sequences $(n)_{n=1}^{\infty}$ and $(n+1)_{n=1}^{\infty}$ are increasing. Hence, by Theorem 7.3(c), the sequences $(n \cdot n)_{n=1}^{\infty} = (n^2)_{n=1}^{\infty}$, $(n^2 \cdot n^2)_{n=1}^{\infty} = (n^4)_{n=1}^{\infty}$ and $((n+1)^{15})_{n=1}^{\infty}$ are increasing. Since $2^{n+1} = 2 \cdot 2^n \geq 2^n$, the sequence $(2^n)_{n=1}^{\infty}$ is increasing and again, by Theorem 7.3(c), the sequence $(n^4 \cdot 2^n)_{n=1}^{\infty}$ is increasing. By Theorem 7.1(a) and (d), the sequence (7.1) is decreasing.

Definition 7.5. If $(a_n)_{n=1}^{\infty}$ is an increasing sequence of real numbers and $\{a_n\}_{n=1}^{\infty}$ is bounded above, we let

$$\lim_{n \to \infty} a_n = lub(\{a_n\}_{n=1}^{\infty}).$$

If $(a_n)_{n=1}^{\infty}$ is a decreasing sequence of real numbers and $\{a_n\}_{n=1}^{\infty}$ is bounded below, we let

$$\lim_{n \to \infty} a_n = glb(\{a_n\}_{n=1}^{\infty}).$$

When $\lim_{n\to\infty} a_n = a$ we also write $a_n \longrightarrow a$ as $n \longrightarrow \infty$ and say that the sequence converges. If the increasing (respectively decreasing) sequence is not bounded above (respectively below), we say that the sequence diverges and write $\lim_{n\to\infty} a_n = \infty$ in either case.

We pause to see why Definition 7.5 is plausible. Let $l = \lim_{n\to\infty} a_n = lub(\{a_n\}_{n=1}^{\infty})$ where $(a_n)_{n=1}^{\infty}$ is increasing and bounded above. By the definition of upper bound, all a_n are to the left of l. If l' is *any* number to the *left* of l (see Figure 7.2) and the closed interval $[l', l]$ does not contain any a_n, then all a_n are to the left of l'. Hence, $a_n \leq l'$ for all n and we have

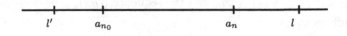

$$l' \qquad a_{n_0} \qquad\qquad\qquad a_n \qquad l$$

Fig. 7.2

found an upper bound less than the least upper bound. This is impossible.

Hence there is some positive integer n_0 such that $l' \leq a_{n_0} \leq l$. This implies $l' \leq a_{n_0} \leq a_n \leq l$ for all $n \geq n_0$ and shows, as expected, that a_n is getting closer to l as n gets larger.

Clearly, a decreasing sequence of *positive* numbers is bounded below by 0 and hence converges to a non-negative limit. Our next theorem refines this comment. This result requires increasing and decreasing sequences and sets that are bounded above and sets that are bounded below.

Theorem 7.6. *If $(a_n)_{n=1}^{\infty}$ is a decreasing sequence of positive real numbers then $\lim_{n \to \infty} a_n = 0$ if and only if $\left(\frac{1}{a_n}\right)_{n=1}^{\infty}$ is not bounded above.*[4]

Proof. Let $\mathbf{P}$ denote the statement: $\lim_{n \to \infty} a_n = 0$ and let $\mathbf{Q}$ denote the statement: $\left(\frac{1}{a_n}\right)_{n=1}^{\infty}$ is not bounded above. To show that $\mathbf{P}$ and $\mathbf{Q}$ are equivalent we show $\mathbf{NQ}$ implies $\mathbf{NP}$ and then that $\mathbf{NP}$ implies $\mathbf{NQ}$.

If $\mathbf{NQ}$ holds, then $\left(\frac{1}{a_n}\right)_{n=1}^{\infty}$ is bounded above and so there exists $c \in \mathbf{R}$ such that $\frac{1}{a_n} \leq c$ for all $n \in \mathbf{N}$. Hence, $a_n \geq 1/c$ for all n and $\lim_{n \to \infty} a_n \geq 1/c > 0$, that is $\mathbf{NP}$ holds.

If $\mathbf{NP}$ holds, then $\lim_{n \to \infty} a_n = glb(\{a_n\}_{n=1}^{\infty}) = a > 0$. Hence, $a_n \geq a$ and $\frac{1}{a_n} \leq \frac{1}{a}$ for all n, and $\left(\frac{1}{a_n}\right)_{n=1}^{\infty}$ is bounded above. This means that $\mathbf{NQ}$ holds and the proof is complete. $\square$

Example 7.7.

$$\lim_{n \to \infty} \frac{1}{n^2 + n^3} = 0. \tag{7.2}$$

Proof. The sequences $(n^2)_{n=1}^{\infty}$ and $(n^3)_{n=1}^{\infty}$ are increasing, so $(n^2 + n^3)_{n=1}^{\infty}$ is increasing and the sequence in (7.2) is decreasing and positive. Hence $\lim_{n \to \infty} \frac{1}{n^2 + n^3} \geq 0$. By Example 6.8, the sequence $(n)_{n=1}^{\infty}$ is not bounded above and, as $n^2 + n^3 \geq n$, the sequence $(n^2 + n^3)_{n=1}^{\infty}$ is not bounded above. By Theorem 7.6, $\lim_{n \to \infty} \frac{1}{n^2 + n^3} = 0$. $\square$

The simple rules in the next theorem follow easily from Theorems 6.11, 6.12, 7.3 and Definition 7.5.

---

[4]It is worth remembering this result in reverse: *For any decreasing sequence of positive real numbers $(a_n)_{n=1}^{\infty}$, we have $\lim_{n \to \infty} a_n > 0 \iff \left(\frac{1}{a_n}\right)_{n=1}^{\infty}$ is bounded above.*

Theorem 7.8. *Let $(a_n)_{n=1}^\infty$ and $(b_n)_{n=1}^\infty$ denote positive increasing convergent sequences and suppose $c \in \mathbf{R}$. Then*

(a) $\lim_{n\to\infty}(a_n + b_n) = \lim_{n\to\infty} a_n + \lim_{n\to\infty} b_n$,

(b) $\lim_{n\to\infty} ca_n = c\lim_{n\to\infty} a_n$,

(c) $\lim_{n\to\infty} a_n \cdot \lim_{n\to\infty} b_n = \lim_{n\to\infty}(a_n \cdot b_n)$,

(d) *If $a_n \le b_n$ for all n then $\lim_{n\to\infty} a_n \le \lim_{n\to\infty} b_n$. In particular, if $(b_n)_{n=1}^\infty$ converges then $(a_n)_{n=1}^\infty$ converges and if $(a_n)_{n=1}^\infty$ diverges then $(b_n)_{n=1}^\infty$ diverges.*

Proof. We prove (a). The remaining proofs are similar and left as an exercise. If $\lim_{n\to\infty} a_n = a$ and $\lim_{n\to\infty} b_n = b$, then $a_n \le a$ and $b_n \le b$ for all n. Hence, $(a_n + b_n)_{n=1}^\infty$ is an increasing sequence which is bounded above by $a + b$. This implies that the sequence $(a_n + b_n)_{n=1}^\infty$ converges and $\lim_{n\to\infty}(a_n + b_n) \le a + b$.

If $m < a+b$ then $a+b-m =: 2c > 0$. Since $a - c < a$ and $b - c < b$, we can find positive integers n_1 and n_2 such that $a_{n_1} > a - c$ and $b_{n_2} > b - c$. If $n > n_1 + n_2$, then $a_n + b_n \ge a_{n_1} + b_{n_2} > a - c + b - c = a + b - 2c = m$. This implies $\lim_{n\to\infty}(a_n + b_n) \ge m$. Hence $\lim_{n\to\infty}(a_n + b_n)$ is greater than any number less than $a + b$. By Example 6.10, $\lim_{n\to\infty}(a_n + b_n) = lub(\{a_n + b_n\}_{n=1}^\infty) \ge a + b$. Combining the above estimates we obtain $\lim_{n\to\infty}(a_n + b_n) = a + b$. $\qquad\square$

Similar results for decreasing sequences are easily stated and proved.

7.2   Series with Positive Entries

One of the most fertile sources for increasing sequences are series with *positive terms*. If $(a_n)_{n=1}^\infty$ is a sequence of real numbers we let

$$s_n = a_1 + a_2 + \cdots + a_n$$

and call s_n the nth partial sum of the series $\sum_{n=1}^\infty a_n$. If $a_n \ge 0$ for all n, then $(s_n)_{n=1}^\infty$ is an increasing sequence of positive real numbers and the sequence $(s_n)_{n=1}^\infty$ converges if and only if it is bounded above.

Definition 7.9. The *series* $\sum_{n=1}^{\infty} a_n$, where $a_n \geq 0$ for all n, converges when the increasing *sequence* $(s_n)_{n=1}^{\infty}$ converges. In this case, we let

$$\sum_{n=1}^{\infty} a_n = \lim_{n \to \infty} s_n =: s$$

and call s the sum of the series.

Our next example is very important. We get an explicit formula for the sum and, moreover, the main tests for absolute convergence are based on this series.

Example 7.10. (The Geometric Series) If $0 \leq r < 1$, then the series $\sum_{n=0}^{\infty} r^n = 1 + r + r^2 + \cdots$ has all positive entries. If $s_n = 1 + r + r^2 + \cdots + r^n$ then $r s_n = r + r^2 + \cdots + r^{n+1}$ and $(1 - r)s_n = 1 - r^{n+1}$. This implies[5]

$$s_n = \frac{1 - r^{n+1}}{1 - r} \leq \frac{1}{1 - r}$$

for all n. Hence $(s_n)_{n=1}^{\infty}$ is increasing and bounded above and thus the series $\sum_{n=0}^{\infty} r^n$ converges. We denote its limit by $\phi(r)$. Since

$$s_{n+1} = 1 + r + \cdots + r^{n+1} = 1 + r \cdot s_n$$

Theorem 7.8 (a) and (b) imply

$$\phi(r) = \lim_{n \to \infty} s_{n+1} = 1 + r \cdot \lim_{n \to \infty} s_n = 1 + r\phi(r)$$

and we obtain a formula, first proved by François Vièta in 1593,

$$\phi(r) = \frac{1}{1 - r} = \sum_{n=0}^{\infty} r^n.$$

A useful method for checking the convergence of series with positive entries is the following comparison test whose proof is clear.

Theorem 7.11. *If $0 \leq a_n \leq b_n$ for all n sufficiently large, then* (a) $\sum_{n=1}^{\infty} a_n$ *converges if* $\sum_{n=1}^{\infty} b_n$ *converges and* $\sum_{n=1}^{\infty} a_n \leq \sum_{n=1}^{\infty} b_n$ *and* (b) $\sum_{n=1}^{\infty} b_n$ *diverges if* $\sum_{n=1}^{\infty} a_n$ *diverges.*

---

[5] This identity also follows, after n steps, on dividing $1 - r^{n+1}$ by $1 - r$ using the classical method of *long division*. One may also use Exercise 5.2 with $a = 1$ and $b = r$.

Unfortunately, not all series have positive terms and not all sequences are monotonic. We shall soon see how to deal with both but, in the meantime, we look at another simple situation.

Definition 7.12. The series $\sum_{n=1}^{\infty} a_n$ is absolutely convergent if $\sum_{n=1}^{\infty} |a_n|$ converges.

Let

$$a^+ = \begin{cases} a & \text{if } a \geq 0, \\ 0 & \text{if } a < 0 \end{cases}$$

and

$$a^- = \begin{cases} -a & \text{if } a \leq 0, \\ 0 & \text{if } a > 0. \end{cases}$$

We call a^+ the positive part of a and a^- the negative part of a. Clearly, $a = a^+ - a^-$, $|a| = a^+ + a^-$, $0 \leq a^+ \leq |a|$ and $0 \leq a^- \leq |a|$. If $\sum_{n=1}^{\infty} a_n$ is absolutely convergent then, by Theorem 7.11(a), $\sum_{n=1}^{\infty} a_n^+$ and $\sum_{n=1}^{\infty} a_n^-$ both converge. We denote the sums by s^+ and s^-, respectively, and let $\sum_{n=1}^{\infty} a_n = s^+ - s^-$.

Example 7.13. Let $-1 < r < 0$. We consider the series

$$\sum_{n=0}^{\infty} r^n = 1 + r + r^2 + \cdots .$$

Since $|r^n| = |r|^n$ and $0 \leq |r| < 1$ we see, by Example 7.10, that $\sum_{n=0}^{\infty} r^n$ is absolutely convergent. If we let $t = -r$, then $0 \leq t < 1$ and

$$\sum_{n=0}^{\infty} r^n = 1 - t + t^2 - t^3 + \cdots .$$

By Example 7.10, we obtain the sum of the positive terms

$$1 + t^2 + t^4 + \cdots = \frac{1}{1 - t^2}$$

and the sum of the negative terms

$$t + t^3 + t^5 + \cdots = t(1 + t^2 + t^4 \cdots) = \frac{t}{1 - t^2}.$$

Hence

$$\sum_{n=0}^{\infty} r^n = \frac{1}{1 - t^2} - \frac{t}{1 - t^2} = \frac{1 - t}{1 - t^2} = \frac{(1 - t)}{(1 - t)(1 + t)} = \frac{1}{1 + t} = \frac{1}{1 - r}$$

and we obtain, now for $-1 < r < 1$, the formula in Example 7.10.

Example 7.14. Consider the decimal expansion of the real number $a = b.c_1c_2\ldots$ where $b \in \mathbf{Z}$ and $0 \le c_i \le 9$. We have

$$.c_1c_2c_3\cdots\cdots = \sum_{n=1}^{\infty} \frac{c_n}{10^n}.$$

Since $0 \le c_n \le 9$,

$$0 \le \frac{c_n}{10^n} \le \frac{1}{10^{n-1}}$$

and, as $\sum_{n=1}^{\infty} \frac{1}{10^{n-1}} < \infty$, the comparison principle in Theorem 7.11(a), implies that $\sum_{n=1}^{\infty} \frac{c_n}{10^n}$ converges.

What happens if we change the order in which we sum a convergent series with positive terms, that is if we rearrange the terms of the series $\sum_{n=1}^{\infty} a_n$ to get $\sum_{n=1}^{\infty} b_n$? This means that $b_1 = a_{\phi(1)}$, $b_2 = a_{\phi(2)}$, $b_3 = a_{\phi(3)}$, etc. and we obtain a sequence of integers $(\phi(n))_{n=1}^{\infty}$. Since each term in the original series must be included once and only once, the mapping $\phi : \mathbf{N} \longrightarrow \mathbf{N}$ is bijective. Thus, there is a one to one correspondence between rearrangements of series and bijective functions from $\mathbf{N}$ to $\mathbf{N}$. Suppose $\sum_{n=1}^{\infty} a_n = s$. Let $s'_n = \sum_{j=1}^{n} a_{\phi(j)}$. If $l = \max(\phi(1), \phi(2), \ldots, \phi(n))$, then,

$$s'_n = \sum_{j=1}^{n} a_{\phi(j)} \le \sum_{j=1}^{l} a_j \le s$$

and $(s'_n)_{n=1}^{\infty}$ is an increasing sequence which is bounded above by s. Hence $\sum_{j=1}^{\infty} a_{\phi(j)}$ converges and $\sum_{j=1}^{\infty} a_{\phi(j)} =: s' \le s$. However, we can rearrange the series $\sum_{j=1}^{\infty} a_{\phi(j)}$ to get $\sum_{j=1}^{\infty} a_j$ and the above analysis shows that $s \le s'$. Hence $s = s'$.

If $\sum_{n=1}^{\infty} a_n$ is an absolutely convergent series and $\phi : \mathbf{N} \longrightarrow \mathbf{N}$ is bijective then, as $\sum_{n=1}^{\infty} |a_{\phi(n)}|$ is a rearrangement of $\sum_{n=1}^{\infty} |a_n|$, the series $\sum_{n=1}^{\infty} a_{\phi(n)}$ is absolutely convergent. Since $\sum_{n=1}^{\infty} a_n$ and $\sum_{n=1}^{\infty} a_{\phi(n)}$ have the same positive and negative terms, this proves the following result of Dirichlet.[6]

---

[6]Johan Peter Gustav Lejeune Dirichlet, 1805-1859, was born in Düren in Germany. He was a student of Fourier at the University of Paris and was a professor at Breslau, Berlin and Göttingen. He worked on number theory, quadratic forms, mechanics, potential theory and Fourier series and defined a class of series now called Dirichlet Series. The most well known Dirichlet series is $\zeta(s)$ for $s > 1$. This is, as we have mentioned earlier, connected to *The Riemann Hypothesis*. Bernard Riemann was a student of Dirichlet.

Theorem 7.15. *If $\sum_{n=1}^{\infty} a_n$ is an absolutely convergent series, then any rearrangement of the terms gives an absolutely convergent series with the same sum as the original series.*

7.3 The Exponential Function

In this section we discuss the fundamental properties of the exponential function.

Definition 7.16. We define $\exp : \mathbf{R} \longrightarrow \mathbf{R}^+$ by letting

$$\exp x = \sum_{n=0}^{\infty} \frac{x^n}{n!} = 1 + x + \frac{x^2}{2!} + \cdots . \tag{7.3}$$

The notation $n! = 1 \times 2 \times 3 \times \cdots \times n$ was introduced by Christian Kramp, 1760-1826, from Strasbourg, to circumvent printing difficulties and is now standard. However, not everyone was favorably impressed when it first appeared. According to Augustus De Morgan: *Writers have borrowed from the Germans the abbreviation n! to signify $1 \cdot 2 \cdot 3 \cdots (n-1) \cdot n$ which gives their pages the appearance of expressing surprise and admiration that 2,3,4, etc. should be found in mathematical results.*[7] Gauss used the notation $\Pi(n)$ in 1816 with Π denoting the product.

We first show that the series converges absolutely. Since $\left| \frac{x^n}{n!} \right| = \frac{|x|^n}{n!}$, it suffices to show that (7.3) converges for all positive x. Fix $x \in \mathbf{R}^+$. By Example 6.8, we can choose a positive integer n_0 such that $n_0 > x$. Then $0 < \frac{x}{n_0} < 1$. For $n > n_0$, we have

$$0 < \frac{x^n}{n!} \leq \frac{x}{1} \cdot \frac{x}{2} \cdot \frac{x}{3} \cdots \frac{x}{n_0} \cdot \frac{x}{n_0} \cdots \frac{x}{n_0} = \frac{x^{n_0}}{n_0!} \left(\frac{x}{n_0} \right)^{n-n_0}$$

$$= \frac{x^{n_0}}{n_0!} \left(\frac{n_0}{x} \right)^{n_0} \left(\frac{x}{n_0} \right)^n = C r^n$$

where $C := \frac{x^{n_0}}{n_0!} \left(\frac{n_0}{x} \right)^{n_0}$ and $0 < r := x/n_0 < 1$. Comparison with the geometric series and Theorem 7.11(a) imply convergence. The method

---

[7] Augustus De Morgan, 1806-1871, was born in India to English parents. He was first professor of mathematics at University College London and first president of the London Mathematical Society. De Morgan wrote a number of successful textbooks, worked on mathematical logic, and resigned his professorship twice on matters of principle.

used in this example can be slightly modified to give the ratio test (see Chapter 8).

The most remarkable property[8] of the exponential function is given in Theorem 7.17: *it transforms addition into multiplication.* We prove this later but first see how useful it is in deriving basic properties of the exponential function.

Theorem 7.17. *If* $x, y \in \mathbf{R}$, *then*

$$\sum_{n=0}^{\infty} \frac{(x+y)^n}{n!} = \exp(x+y) = \exp x \cdot \exp y = \sum_{n=0}^{\infty} \frac{x^n}{n!} \cdot \sum_{n=0}^{\infty} \frac{y^n}{n!}. \quad (7.4)$$

Theorem 7.18.

(a) $\exp(0) = 1$,

(b) $\exp(-x) = 1/\exp(x)$ *for all* $x \in \mathbf{R}$,

(c) *if* $x < 0 < y$ *then* $0 < \exp(x) < 1 < \exp(y)$,

(d) $\exp(x) \longrightarrow +\infty$ *as* $x \longrightarrow +\infty$,

(e) $\exp(x) \longrightarrow 0$ *as* $x \longrightarrow -\infty$,

(f) *the exponential function is strictly increasing (and hence injective)*,

(g) $\exp(\mathbf{R}) = \mathbf{R}^+$,

(h) $\exp : \mathbf{R} \longrightarrow \mathbf{R}^+$ *is a bijective function.*

Proof.   (a) Let $x = 0$ in (7.3).

(b) The result follows from (7.4) and (a), which imply

$$\exp(x) \cdot \exp(-x) = \exp(0) = 1.$$

(c) By (7.3), $\exp(y) > 1$ when $y > 0$. If $x < 0$ then $-x > 0$ and $\exp(-x) > 1$. Hence, by (b), $0 < \exp(x) = [\exp(-x)]^{-1} < 1$.

(d) For $x > 0$ we have, by (7.3), $\exp(x) > 1 + x \longrightarrow +\infty$ as $x \longrightarrow +\infty$.

(e) By (b) and (d), $\exp(x) = (1/\exp(-x)) \longrightarrow 0$ as $x \longrightarrow -\infty$.

(f) If $0 < x < y$ then, by (7.3), $\exp(y) - \exp(x) \geq y - x > 0$ and $\exp(y) > \exp(x)$. On applying this result and (b) to $x < y < 0$ we see that $\exp(y) >$

---

[8]Another remarkable connection between quadratic functions and the exponential function occurs in the *normal density function* (see Example 14.9). The general form of the one dimensional normal density is given by $f(x) = ae^{-q(x)}$ where q is a positive quadratic function and a is a suitably chosen positive real number (see Exercise 14.12).

$\exp(x)$ for $x < y < 0$. If $x < 0 < y$, (c) implies $\exp(x) < 1 < \exp(y)$. This covers all possibilities and proves (f).

The proof of (g) requires the *Intermediate Value Theorem* and continuity of the exponential function. We return to these later.

(h) This follows from (f) and (g).   □

7.4   Multiplication of Series

To prove Theorem 7.17 we need to multiply, term by term, the two infinite series on the right-hand side of (7.4) and then add the resulting terms together. It is more convenient to consider two arbitrary absolutely convergent series, $\sum_{n=1}^{\infty} a_n$ and $\sum_{n=1}^{\infty} b_n$. If we multiply each term in $\sum_{n=1}^{\infty} a_n$ by each term in $\sum_{n=1}^{\infty} b_n$ and add the resulting terms together we obtain $\sum_{n=1}^{\infty} \sum_{m=1}^{\infty} a_n b_m$. This is *not* a series. So, we must first ask if it can be re-arranged to form a sequence. Since the terms in the product are indexed by pairs (n, m) in $\mathbf{N} \times \mathbf{N} = \mathbf{N}^2$, this will be the case if there is a bijective map from $\mathbf{N}$ onto $\mathbf{N}^2$. By Example 6.4, $\mathbf{N}^2$ is countable and the mapping $\phi(n^2, m^2) = (n, m)$ shows that $\mathbf{N}^2$ is infinite. By Theorem 6.3 there exists a bijective mapping $\phi : \mathbf{N} \longrightarrow \mathbf{N}^2$. If we let $c_n = a_i \cdot b_j$ when $\phi(n) = (i, j)$, we obtain the series $\sum_{n=1}^{\infty} c_n$. Any other bijective mapping gives a rearrangement of this series and hence all bijective mappings give rise to convergent series with the same sum if one arrangement is absolutely convergent. Let $\sum_{n=1}^{\infty} |a_n| = s$ and $\sum_{m=1}^{\infty} |b_m| = s'$. If $J := \{\phi(n)\}_{n=1}^{j}$ is a finite subset of $\mathbf{N} \times \mathbf{N}$, then there exists a positive integer n_0 such that $J \subset \{1, 2, \ldots, n_0\} \times \{1, 2, \ldots, n_0\}$. Hence

$$\sum_{n=1}^{j} |c_n| = \sum_{(n,m) \in J} |a_n| \cdot |b_m| \leq \sum_{n=1}^{n_0} |a_n| \cdot \sum_{m=1}^{n_0} |b_m| \leq s \cdot s'.$$

This implies $\sum_{n=1}^{\infty} |c_n| < \infty$ for any bijective mapping $\phi : \mathbf{N} \longrightarrow \mathbf{N}^2$, and there is no ambiguity if we write $\sum_{n=1}^{\infty} \sum_{m=1}^{\infty} a_n \cdot b_m$ in place of $\sum_{n=1}^{\infty} c_n$. Moreover, we may sum the double series in *any* order we wish and obtain the following result of Cauchy:

$$\sum_{n=1}^{\infty}\sum_{m=1}^{\infty} a_n \cdot b_m = \lim_{n \to \infty}\sum_{j=1}^{n}\sum_{k=1}^{n} a_j \cdot b_k$$

$$= \lim_{n \to \infty}\sum_{j=1}^{n} a_j \cdot \sum_{k=1}^{n} b_k$$

$$= \sum_{j=1}^{\infty} a_j \cdot \sum_{k=1}^{\infty} b_k.$$

Since the series $\sum_{n=0}^{\infty} x^n/n!$ is absolutely convergent for all $x \in \mathbf{R}$ the Binomial Theorem (see Exercise 5.16) implies

$$\exp x \cdot \exp y = \Big(\sum_{n=0}^{\infty}\frac{x^n}{n!}\Big) \cdot \Big(\sum_{m=0}^{\infty}\frac{y^m}{m!}\Big)$$

$$= \sum_{n=0}^{\infty}\Big\{\sum_{j=0}^{n}\frac{x^j}{j!} \cdot \frac{y^{n-j}}{(n-j)!}\Big\}$$

$$= \sum_{n=0}^{\infty}\frac{1}{n!}\Big\{\sum_{j=0}^{n}\frac{n!}{j!(n-j)!}x^j y^{n-j}\Big\}$$

$$= \sum_{n=0}^{\infty}\frac{(x+y)^n}{n!}$$

$$= \exp(x+y).$$

This completes the proof of Theorem 7.17.

7.5   The Log Function

By Theorem 7.18(h), the exponential function has an inverse. We call this inverse the *log function*. The expressions *logarithms, natural logs* and *logs to the base e* (see Definition 7.21) also appear in the literature and the notation *ln* is common. In this book we use only one log function, the inverse function to the exponential function (Definition 7.16). Logs replace multiplication by addition and, for this reason, are a practical tool in certain disciplines, e.g. astronomy. But, to be of use, tables were required and these were produced, after twenty years work, by John Napier, 1550-1617, and published in 1614 and (posthumously) in 1619. The possibility of using machines to produce log tables motivated Charles Babbage to construct his *Difference Engine* (see Chapters 11 and 12).

Formally. we have

$$\log : \mathbf{R}^+ \longrightarrow \mathbf{R}$$

and have already sketched, in Chapter 3 (Figure 3.11), the graphs of both the exp and log functions. It is easily seen that the properties in Theorem 7.18 can be used to produce these sketches. From Definition 4.8 we obtain the two fundamental formula:

$$\log(\exp x) = x \text{ for all } x \in \mathbf{R} \tag{7.5}$$

and

$$\exp(\log x) = x \text{ for all } x \in \mathbf{R}^+. \tag{7.6}$$

Every property of the exponential function gives rise to a corresponding property of the log function. For example if $a, b \in \mathbf{R}^+$ then $\exp(\log a) = a$ and $\exp(\log b) = b$. Hence

$$ab = \exp(\log a) \cdot \exp(\log b) = \exp(\log a + \log b) \tag{7.7}$$

and applying the log function to both sides of (7.7) we obtain

$$\log(ab) = \log(\exp(\log a + \log b)) = \log a + \log b. \tag{7.8}$$

One can show similarly that $\log(1/a) = -\log(a)$ and $\log(a/b) = \log(a) - \log(b)$ for $a, b \in \mathbf{R}^+$ and, from (7.5) and Theorem 7.18(a), we see that $\log(1) = \log(\exp(0)) = 0$.

Using the inverse of the function $x \longrightarrow x^n, x \in \mathbf{R}^+$, we define $x^{1/n}$ for all $x > 0$ and all positive integers n. If $n, m \in \mathbf{N}$ and $x \in \mathbf{R}^+$, we let $x^{m/n} := (x^{1/n})^m$, $x^0 = 1$ and $x^{-m/n} = (1/x)^{m/n}$. This defines x^r for all $r \in \mathbf{Q}^+$ and all $x \in \mathbf{R}^+$.

We now extend this definition using the exponential and log functions and define a^b for all $a > 0$ and all $b \in \mathbf{R}$. In fact, the exponential function arose from considering powers such as a^x and its basic properties, given in Section 7.4, are due to Euler.

Definition 7.19. If $a \in \mathbf{R}^+$ and $b \in \mathbf{R}$ we let

$$a^b = \exp(b \log a). \tag{7.9}$$

If n is a positive integer and $a \in \mathbf{R}$, then $[\exp(\frac{1}{n} \log a)]^n = \exp(\log a) = a$ and $a^{1/n} = \exp(\frac{1}{n} \log a)$. If $m \in \mathbf{N}$ and $a \in \mathbf{R}$ then

$$a^m = a \times a \times \cdots \times a(m \text{ times}) = \exp(\log a) \times \exp(\log a) \times \cdots \times \exp(\log a)$$
$$= \exp(\log a + \cdots + \log a)$$
$$= \exp(m \log a).$$

The two definitions coincide since

$$a^{m/n} = [a^{1/n}]^m = a^{1/n} \times a^{1/n} \times \cdots \times a^{1/n}(m \text{ times})$$
$$= \exp\left(\frac{\log a}{n}\right) \times \exp\left(\frac{\log a}{n}\right) \times \cdots \times \exp\left(\frac{\log a}{n}\right)$$
$$= \exp\left(\frac{\log a}{n} + \cdots + \frac{\log a}{n}\right)$$
$$= \exp\left(\frac{m}{n} \log a\right)$$

and all the laws of indices can now be obtained from (7.9). In the following theorem we prove a number of these and leave the remainder as exercises.

Theorem 7.20. *For positive real numbers a, b and real numbers c and d the following hold:*

(a) $a^0 = 1, \quad a^{-c} = 1/a^c = (1/a)^c,$

(b) $a^c \cdot b^c = (ab)^c,$

(c) $a^c/b^c = (a/b)^c,$

(d) $a^c \cdot a^d = a^{c+d},$

(e) $a^c/a^d = a^{c-d},$

(f) $(a^c)^d = a^{cd}.$

Proof. (a) $a^0 = \exp(0 \log a) = \exp(0) = 1$, $(1/a)^c = \exp(c \log(1/a)) = \exp(-c \log a) = a^{-c} = 1/\exp(c \log a) = 1/a^c$. For (b), (7.8) implies

$$a^c \cdot b^c = \exp(c \log a) \exp(c \log b) = \exp(c \log a + c \log b)$$
$$= \exp(c(\log a + \log b)) = \exp(c \log ab) = (ab)^c.$$

By (a) and (b), $a^c/b^c = a^c(1/b)^c = (a/b)^c$ and we obtain (c). Since

$$a^c \cdot a^d = \exp(c \log a) \cdot \exp(d \log a) = \exp((c + d) \log a) = a^{c+d}$$

we obtain (d). By (a) and (d), $a^c/a^d = a^c \cdot a^{-d} = a^{c-d}$, that is (e) holds. By (7.9), $\log(a^b) = b \log a$, and hence

$$(a^c)^d = \exp(d \log a^c) = \exp(dc \log a) = a^{cd}.$$

This proves (f).   □

Definition 7.21. $e = \exp(1)$.

Theorem 7.22. *If $x \in \mathbf{R}^+$, then $\exp(x) = e^x$.*

Proof.   If $x \in \mathbf{R}$ then

$$e^x = \exp(x \log e) = \exp(x).$$ □

If m is a positive integer and $x \in \mathbf{R}^+$ then, by the Binomial Theorem,

$$a_m := (1 + \frac{x}{m})^m = 1 + x + \frac{m(m-1)}{2}\left(\frac{x}{m}\right)^2 + \cdots$$

$$= 1 + x + \left(1 - \frac{1}{m}\right)\frac{x^2}{2!} + \left(1 - \frac{1}{m}\right)\left(1 - \frac{2}{m}\right)\frac{x^3}{3!} + \cdots$$

$$\leq a_{m+1} \leq \sum_{n=0}^{\infty} \frac{x^n}{n!} = e^x.$$

This implies that the sequence $(a_m)_{m=1}^{\infty}$ is increasing and bounded above and hence converges. Since the terms in the series above are positive and $\frac{1}{m} \longrightarrow 0$ as $m \longrightarrow \infty$ the above also shows that for any positive integer k we have

$$\sum_{n=0}^{k} \frac{x^n}{n!} \leq \lim_{m \to \infty} \left(1 + \frac{x}{m}\right)^m \leq e^x.$$

On letting k tend to infinity we obtain the following result.

Theorem 7.23. *If $x \in \mathbf{R}$ then*

$$\lim_{n \longrightarrow \infty} \left(1 + \frac{x}{n}\right)^n = e^x.$$

Finally we note that it is possible to define a^b for any real number a and certain rational numbers. If q is an *odd* positive integer, then $f : \mathbf{R} \longrightarrow \mathbf{R}, f(x) = x^q$, is easily seen to be bijective (see Example 11.6). We let $x^{1/q} = f^{-1}(x)$ for *all* $x \in \mathbf{R}$. Since $(-a)^{1/q} = -|a|^{1/q}$ for $a \in \mathbf{R}^+$, for example $(-27)^{1/3} = -(27^{1/3}) = -3$, the laws of indices, as in Theorem 7.20, follow from the corresponding results for $a > 0$.

7.6 Exercises

(7.1) Show that the only sequences which are both increasing and decreasing are the constant sequences.

(7.2) If $(a_n)_{n=1}^{\infty}$ and $(b_n)_{n=1}^{\infty}$ are increasing sequences of real numbers, is $(a_n b_n)_{n=1}^{\infty}$ also increasing? Justify your assertions.

(7.3) Show that the function $f : \mathbf{R} \longrightarrow \mathbf{R}, f(x) = \exp\{-e^{-x}\}$ is increasing.

(7.4) Find from first principles, that is using the upper bounds or lower bounds, the following limits

$$(i) \ \lim_{n \longrightarrow \infty} (\sqrt{n+1} - \sqrt{n}), \qquad (ii) \ \lim_{n \to \infty} n^3/3^n.$$

(7.5) If $a_1 = 1$ and $a_n = +\sqrt{2 + a_{n-1}}$ for $n \geq 2$, show that $(a_n)_{n=1}^{\infty}$ converges and find $\lim_{n \to \infty} a_n$.

(7.6) Find all real numbers x such that the series

$$\sum_{n=1}^{\infty} \left(\frac{x}{1-x}\right)^n$$

converges.

(7.7) Sketch the graphs of the functions $f, g, f \pm g : \mathbf{R} \longrightarrow \mathbf{R}$ where $f(x) = x^+$ and $g(x) = x^-$.

(7.8) Find all real solutions of the equation $2(\log x)^2 - 3\log(x^2) = 8$.

(7.9) Find all positive real numbers x such that the series

$$\sum_{n=1}^{\infty} \left(\frac{x^2}{1+x}\right)^n$$

converges. Sum the series when it converges.

(7.10) For which positive real numbers does the series

$$\sum_{n=1}^{\infty} \frac{2^n (x + 2\log(x^2))^n}{n!}$$

converge absolutely. Sum the series when it converges.

(7.11) Find all real numbers x which satisfy the inequality $|\, 1 - 2\log x \,| < 1$.

(7.12) If n is a positive integer, let $s_n = \sum_{i=0}^{n} \frac{1}{i!}$. Show, using geometric series, that

$$0 < e - s_n < \frac{1}{n!n}.$$

Hence show that $n!(e - s_n)$ is not an integer and deduce that e is irrational. Show that $\sum_{n=2}^{\infty}(n!)^{-2}$ is irrational.

(7.13) Show that the series

$$\sum_{n=1}^{\infty} \left(\frac{1}{\sqrt{2}}\right)^n$$

converges to an irrational number.

(7.14) Find the least upper bound of the set $(\frac{n^4+1}{n^4+2} : n = 1, 2, 3, \ldots)$.

(7.15) Find a series expansion, in powers of x, for $\exp(x^2 - 1)$.

(7.16) If a real number $a, 0 \le a < 1$, has two different decimal expansions, show that one of them consists of 9's after some point.

(7.17) Find all real x which satisfy $4(3^{2x+1}) + 17(3^x) - 7 = 0$.

(7.18) If $a_\alpha \ge 0$ for all $\alpha \in \Gamma$ and $\sum_{\alpha \in \Gamma} a_\alpha < \infty$ show that $\{\alpha : a_\alpha > 0\}$ is countable.

(7.19) If Ω is a countable infinite set, show that the set of all finite subsets of Ω is countable and that the set of all subsets of Ω is uncountable.

(7.20) If $A_\alpha \subset \mathbf{R}$ for all $\alpha \in \Gamma$, show that $lub\left(\bigcup_{\alpha \in \Gamma} A_\alpha\right) = lub(\{lub(A_\alpha)\}_{\alpha \in \Gamma})$. If $A, B \subset \mathbf{R}^+$ use this result and $A \cdot B = \bigcup_{b \in B} A \cdot \{b\}$ to prove Theorem 6.12.

(7.21) Use Theorems 6.11 and 7.17 to give another proof of Theorem 6.12.

(7.22) Let $r \in \mathbf{R}$. Show there exists an increasing sequence of rational numbers $(p_n)_{n=1}^{\infty}$ and a decreasing sequence of rational numbers $(q_n)_{n=1}^{\infty}$ such that $\lim_{n \to \infty} p_n = \lim_{n \to \infty} q_n = r$.

(7.23) If $A \subset \mathbf{R}$ and $lub(A) < \infty$ show that A contains an increasing sequence $(x_n)_{n=1}^{\infty}$ such that $\lim_{n \to \infty} x_n = lub(A)$. State and prove a similar result for the greatest lower bound.

References [9; 16]

Chapter 8

REVIEW AND TRANSITION

If a new result has value it is when, by binding together long known elements, until now scattered and appearing unrelated to each other, it suddenly brings order where there reigned apparent disorder.

Henri Poincaré, 1854-1912

Summary

We give a general definition of convergent sequence using increasing and decreasing sequences. We define power series and apply the ratio test to find the radius of convergence of certain power series.

8.1 Review and Preview

We pause in this chapter to review our progress, to highlight aspects of different ideas that we have encountered, and to consider how we might proceed.

So far, we have looked at *Sets, Functions, Logic, and Number Systems*. The starting point was counting and the natural numbers. The number systems $\mathbf{Z}$ and $\mathbf{Q}$ were developed from $\mathbf{N}$ in response to different demands. Logic and different methods of proof allowed us to proceed in an agreed fashion. Without proofs and logic we would have descended long ago into argument about what was true, what was acceptable, what could be used in the development of the subject, and would not have been able to separate intuition from fact. Set theory provided the language in which it was possible to phrase different statements and classify different collections. Functions allowed us to pass from one set to another.

The biggest gap in our development is the fact that we have not yet constructed the real numbers $\mathbf{R}$, a task that was first satisfactorily completed in the second half of the 19th century and that we postpone until Chapter 12. Starting from $\mathbf{N}$ we were motivated to construct different collections of numbers by attempting to solve certain equations:

   (a) $n + x = m$, where $n, m \in \mathbf{N}$,
   (b) $n \cdot x = m$, where $n, m \in \mathbf{N}$,
   (c) $x^2 = n$, where $n \in \mathbf{N}$.

The equations (a) and (b) led, via equivalence classes in Chapter 5, to the rational numbers $\mathbf{Q}$ while (c) showed that the number line $\mathbf{R}$ was larger than $\mathbf{Q}$. In fact, we proved afterwards that $\mathbf{R}$ was much larger than $\mathbf{Q}$. These show that both necessity and curiosity motivated the introduction of new number systems.[1] However, the main reason for constructing the real numbers is to provide a collection of domains, the *closed bounded intervals*, on which a sufficiently rich class of functions, the *continuous functions*, achieve their maxima and minima. Essentially, this means that the set of real numbers, when looked upon as the points on a straight line *contains no gaps*. This, even though it appears unrelated, is equivalent to showing that the real numbers satisfy an *Upper Bound Principle*. There are several approaches to this problem. One could consider equivalence classes of Cauchy sequences[2] or Dedekind cuts of $\mathbf{Q}$ or one could, as we do in Chapter 12, use dyadic sections and the Bisection Principle. The construction is ingenious

---

[1] If we considered all solution to equations similar to (c) we would be led to the *algebraic numbers* and on pursuing it further we would be led to *complex numbers*.

[2] A sequence $(a_n)_{n=1}^{\infty}$ is a Cauchy sequence if $\lim_{n,m\to\infty} |a_n - a_m| = 0$. Cauchy sequences were introduced independently by Bolzano and Cauchy.

and, apart from the routine task of verifying that certain basic properties are satisfied, it is short and direct. The consequences of having the real numbers at our disposal are immense. Delaying their rigorous introduction until we have observed their effectiveness allows us to more fully appreciate the almost inevitable steps in their construction.

Very few concepts or structures in mathematics appear out of the void. They are usually inspired by simpler notions, and the more sophisticated ideas retain, sometimes disguised and perhaps with modifications, traces of their humble origin. Insight may follow from observing such progressions; for example, it is interesting to compare the following properties of **R** and **N**:

(i) every non-empty subset of **N** has a *smallest element*,

(ii) every non-empty subset of **R** which is bounded below has a *greatest lower bound*.

To study continuous and differentiable functions we require *convergence of an arbitrary sequence* of real numbers. This we define using, from the previous chapter, convergence of increasing and decreasing sequences. We then introduce an important notion, that of a *subsequence of a given sequence*. The main result in this chapter is Theorem 8.6, proved using a diagonal process. We extend our repertoire by considering alternating series and power series, neither of which could be investigated by the methods of Chapter 7.

8.2 Convergent Sequences

In this section we discuss convergence of an arbitrary sequence of real numbers.

Definition 8.1. A sequence of real numbers $(a_n)_{n=1}^{\infty}$ converges to the real number a if there exists an increasing sequence $(b_n)_{n=1}^{\infty}$ and a decreasing sequence $(c_n)_{n=1}^{\infty}$ both of which converge to a such that

$$b_n \leq a_n \leq c_n$$

for all n.

Suppose $(a_n)_{n=1}^{\infty}$ is an increasing sequence and $a = lub(\{a_n\}_{n=1}^{\infty})$ is finite. Let $b_n = a_n$ for all n and let $c_n = a$ for all n. Then $(b_n)_{n=1}^{\infty}$ is increasing and $(c_n)_{n=1}^{\infty}$ is decreasing and both converge to a. Since $b_n \leq a_n \leq c_n$ this shows that $(a_n)_{n=1}^{\infty}$ converges to a according to Definition 8.1 and we have extended the definition of convergent sequence given in Definition 7.5. A similar argument works for decreasing sequences. Since an increasing sequence is convergent if and only if it is bounded above and a decreasing sequence is convergent if and only if it is bounded below, Definition 8.1 implies that a convergent sequence is bounded, that is bounded above and below. Our next result follows immediately from Theorem 7.8.

Theorem 8.2. *If $(a_n)_{n=1}^{\infty}$ and $(b_n)_{n=1}^{\infty}$ are convergent sequences of real numbers and c is a real number, then $(a_n+b_n)_{n=1}^{\infty}$, $(c \cdot a_n)_{n=1}^{\infty}$ and $(a_n \cdot b_n)_{n=1}^{\infty}$ converge and*

$$\lim_{n \to \infty} (a_n \pm b_n) = \lim_{n \to \infty} (a_n) \pm \lim_{n \to \infty} (b_n),$$
$$\lim_{n \to \infty} (c \cdot a_n) = c \lim_{n \to \infty} a_n,$$
$$\lim_{n \to \infty} (a_n \cdot b_n) = \lim_{n \to \infty} a_n \cdot \lim_{n \to \infty} b_n.$$

If, additionally, $\lim_{n \to \infty} b_n \neq 0$, then

$$\lim_{n \to \infty} \frac{a_n}{b_n} = \frac{\lim_{n \to \infty} a_n}{\lim_{n \to \infty} b_n}.$$

Example 8.3. Using Theorem 8.2 and $\lim_{n \to \infty}(1/n) = 0$ we obtain:

$$\lim_{n \to \infty} \frac{n^2+1}{2n^2-n} = \lim_{n \to \infty} \frac{1+\frac{1}{n^2}}{2-\frac{1}{n}} = \frac{1+\lim_{n \to \infty}\frac{1}{n^2}}{2-\lim_{n \to \infty}\frac{1}{n}} = \frac{1+\left(\lim_{n \to \infty}\frac{1}{n}\right)^2}{2-\lim_{n \to \infty}\frac{1}{n}} = \frac{1}{2}.$$

Theorem 8.4. *The sequence of real numbers $(a_n)_{n=1}^{\infty}$ converges to a if and only if, for any open interval (r,s) containing a, there exists a positive integer n_0 such that $a_n \in (r,s)$ for all $n \geq n_0$.*

Proof. Suppose $\lim_{n \to \infty} a_n = a$. Let $(b_n)_{n=1}^{\infty}$ and $(c_n)_{n=1}^{\infty}$ denote increasing and decreasing sequences respectively both converging to a and satisfying $b_n \leq a_n \leq c_n$ for all n. Since $lub(\{b_n\}_{n=1}^{\infty}) = glb(\{c_n\}_{n=1}^{\infty}) = a$ and $r < a < s$, there exist positive integers i and j such that $r < b_i$ and $c_j < s$. Let $n_0 = i + j$. If $n \geq n_0$, then $r < b_i \leq b_n \leq a_n \leq c_n \leq c_j < s$.

Conversely, suppose that for every interval (r, s) containing a, we can choose an integer n_0 such that $a_n \in (r, s)$ for all $n \geq n_0$. For each positive integer j choose a positive integer m_j such that $a_n \in (a - \frac{1}{j}, a + \frac{1}{j})$ for all $n \geq m_j$. For all j, let $n_j = m_1 + \cdots + m_j$. Then $(n_j)_{j=1}^{\infty}$ is a strictly increasing sequence of positive integers. Let $b_n = a + 1$ and $c_n = a - 1$ for $n \leq n_2$ and for all $j > 1$ let $b_n = a + \frac{1}{j}$ and $c_n = a - \frac{1}{j}$ for $n_j < n \leq n_{j+1}$. Then $(b_n)_{n=1}^{\infty}$ is decreasing to a, $(c_n)_{n=1}^{\infty}$ is increasing to a, and $c_n \leq a_n \leq b_n$ for all n. This completes the proof. $\square$

Theorem 8.5. *(Squeezing Principle) If* $(a_n)_{n=1}^{\infty}$, $(b_n)_{n=1}^{\infty}$ *and* $(c_n)_{n=1}^{\infty}$ *are sequences of real numbers with* $\lim_{n \to \infty} a_n = \lim_{n \to \infty} c_n = a \in \mathbf{R}$ *and*

$$a_n \leq b_n \leq c_n$$

for all n*, then* $\lim_{n \to \infty} b_n = a$*.*

Proof. By Definition 8.1, there exists an increasing sequence $(d_n)_{n=1}^{\infty}$ that converges to a such that $d_n \leq a_n$ for all n and a decreasing sequence $(f_n)_{n=1}^{\infty}$ that converges to a such that $c_n \leq f_n$ for all n. Since

$$d_n \leq a_n \leq b_n \leq c_n \leq f_n$$

for all n an application of Definition 8.1 completes the proof. $\square$

By dropping some terms but keeping an infinite number from a given sequence $(a_n)_{n=1}^{\infty}$ we obtain a new sequence, a *subsequence* of the original sequence. We often write this sequence as $(a_{n_j})_{j=1}^{\infty}$. More formally every subsequence of the sequence $f : \mathbf{N} \longrightarrow A$ has the form $f \circ \varphi$ where $\varphi : \mathbf{N} \longrightarrow \mathbf{N}$ is a strictly increasing mapping. Clearly, a subsequence of a convergent increasing or decreasing sequence converges to the same limit as the original sequence and hence, by Definition 8.1, $\lim_{j \to \infty} a_{n_j} = a$ whenever $\lim_{n \to \infty} a_n = a$. We now use a *diagonal process* to prove a fundamental result first mentioned by Bolzano in 1817 and first rigorously proved by Weierstrass in his Berlin lectures around 1860. The result is now known as the Bolzano-Weierstrass Theorem.

Theorem 8.6. *Every bounded sequence of real numbers contains a convergent subsequence.*

Proof. Since a sequence is bounded if and only if it is contained in a finite interval, we may suppose that $(x_n^1)_{n=1}^\infty$ is a sequence in $[a_1, b_1]$. At least one of the intervals $[a_1, \frac{a_1+b_1}{2}]$ or $[\frac{a_1+b_1}{2}, b_1]$ contains a subsequence of $(x_n^1)_{n=1}^\infty$. We denote an interval that does by $[a_2, b_2]$ and a chosen subsequence by $(x_n^2)_{n=1}^\infty$. Repeating this process we generate a sequence of intervals $([a_n, b_n])_{n=1}^\infty$, where $(a_n)_{n=1}^\infty$ is an increasing sequence and $(b_n)_{n=1}^\infty$ is decreasing, and a sequence of sequences $(x_n^m)_{n=1}^\infty, m = 1, 2, \ldots$ where $(x_n^{m+1})_{n=1}^\infty$ is a sequence in $[a_{m+1}, b_{m+1}]$ and a subsequence of $(x_n^m)_{n=1}^\infty$. Since $a_1 \le a_n \le b_n \le b_1$ for all n, the sequences $(a_n)_{n=1}^\infty$ and $(b_n)_{n=1}^\infty$ converge and $\lim_{n\to\infty} a_n \le \lim_{n\to\infty} b_n$. We now have the following display:

$$[x_1^1], x_2^1, x_3^1, \ldots, x_n^1, \ldots$$
$$x_1^2, [x_2^2], x_3^2, \ldots, x_n^2, \ldots$$
$$x_1^3, x_2^3, [x_3^3], \ldots, x_n^3, \ldots$$

$$\cdots\cdots\cdots\cdots\cdots\cdots\cdots\cdots$$
$$\cdots\cdots\cdots\cdots\cdots\cdots [x_n^n] \cdots$$

$$\cdots\cdots\cdots\cdots\cdots\cdots$$

and we claim that the diagonal sequence $(x_n^n)_{n=1}^\infty$ has the properties we seek. Clearly, it is subsequence of the original sequence and thus it suffices to show it converges. Since $a_n \le x_n^n \le b_n$ for all n, we must prove $\lim_{n\to\infty} a_n = \lim_{n\to\infty} b_n$. For all n

$$b_n - a_n = \frac{b_1 - a_1}{2^{n-1}} \longrightarrow 0$$

as $n \longrightarrow \infty$ and hence,

$$\lim_{n\longrightarrow\infty} b_n = \lim_{n\longrightarrow\infty} a_n + \lim_{n\longrightarrow\infty} \frac{b_1 - a_1}{2^{n-1}} = \lim_{n\longrightarrow\infty} a_n.$$

This completes the proof. □

8.3 Convergent Series

The crucial concepts in this chapter are now in place. In this section we provide practical tests for convergence of series and in the following section we consider power series.

Definition 8.7. Let $(a_n)_{n=1}^{\infty}$ denote a sequence of real numbers. If $s_n = \sum_{i=1}^{n} a_i$ and the sequence $(s_n)_{n=1}^{\infty}$ converges, we say that the series $\sum_{n=1}^{\infty} a_n$ converges and write $\sum_{n=1}^{\infty} a_n < \infty$. If $s_n \longrightarrow s$ as $n \to \infty$ we say that the series $\sum_{n=1}^{\infty} a_n$ converges to s and write $\sum_{n=1}^{\infty} a_n = s$.

We call a_n the nth term and s_n the nth partial sum of the series $\sum_{n=1}^{\infty} a_n$.

Theorem 8.8. *If the series $\sum_{n=1}^{\infty} a_n$ converges, then $\lim_{n \to \infty} a_n = 0$.*

Proof. Suppose $\sum_{n=1}^{\infty} a_n = s$. If $s_n = \sum_{i=1}^{n} a_i$ then $\lim_{n \to \infty} s_n = \lim_{n \to \infty} s_{n+1} = s$ and

$$\lim_{n \to \infty} a_n = \lim_{n \to \infty} (s_{n+1} - s_n) = \lim_{n \to \infty} s_{n+1} - \lim_{n \to \infty} s_n = s - s = 0. \qquad \square$$

Our remarks in Section 8.1 show that Definition 8.7 extends Definition 7.9. The following result shows that absolutely convergent series, defined in the previous chapter, converge according to Definition 8.7 to the sum assigned to them in Chapter 7.

Theorem 8.9. *If $\sum_{n=1}^{\infty} |a_n| < \infty$, then $\sum_{n=1}^{\infty} a_n < \infty$ and*

$$\sum_{n=1}^{\infty} a_n = \sum_{n=1}^{\infty} a_n^+ - \sum_{n=1}^{\infty} a_n^-.$$

Proof. Let $s_n^+ = \sum_{i=1}^{n} a_i^+$ and $s_n^- = \sum_{i=1}^{n} a_i^-$. Since $s_n^+ \le \sum_{i=1}^{n} |a_i| \le s$, where $s = \sum_{i=1}^{\infty} |a_i|$, and $s_n^- \le \sum_{i=1}^{n} |a_i| \le s$, it follows that $s^+ := \sum_{i=1}^{\infty} a_i^+ < \infty$ and $s^- := \sum_{i=1}^{\infty} a_i^- < \infty$. The sequences $(s_n^+)_{n=1}^{\infty}$ and $(s_n^-)_{n=1}^{\infty}$ are both increasing and

$$s_n^+ - s^- \le s_n \le s^+ - s_n^-.$$

Since $\lim_{n \to \infty}(s_n^+ - s^-) = \lim_{n \to \infty}(s^+ - s_n^-) = s^+ - s^-$, this implies $s_n \longrightarrow s^+ - s^-$ as $n \longrightarrow \infty$ and the series $\sum_{n=1}^{\infty} a_n$ converges. This completes the proof. $\qquad \square$

The general converses to Theorems 8.8 and 8.9 are both false. To give one counter-example to both, in Example 8.11, we look at another type of series, alternating series. A series in which alternate terms are positive and negative is called an *alternating series*. The classical example is the series with n^{th} term $(-1)^{n+1}$,

$$\sum_{n=1}^{\infty}(-1)^{n+1} = +1 - 1 + 1 - 1 \cdots .$$

We confine our proofs to alternating series in which the first term is positive. If $(a_n)_{n=1}^{\infty}$ is an alternating sequence with $a_1 > 0$, then $b_n := (-1)^{n+1}a_n > 0$ and $(-1)^{n+1}b_n = (-1)^{2n+2}a_n = a_n$ for all n. Hence

$$\sum_{n=1}^{\infty} a_n = \sum_{n=1}^{\infty}(-1)^{n+1}b_n.$$

Theorem 8.10. *(Alternating Series Test) If $\sum_{n=1}^{\infty}(-1)^{n+1}a_n$ is an alternating series with $a_n > 0$ for all n and $(a_n)_{n=1}^{\infty}$ is a decreasing sequence then $\sum_{n=1}^{\infty}(-1)^{n+1}a_n$ converges if and only if $a_n \longrightarrow 0$ as $n \longrightarrow \infty$.*

Proof. By Theorem 8.8, the nth term in a convergent series tends to 0. This proves one half of the theorem.

Suppose $a_n \longrightarrow 0$ as $n \longrightarrow \infty$. Let

$$b_{2n-1} := b_{2n} := \sum_{j=1}^{2n}(-1)^{j+1}a_j = \sum_{j=1}^{n}(a_{2j-1} - a_{2j})$$

and

$$c_{2n-1} := c_{2n} := \sum_{j=1}^{2n-1}(-1)^{j+1}a_j = a_1 - \sum_{j=1}^{n-1}(a_{2j} - a_{2j+1})$$

for $n \geq 1$. Since $a_{2j-1} - a_{2j} \geq 0$ for all j, $(b_n)_{n=1}^{\infty}$ is an increasing sequence and $(c_n)_{n=1}^{\infty}$ is an decreasing sequence. Moreover,

$$b_{2n-1} = b_{2n} = \sum_{j=1}^{2n}(-1)^{j+1}a_j \leq \sum_{j=1}^{2n-1}(-1)^{j+1}a_j = c_{2n-1} = c_{2n}.$$

Hence

$$b_n \leq \sum_{j=1}^{n}(-1)^{j+1}a_n \leq c_n$$

for all n and, by Exercise 8.3, the sequences $(b_n)_{n=1}^{\infty}$ and $(c_n)_{n=1}^{\infty}$ converge. Since

$$c_{2n} - b_{2n} = c_{2n-1} - b_{2n-1} = -a_{2n} \longrightarrow 0$$

as $n \longrightarrow \infty$, $\lim_{n\to\infty} b_n = \lim_{n\to\infty} c_n$ and, by Definition 8.1, the series converges. This completes the proof. $\qquad\square$

In this book we do not examine the *rate of convergence* of partial sums to their limit, an important consideration in applied mathematics. Exercise 8.21 shows that the rate of convergence may be extremely slow, for instance by Exercise 8.12 one must add the first 1000 terms from the series $\sum_{n=1}^{\infty} \frac{(-1)^{n+1}}{n}$ to approximate $\log 2$ to three places of decimals. The inequality implicit in Definition 8.1, $|a_n - a| \le c_n - b_n$, estimates the rate of convergence of a sequence. James Stirling, in his book on infinite series, obtained various transformations of series which speeded up the rate of convergence.[3]

Example 8.11. Since $(\frac{1}{n})_{n=1}^{\infty}$ is a decreasing sequence which converges to 0 the series $\sum_{n=1}^{\infty} (-1)^{n+1}/n$ is convergent by Theorem 8.10. If $\sum_{n=1}^{\infty} 1/n = s < \infty$ then $s_{2^n} = \sum_{j=1}^{2^n} 1/j \longrightarrow s$ and $s_{2^{n+1}} = \sum_{j=1}^{2^{n+1}} 1/j \longrightarrow s$ as $n \longrightarrow \infty$. Hence $s_{2^{n+1}} - s_{2^n} \longrightarrow 0$ as $n \longrightarrow \infty$. This contradicts

$$s_{2^{n+1}} - s_{2^n} = \sum_{j=2^n+1}^{2^{n+1}} 1/j \ge 2^n \cdot \frac{1}{2^{n+1}} = \frac{1}{2}$$

and shows that the series $\sum_{n=1}^{\infty} 1/n$ diverges, a result from the 14th century due to Bishop Nicolas Oresme. Hence $\sum_{n=1}^{\infty} (-1)^{n+1}/n$ is not absolutely convergent (see Exercise 8.12).

The following theorem contains two tests for convergence. The *comparison test* in (a) is just Theorem 7.11, while the *ratio test* in (b) follows from using geometric series and the comparison test. Note that by Theorem 8.9 and Examples 7.10 and 7.13, the series $\sum_{n=0}^{\infty} r^n$ converges absolutely if and only if $|r| < 1$.

---

[3] James Stirling, 1692-1770, was born near Stirling in Scotland. He was educated at Oxford and spent a number of years in Italy where he became friendly with the first Nicolaus Bernoulli. He settled in London in 1724 for ten years where he was friendly with Newton and Swift. Afterwards he returned to Scotland where he managed a mine for over thirty years. He was a progressive manager, a model for social reformers, and an influential figure in the opening stages of the Scottish Industrial Revolution. Stirling's main mathematical contributions were to the calculus and to the theory of infinite series and these appeared in his 1730 book *Methodus Differentialis: sive Tractatus de Summatione et Interpolatione Serierum Infinitarum.*

Theorem 8.12. (a) *If $|a_n| \le |b_n|$ for all n, then $\sum_{n=1}^{\infty} |a_n|$ converges when $\sum_{n=1}^{\infty} |b_n|$ converges, and $\sum_{n=1}^{\infty} |b_n|$ diverges when $\sum_{n=1}^{\infty} |a_n|$ diverges.*

(b) *If $a_n \ne 0$ for all n and $\lim_{n \longrightarrow \infty} \left| \frac{a_{n+1}}{a_n} \right| = l$, then the series $\sum_{n=1}^{\infty} a_n$ converges if $l < 1$ and diverges if $l > 1$.*

Proof. (b) If $l < 1$, choose r such that $l < r < 1$. By Theorem 8.4, there exists a positive integer n_0 such that $|a_{n+1}| \le r|a_n|$ for all $n \ge n_0$. For every positive integer m we have

$$|a_{n_0+m}| \le r|a_{n_0+m-1}| \le r^2|a_{n_0+m-2}| \le \cdots \le r^m|a_{n_0}|.$$

By part (a), $\sum_{n=n_0}^{\infty} |a_n| \le |a_{n_0}| \sum_{m=1}^{\infty} r^m < \infty$, and $\sum_{n=1}^{\infty} a_n$ converges.

If $l > 1$, choose r such that $1 < r < l$. By Theorem 8.4, there exists a positive integer n_0 such that $|a_{n+1}| \ge r|a_n|$ for all $n \ge n_0$. For every positive integer m we have

$$|a_{n_0+m}| \ge r|a_{n_0+m-1}| \ge r^2|a_{n_0+m-2}| \ge \cdots \ge r^m|a_{n_0}| \ge |a_{n_0}|.$$

Hence $|a_n| \ge |a_{n_0}|$ for all $n \ge n_0$ and $\lim_{n \to \infty} a_n \ne 0$. By Theorem 8.8, the series $\sum_{n=1}^{\infty} a_n$ does not converge. This completes the proof. $\square$

Example 8.13. If $a_n = 2^n \cdot n^2/(n+1)^3$ then

$$
\begin{aligned}
\left| \frac{a_{n+1}}{a_n} \right| &= \left| \frac{\frac{2^{n+1} \cdot (n+1)^2}{(n+2)^3}}{\frac{2^n \cdot n^2}{(n+1)^3}} \right| \\
&= \left| \frac{2^{n+1}}{2^n} \cdot \frac{(n+1)^2}{n^2} \cdot \frac{(n+1)^3}{(n+2)^3} \right| \\
&= \left| 2 \cdot \left(\frac{n+1}{n} \right)^2 \cdot \left(\frac{n+1}{n+2} \right)^3 \right| \\
&= \left| 2 \cdot \left(1 + \frac{1}{n} \right)^2 \cdot \left(1 - \frac{1}{n+2} \right)^3 \right| \\
&\longrightarrow 2 \text{ as } n \longrightarrow \infty
\end{aligned}
$$

and the series $\sum_{n=0}^{\infty} \frac{2^n \cdot n^2}{(n+1)^3}$ diverges.

Example 8.14. Let $a_n = \frac{n^4}{10^n\sqrt{n!}}$. Then

$$\left|\frac{a_{n+1}}{a_n}\right| = \left|\frac{\frac{(n+1)^4}{10^{n+1}\cdot\sqrt{(n+1)!}}}{\frac{n^4}{10^n\cdot\sqrt{n!}}}\right|$$

$$= \left|\frac{10^n}{10^{n+1}}\cdot\frac{(n+1)^4}{n^4}\cdot\frac{(n!)^{1/2}}{((n+1)!)^{1/2}}\right|$$

$$= \left|\frac{1}{10}\cdot\left(\frac{n+1}{n}\right)^4\cdot\left(\frac{n!}{(n+1)!}\right)^{1/2}\right|$$

$$= \left|\frac{1}{10}\cdot\left(1+\frac{1}{n}\right)^4\cdot\frac{1}{\sqrt{n+1}}\right|$$

$$\longrightarrow 0 \text{ as } n \longrightarrow \infty$$

and the series $\sum_{n=1}^{\infty}\frac{n^4}{10^n\sqrt{n!}}$ converges. By Theorem 8.8, $\lim_{n\to\infty}\frac{n^4}{10^n\sqrt{n!}} = 0$.

It is useful to compare, using the exponential function, the growth rates of sequences that regularly occur in analysis, for instance

$$(n!)^a, \quad b^n, \quad n^c, \quad (\log n)^d$$

where a, b, c and d are *positive* real numbers with $b > 1$. These can be rewritten as

$$\exp\left(a\sum_{j=1}^{n}\log j\right), \quad \exp(n\log b), \quad \exp(c\log n), \quad \exp(d\log(\log n)).$$

By Exercise 8.6,

$$\lim_{n\to\infty}\frac{n}{\log n} = +\infty$$

and, moreover, for $n > 1$

$$4n\log(n) \geq \log(2n!) = \sum_{j=1}^{2n}\log j \geq \sum_{j=n}^{2n}\log j \geq n\log n.$$

Since

$$\frac{(n!)^a}{b^n} \geq \exp\left(a\frac{n}{2}\log\frac{n}{2} - n\log b\right) \geq \exp\left(n\left(\frac{a}{2}\log n - \frac{a}{2}\log 2 - b\right)\right),$$

$$\frac{b^n}{n^c} \geq \exp\left(n\log b - c\log n\right) \geq \exp\left(n\left(\log b - \frac{c\log n}{n}\right)\right),$$

and

$$\frac{n^c}{(\log n)^d} \geq \exp(c \log n - d \log(\log n)) \geq \exp\left(\log n \left(c - \frac{\log(\log n)}{\log n}\right)\right),$$

the sequences

$$\left(\frac{(n!)^a}{b^n}\right)_{n=1}^{\infty}, \quad \left(\frac{b^n}{n^c}\right)_{n=1}^{\infty}, \quad \left(\frac{n^c}{(\log n)^d}\right)_{n=1}^{\infty},$$

all tend to $+\infty$ as $n \longrightarrow \infty$ (see Figure 8.1).[4]

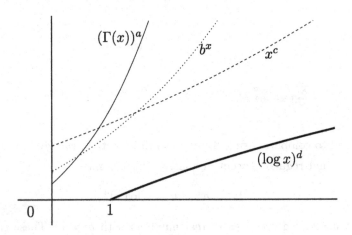

Fig. 8.1

8.4   Power Series

There is an informal hierarchy of increasing complexity within the world of functions. The three simplest kinds are the constant, linear, and quadratic functions. After that, we have the polynomial functions and, in particular, functions such as $f : \mathbf{R} \longrightarrow \mathbf{R}$, $f(x) = (1+x)^n$. Motivated by the *Binomial Theorem* (see Exercise 5.16):

$$(1+x)^n = \sum_{k=0}^{n} \frac{n!}{k!(n-k)!} x^k = 1 + nx + \frac{n(n-1)}{2} x^2 + \cdots + x^n$$

---

[4]The Gamma function or Γ function was introduced by Euler (see Chapter 14). We have $\Gamma(n+1) = n!$ for every positive integer n.

Newton considered

$$f : \mathbf{R} \longrightarrow \mathbf{R}, f(x) = (1+x)^q$$

where $q \in \mathbf{Q}$ and obtained series representations such as

$$\frac{1}{(1-x^2)^{1/2}} = (1-x^2)^{-1/2} = 1 + \frac{1}{2}x^2 + \frac{3}{8}x^4 + \frac{5}{16}x^6 + \frac{35}{128}x^8 + \cdots$$

and verified his results by multiplication of series and using the known formula for the sum of a geometric series. The next step is to consider series of the form

$$\sum_{n=0}^{\infty} a_n x^n. \tag{8.1}$$

We call these *power series*[5] as they are constructed by combining an infinite set of powers of x. In the 17th and 18th centuries the generally accepted concept of function was rather vague and many, including Euler, assumed it was equivalent to being expressible as a power series. Indeed, Lagrange[6] managed to convince himself, by virtue of Taylor's Theorem (see Chapter 14), that the differential calculus could be based on power series and thus made rigorous without the aid of either infinitesimals or limits.

Let

$$A := \left\{ x \in \mathbf{R} : \sum_{n=0}^{\infty} |a_n x^n| < \infty \right\}.$$

If $x \in A, y \in \mathbf{R}$ and $|y| \leq |x|$, then $\pm y \in A$. Hence A is an interval centered at the origin. If $A = \mathbf{R}$, then (8.1) converges for all real numbers

---

[5]We consider power series about the origin. Power series about the point a have the form $\sum_{n=0}^{\infty} a_n(x-a)^n$.

[6]Joseph Louis Lagrange, 1736-1813, was a self-taught mathematician from Turin. He was a professor of mathematics in Turin from 1755 until 1766, when he succeeded Euler as Director of Mathematics at the Berlin Academy. In 1778 he left Berlin to become a member of the Académie des Sciences in Paris and he remained there until he died in 1813. Lagrange was a prolific mathematician who made important contributions to the calculus of variations, mechanics, differential equations, probability theory, number theory, astronomy, fluid mechanics and dynamics. He worked with Legendre and Laplace on the committee standardizing weights and measures. The Lagrangian in mechanics, the Lagrange Interpolation Formula, Lagrange Multipliers and the notation f' for the derivative of f are part of his legacy. His research was always of a high standard but his intuition regarding a power series approach to the differential calculus was mistaken.

and the power series has infinite radius of convergence. If $A \neq \mathbf{R}$, then $A = (-r, r)$ or $A = [-r, r]$ and we call r the radius of convergence of the power series. Suppose $r < \rho < +\infty$ and $\lim_{n \to \infty} |a_n \rho^n| = 0$. If $r < |x| < \rho$, $C = \max_n |a_n \rho^n|$ and $\sigma := (|x|/\rho) < 1$, then

$$|a_n x^n| \leq |a_n \rho^n| \cdot \left(\frac{|x|}{\rho}\right)^n \leq C \cdot \sigma^n.$$

By the comparison test, $x \in A$, that is $|x| \leq r$, and this contradicts our choice of x. Hence $\lim_{n \to \infty} |a_n x^n| \neq 0$ and, by Theorem 8.8, the series $\sum_{n=0}^{\infty} a_n x^n$ diverges when $|x| > r$. We have shown that the series (8.1) converges absolutely for $|x| < r$ and diverges when $|x| > r$ and essentially proved the following result.

Theorem 8.15. *If $\sum_{n=0}^{\infty} a_n x^n$ is a power series, $A = \{|x| : \sum_{n=0}^{\infty} |a_n x^n| < \infty\}$, $B = \{|x| : \{|a_n x^n|\}_{n=0}^{\infty}$ is bounded above$\}$, $C = \{|x| : \lim_{n \to \infty} |a_n x^n| = 0\}$, $D = \{|x| : \sum_{n=0}^{\infty} |a_n x^n| = \infty\}$, $E = \{|x| : \{|a_n x^n|\}_{n=0}^{\infty}$ is not bounded above$\}$, $F = \{|x| : \lim_{n \to \infty} |a_n x^n| \neq 0\}$, then the common value[7] in (8.2)*

$$lub(A) = lub(B) = lub(C) = glb(D) = glb(E) = glb(F) \tag{8.2}$$

is the radius of convergence of $\sum_{n=0}^{\infty} a_n x^n$.

Theorem 8.15 together with Theorems 8.8 and 8.12 enable us to find this interval for many different power series (see Exercises 8.17 and 8.20).

We have already shown that

$$\frac{1}{1-x} = \sum_{n=0}^{\infty} x^n, \quad \text{when } |x| < 1, \tag{8.3}$$

and

$$\exp(x) = \sum_{n=0}^{\infty} \frac{x^n}{n!}, \quad \text{for all } x \in \mathbf{R}. \tag{8.4}$$

---

[7]If $\emptyset$ is the empty set, we let $glb(\emptyset) = +\infty$ and, if A is not bounded above, we let $lub(A) = +\infty$.

When a power series converges for all x in a given set we can use this to define new power series expansions. For example, since $|x^2| < 1$ if and only if $|x| < 1$, we have, by (8.3),

$$\frac{1}{1+x^2} = \frac{1}{1-(-x^2)} = \sum_{n=0}^{\infty}(-x^2)^n = \sum_{n=0}^{\infty}(-1)^n x^{2n}$$

when $|x| < 1$ and, by (8.4), we have

$$\exp(x^3) = \sum_{n=0}^{\infty} \frac{x^{3n}}{n!}$$

when $x \in \mathbf{R}$.

Example 8.16. Consider the power series $\sum_{n=1}^{\infty} \frac{x^{2n}}{n2^{n+1}}$. Since

$$\left| \frac{\frac{x^{2(n+1)}}{(n+1)2^{n+2}}}{\frac{x^{2n}}{n2^{n+1}}} \right| = \left| \frac{n}{n+1} \cdot \frac{x^{2n+2}}{x^{2n}} \cdot \frac{2^{n+1}}{2^{n+2}} \right|$$

$$= \left| \frac{1}{1+\frac{1}{n}} \frac{x^2}{2} \right|$$

$$\longrightarrow \left| \frac{x^2}{2} \right|$$

the series converges absolutely when $|x^2/2| < 1$, that is when $|x| < \sqrt{2}$, and diverges when $|x| > \sqrt{2}$. When $x = \sqrt{2}$, we obtain the series $\sum_{n=1}^{\infty} \frac{2}{n}$, which diverges by Example 8.11 and, if $x = -\sqrt{2}$, we obtain the series $\sum_{n=1}^{\infty} \frac{2(-1)^{n+1}}{n}$, which converges by Theorem 8.10.

8.5 Exercises

(8.1) Is the following statement true or false: a sequence of real numbers $(x_n)_{n=1}^{\infty}$ converges to the real number x if and only if every subsequence of $(x_n)_{n=1}^{\infty}$ contains a subsequence which converges to x. Justify your assertions.

(8.2) Let $(x_n)_{n=1}^{\infty}$ denote a sequence of real numbers. Decide whether or not the following holds: $\lim_{n \to \infty} x_n = 0 \iff \lim_{n \to \infty} |x_n| = 0$.

(8.3) If $(a_n)_{n=1}^{\infty}$ is an increasing sequence and $(b_n)_{n=1}^{\infty}$ is a decreasing sequence and $a_n \le b_n$ for all n, show that both sequences converge and

$\lim_{n\to\infty} a_n \le \lim_{n\to\infty} b_n$. Show that $\lim_{n\to\infty} a_n = \lim_{n\to\infty} b_n$ if and only if $\lim_{n\to\infty}(b_n - a_n) = 0$.

(8.4) Use the alternating series test to show that $\sum_{n=0}^{\infty} \frac{x^n}{n!}$ converges when x is negative.

(8.5) Show that the bounded sequence of real numbers $(x_n)_{n=1}^{\infty}$ converges if and only if any two subsequences of $(x_n)_{n=1}^{\infty}$ contain subsequences which converge to the same limit. Justify your assertions.[8]

(8.6) Show that $2e^{\sqrt{n}} > n$ for every positive integer n. Hence show, using the squeezing principle, that $\lim_{n\to\infty}(\log n)/n = 0$.

(8.7) Sum the series $\sum_{j=1}^{n} jr^j$ (a) directly and (b) by differentiating $\sum_{j=0}^{n} r^j$. Hence show that $\sum_{j=1}^{\infty} jr^j$ converges when $|r| < 1$. Find $\sum_{j=1}^{\infty} jr^j$.

(8.8) Show that $\sum_{n=1}^{\infty} \frac{(-1)^n}{p_n}$ converges where p_n denotes the nth prime.

(8.9) If $(a_n)_{n=1}^{\infty}$ is a decreasing sequence of positive numbers show that $\sum_{n=1}^{\infty} a_n$ converges if and only if $\sum_{n=1}^{\infty} 2^n a_{2^n}$ converges. Is the result true for non-decreasing sequences? Can we replace 2 by any prime number? by an arbitrary positive integer? Prove your assertions. Find all positive real numbers p such that $\sum_{n=1}^{\infty} \frac{1}{n^p}$ converges.

(8.10) Find all sets $\{a, b, c, d\} \in \mathbf{R}^4$ for which the series

$$\sum_{n=2}^{\infty} (\log n)^a \cdot n^b \cdot 2^{nc} \cdot (n!)^d$$

converges.

(8.11) If $\lim_{n\to\infty} a_n = a$ and $b_n := \frac{1}{n}\sum_{i=1}^{n} a_n$ for all n, show that $\lim_{n\to\infty} b_n = a$.

(8.12) Show that $\log \frac{1}{1-x} > x$ for $0 < x < 1$ and use this result to show that the sequence $(a_n)_{n=1}^{\infty}$ is decreasing where $a_n = \sum_{j=1}^{n} \frac{1}{j} - \log n$. Show[9] that $0 < \gamma := \lim_{n\to\infty} a_n < 1$. If

$$b_n = 1 - \frac{1}{2} + \frac{1}{3} - \frac{1}{4} + \cdots + \frac{(-1)^{n+1}}{n}$$

---

[8]The proofs of Theorem 8.4 and Exercises 8.5, 9.10 and 9.16 are close in spirit to the $\epsilon - \delta$ approach to convergence due to Weierstrass.

[9]γ is known as Euler's constant. It is not known if γ is rational.

show that $b_{2n} = a_{2n} - a_n + \log 2$ and deduce that

$$\sum_{n=1}^{\infty} \frac{(-1)^{n+1}}{n} = \log 2.$$

(8.13) Let $(a_n)_{n=1}^{\infty} \sim (b_n)_{n=1}^{\infty}$ if $(a_n)_{n=1}^{\infty}$ is a rearrangement of $(b_n)_{n=1}^{\infty}$. Show that $\sim$ is an equivalence relationship.

(8.14) If every rearrangement of the series $\sum_{n=1}^{\infty} a_n$ is convergent, show that the series is absolutely convergent. If $\sum_{n=1}^{\infty} a_n$ is convergent and, for a subsequence $(n_j)_{j=1}^{\infty}$ we have $\sum_{j=1}^{\infty} |a_{n_j}| < \infty$, show that $\sum_{n \neq n_j} a_n$ converges.

(8.15) If $p_1 = 0$, and otherwise if p_n is the largest prime divisor of n, show that $\sum_{n=1}^{\infty} \frac{(-1)^{n+p_n}}{n}$ converges. Sum the series.

(8.16) Test for convergence

$$\sum_{n=1}^{\infty} \frac{n}{2^n}, \qquad \sum_{n=1}^{\infty} \frac{1}{n \cdot 2^n}, \qquad \sum_{n=1}^{\infty} \frac{n^2 \cdot 3^n}{\sqrt{n!}}, \qquad \sum_{n=1}^{\infty} \frac{n^n}{n!}.$$

(8.17) Find the radius of convergence of the following series

$$\sum_{n=0}^{\infty} \frac{n^3}{2^n} x^{2n}, \qquad \sum_{n=0}^{\infty} \frac{5^n}{\sqrt{n!}} \cdot \frac{1}{n+1} \cdot x^n.$$

(8.18) For which $x \in \mathbf{R}^+$ does the series

$$\sum_{n=0}^{\infty} \frac{2^n (\exp x)^n}{n^2}$$

converge?

(8.19) Find the coefficient of x^{10} in the power series expansion of $(1-x^2)^{-1/2}$.

(8.20) Find the sets of real numbers where the following power series converge

$$\sum_{n=1}^{\infty} x^{n!}, \quad \sum_{n=1}^{\infty} \frac{x^{3n}}{n^2 4^n}, \quad \sum_{n=1}^{\infty} \frac{x^{n^2}}{10^n}, \quad \sum_{n=1}^{\infty} x^{p_n}, \quad \sum_{n=1}^{\infty} \frac{x^{p_n}}{2^n}, \quad \sum_{n=1}^{\infty} \frac{x^{n!}}{\sqrt{n!}}$$

where p_n is the nth prime. You may assume that

$$An \log n < p_n < Bn \log n$$

for all $n > 1$ where A and B are positive numbers.

(8.21) If $(a_n)_{n=1}^{\infty}$ is a decreasing sequence of positive numbers that converges to 0,

$$s_n = \sum_{j=1}^{n}(-1)^{j+1}a_j \text{ and } s = \sum_{j=1}^{\infty}(-1)^{j+1}a_j$$

show that $|s - s_n| \leq a_{n+1}$ for all n.

(8.22) Write out carefully a complete proof of the following result: if $\sum_{n=1}^{\infty} a_n$ and $\sum_{n=1}^{\infty} b_n$ are convergent series then $\sum_{n=1}^{\infty}(a_n + b_n)$ is also convergent and

$$\sum_{n=1}^{\infty} a_n + \sum_{n=1}^{\infty} b_n = \sum_{n=1}^{\infty}(a_n + b_n).$$

(8.23) Provide a detailed proof of Theorem 8.15.

(8.24) If $(a_n)_{n=1}^{\infty}$ and $(b_n)_{n=1}^{\infty}$ are sequences in $\mathbf{R}\backslash\{0\}$ and $\lim_{n\to\infty}\frac{a_n}{b_n} = l \neq 0$, show that $\sum_{n=1}^{\infty} a_n$ converges if an only if $(b_n)_{n=1}^{\infty}$ converges. Use this result to test for convergence

$$\sum_{n=1}^{\infty}\frac{n^2 - n + 4}{n^4 - n^3 + n^2 + 1}, \qquad \sum_{n=1}^{\infty}\frac{2^n - \log n}{2^n + (\log n)^4}.$$

(8.25) If $\lambda \in \mathbf{R}^{+}, a \in \mathbf{R}_0^{+}$ and $g_a : \mathbf{N} \longrightarrow \mathbf{R}$ is given by $g_a(n) = n^a$, sum the series

$$\sum_{n=0}^{\infty} g_a(n)\frac{e^{-\lambda} \cdot \lambda^n}{n!}$$

for $a = 0, 1, 2$.

(8.26) (Heine-Borel Theorem).[10] Let $\mathcal{A} = (A_\alpha)_{\alpha\in\Gamma}$ denote a collection of open intervals in $\mathbf{R}$ and let $[a, b]$ denote a closed bounded interval. If $[a, b] \subset \bigcup_{\alpha\in\Gamma} A_\alpha$ show that there is a finite subset of Γ, Γ_1, such that $[a, b] \subset \bigcup_{\alpha\in\Gamma_1} A_\alpha$

References [6; 9; 16; 20; 25; 30; 34]

---

[10]This result is equivalent to the existence of least upper bounds for sets of real numbers which are bounded above (see Theorem 12.8). Called after Émile Borel, 1871-1956, and Eduard Heine, 1821-1881.

Chapter 9

CONTINUOUS FUNCTIONS

One of the secrets of analysis consists in the
characteristic, that is, in the art of skilful
employment of the available signs.

Gottfried Wilhelm Leibniz

Summary

We define continuous functions and discuss two fundamental theorems: the Intermediate Value Theorem and the Fundamental Existence Theorem for Maxima and Minima. We show that the exponential and log functions are continuous.

9.1 Continuous Functions

The introduction of suitable algebraic notation by François Viète, at the end of the 16th century, facilitated the development of *coordinate geometry* by René Descartes and Pierre Fermat towards the middle of the 17th century. The representation of curves by algebraic formulae, the method of Fermat for finding maxima and minima and an approach to finding the tangent at a point on a curve all preceded the discovery of the *differential and integral calculus*. During the 18th century the concept of function was

155

still evolving and the calculus was generally based on sound intuitions not always accompanied by rigorous reasoning. For two hundred years after Leibniz and Newton, mathematicians knew *how* to apply the methods of the calculus but could not say, with the logical certainty of the ancient Greeks, precisely *when* and *why* these methods worked. We now know that a solid foundation depends on a rigorous construction of the real numbers, that one class of functions, the *continuous functions on closed bounded intervals* always have a maximum and a minimum and that, with the help of another collection, the *differentiable functions*, these can often be easily located. It should not come as a surprise that the fundamental properties of both classes can be based on convergent sequences and, in particular, on Theorem 8.6, which relies on the *Upper Bound Principle*. Alternatively, we could have based our analysis on the upper bound principle and set theory and dispensed with sequences and series (see Exercises 8.26 and 9.16).

Bolzano, Cauchy[1] and Gauss were responsible for initiating the move towards rigor in analysis that culminated in the work of Dedekind and Weierstrass. In 1818, Cauchy began to teach at the École Polytechnique in Paris and, in 1821, he published the first version of his lectures on the differential and integral calculus *Cours d'Analyse*. These were very important and are the basis of our present approach to the calculus. Cauchy aimed to base everything on limits. He laid out clearly the mathematical

---

[1] Augustin Louis Cauchy, 1789-1857, was an extremely prolific mathematician from Paris who published almost 800 articles on mathematics. He was a strong supporter of the Bourbons and, on refusing to take an oath of loyalty to Louis Philippe in 1830, was excluded from public employment in France during the eighteen years of the July monarchy. He experienced prejudice and discrimination because of his strongly held political and religious beliefs, but he himself also behaved in a rather bigoted fashion. While his lecture notes were well written, Cauchy himself was a poor lecturer. His classes often went on far too long and he lectured at too high a level. At one stage, the Minister of the Interior appointed a commission to make sure that his course was sufficiently practical. Federico Luigi Menabrea, 1809-1896, who later became prime minister of Italy and who was apparently the only student of 30 who completed a course given by Cauchy during his self-imposed exile in Turin, reports that Cauchy's *presentations were obscure clouds illuminated from time to time by flashes of pure genius*. Cauchy made important contributions to differential equations and complex analysis.

environment in which he worked and gave, shortly after Bolzano, clear and almost modern definitions of *limit* and *continuous function*. Cauchy also gave the *first formal definition of the integral* of a bounded function, continuous except perhaps at a finite set of points, and stated conditions under which infinite series converge.

Definition 9.1. If $A \subset \mathbf{R}$ and $f : A \longrightarrow \mathbf{R}$, then f is continuous at $x \in A$, if for all $(x_n)_{n=1}^{\infty} \subset A$ we have $f(x_n) \longrightarrow f(x)$ as $n \longrightarrow \infty$ whenever $x_n \longrightarrow x$ as $n \longrightarrow \infty$. We say the f is continuous on A if f is continuous at every point in A.

Continuity means
$$\lim_{n \to \infty} f(x_n) = f(\lim_{n \to \infty} x_n) \tag{9.1}$$
whenever $\lim_{n \to \infty} x_n \in A$ and continuous functions are precisely those functions which *commute with limits* or *preserve limits*. We may rewrite (9.1) in different ways for instance as:
$$\text{if } x_n \longrightarrow x \text{ then } f(x_n) \longrightarrow f(x). \tag{9.2}$$
If $f : (a, b) \longrightarrow \mathbf{R}$ and $x \in (a, b)$, we let $\lim_{y \to x} f(y) = l$ if $\lim_{n \to \infty} f(x_n) = l$ for *every* sequence $(x_n)_{n=1}^{\infty} \subset (a, b)$ such that $\lim_{n \to \infty} x_n = x$. Continuity at x and (9.1) may now be conveniently rephrased as follows: for all $x \in A$
$$\lim_{y \to x} f(y) = f(x) = f(\lim_{y \to x} y). \tag{9.3}$$
If $\Delta x \in \mathbf{R} \longrightarrow 0$ then $x + \Delta x \longrightarrow x$ and (9.3) is equivalent to
$$\lim_{\Delta x \to 0} f(x + \Delta x) = f(x) \tag{9.4}$$
and to
$$\lim_{\Delta x \to 0} (f(x + \Delta x) - f(x)) = 0 \tag{9.5}$$
and, using the modulus function (see Exercise 8.2),
$$\lim_{\Delta x \to 0} |f(x + \Delta x) - f(x)| = 0. \tag{9.6}$$

Elementary properties of limits, given in Theorem 8.2, imply the following: if $f, g : A \longrightarrow \mathbf{R}$ are continuous and $c \in \mathbf{R}$, then $f \pm g, f \cdot g, cf$ and, if $g(x) \neq 0$ for $x \in A$, f/g are all continuous.

Lemma 9.2. *If $f : A \longrightarrow \mathbf{R}$ and $g : \mathbf{R} \longrightarrow \mathbf{R}$ are continuous, then $g \circ f$ is continuous.*

Proof. Suppose $(x_n)_{n=1}^{\infty} \subset A$, $x \in A$, and $x_n \longrightarrow x$ as $n \longrightarrow \infty$. Since f is continuous we have $y_n := f(x_n) \longrightarrow y := f(x)$ as $n \longrightarrow \infty$. Since g is continuous we have $g(y_n) \longrightarrow g(y)$ as $n \longrightarrow \infty$, that is

$$(g \circ f)(x_n) = g(f(x_n)) = g(y_n) \longrightarrow g(y) = g(f(x)) = (g \circ f)(x)$$

as $n \longrightarrow \infty$. This completes the proof. $\qquad\qquad\qquad\square$

Example 9.3. Let $f(x) = x$ for all $x \in \mathbf{R}$. To show that f is continuous we need to show $\lim_{n \to \infty} f(x_n) = f(x)$ whenever $\lim_{n \to \infty} x_n = x$. Since $f(x_n) = x_n$ and $f(x) = x$, we must show $\lim_{n \to \infty} x_n = x$ whenever $\lim_{n \to \infty} x_n = x$. This is trivially true. Using this result, multiplication and addition we see that polynomial functions such as $f : \mathbf{R} \longrightarrow \mathbf{R}$ where

$$f(x) = 10x^4 - 3x^3 + 5x + 4$$

are continuous.

We next show that the exponential function is continuous. The same general idea can be used to show that a power series is continuous and differentiable within its radius of convergence (see Examples 9.14 and 11.14).

Lemma 9.4. *If* $|x| < 1$ *then*

$$|e^x - 1| < \frac{|x|}{1 - |x|}.$$

Proof. Since

$$e^x = 1 + x + \frac{x^2}{2!} + \cdots$$

$$e^x - 1 = x + \frac{x^2}{2!} + \cdots$$

and

$$|e^x - 1| \le |x| + \frac{|x^2|}{2!} + \cdots$$
$$\le |x|(1 + |x| + |x|^2 + |x|^3 + \cdots)$$
$$= \frac{|x|}{1 - |x|}.$$

$\qquad\qquad\qquad\square$

Theorem 9.5. *The exponential function is continuous.*

Proof. We first show continuity at the origin. Let $(x_n)_{n=1}^\infty$ denote a sequence of real numbers which converges to 0. We may suppose that $|x_n| < 1$ for all n. Since $e^0 = 1$, Lemma 9.4 implies

$$|e^{x_n} - e^0| \le \frac{|x_n|}{1 - |x_n|} \longrightarrow 0 \text{ as } n \longrightarrow \infty \qquad (9.7)$$

and this proves continuity at the origin. Now suppose x is an arbitrary real number and that the sequence of real numbers $(x_n)_{n=1}^\infty$ converges to x. By Theorem 7.17 and (9.7)

$$\begin{aligned}
\lim_{n \longrightarrow \infty} |e^{x_n} - e^x| &= \lim_{n \longrightarrow \infty} |e^{x + x_n - x} - e^x| \\
&= \lim_{n \longrightarrow \infty} |e^x e^{(x_n - x)} - e^x| \\
&= e^x \lim_{n \longrightarrow \infty} |e^{(x_n - x)} - e^0| \\
&= 0
\end{aligned}$$

and the exponential function is everywhere continuous by (9.6). $\qquad \square$

9.2 Intermediate Value Theorem

In Chapter 1 we solved quadratic equations. We now consider the equation

$$f(x) = 0 \qquad (9.8)$$

where $f : [a, b] \longrightarrow \mathbf{R}$ is continuous. *The Intermediate Value Theorem* provides a simple sufficient criterion to show the existence of a solution: f should take positive and negative values at the ends of the interval $[a, b]$. Geometrically (see Figure 9.1) this result is clear and, until Bolzano proposed, in 1817, that it should be proved analytically, mathematicians were happy to accept a non-rigorous intuitive proof.

Theorem 9.6. *(Intermediate Value Theorem) If $f; [a, b] \longrightarrow \mathbf{R}$ is continuous and $f(a)f(b) < 0$, then there exists $c \in [a, b]$ such that $f(c) = 0$.*

Proof. Suppose $f(a) < 0$ and $f(b) > 0$ and that we have chosen $(a_i)_{i=0}^n$ and $(b_i)_{i=0}^n$ where $f(a_i) < 0$, $f(b_i) > 0$,

$$a =: a_0 \le a_1 \le \cdots \le a_{n-1} \le a_n < b_n \le b_{n-1} \le \cdots \le b_1 \le b_0 =: b$$

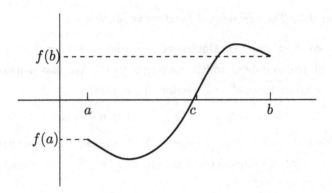

Fig. 9.1

and $2^i(b_i - a_i) = b - a$, for $0 \le i \le n$. Let $c_n = 2^{-1}(a_n + b_n)$. If $f(c_n) = 0$ let $c = c_n$. Otherwise, if $f(c_n) > 0$, let $b_{n+1} = c_n$ and $a_{n+1} = a_n$, and, if $f(c_n) < 0$, let $a_{n+1} = c_n$ and let $b_{n+1} = b_n$. Continuing in this way, we either find a positive integer n with $f(c_n) = 0$ and the proof is complete or we generate sequences such that $f(c_n) \ne 0$ for all n. In the latter situation the sequence $(a_n)_{n=0}^{\infty}$ is an increasing bounded sequence and $f(a_n) < 0$ for all n and $(b_n)_{n=0}^{\infty}$ is a decreasing bounded sequence and $f(b_n) > 0$ for all n. Since $0 \le b_n - a_n \le 2^{-n}(b - a)$, Definition 8.1 implies $\lim_{n \to \infty} a_n = \lim_{n \to \infty} b_n =: c$ (see also Exercise 8.3). Since $f(a_n) < 0$ all n, $f(c) = \lim_{n \to \infty} f(a_n) = lub\{f(a_n)\}_{n=0}^{\infty} \le 0$ and, as $f(b_n) > 0$ all n, $f(c) = \lim_{n \to \infty} f(b_n) = glb\{f(b_n)\}_{n=0}^{\infty} \ge 0$. Hence $f(c) = 0$ (see Figure 9.1). A similar proof works when $f(a) > 0$ and $f(b) < 0$ or, alternatively, one can apply the above proof to the function $g := -f$. This completes the proof.▢

Our first application shows that the *image* of an interval by a continuous function is also an interval and this, in turn, is useful in determining that certain functions are *surjective*.

Corollary 9.7. *If I is an interval and $f : I \mapsto \mathbf{R}$ is continuous, then $f(I) := \{f(x) : x \in I\}$ is an interval.*

Proof. Let $a, b \in I, a < b$, and suppose $f(a) < f(b)$. Let $f(a) < \alpha < f(b)$. The function $g := f - \alpha$, restricted to $[a, b]$, is continuous and

$$g(a)g(b) = (f(a) - \alpha) \cdot (f(b) - \alpha) < 0.$$

By the Intermediate Value Theorem, there exists $c, a < c < b$, such that $g(c) = 0$. Hence $f(c) = \alpha$. A similar proof works when $f(a) > f(b)$. By Definition 6.13, $f(I)$ is an interval. This completes the proof. $\square$

Example 9.8. Since $\exp x \longrightarrow +\infty$ as $x \longrightarrow +\infty$ and $\exp x \longrightarrow 0$ as $x \longrightarrow -\infty$ the Intermediate Value Theorem, implies that $\mathbf{R}^+$ belongs to the image of the exponential function. By Theorem 7.18(c), the image of the exponential function is contained in $\mathbf{R}^+$. This proves Theorem 7.18(g) and completes the proof of Theorem 7.18(h).

Example 9.9. Let $P : \mathbf{R} \longrightarrow \mathbf{R}$ denote the polynomial of degree n

$$P(x) = a_0 + a_1 x + a_2 x^2 + \cdots + a_n x^n$$

where $a_n \neq 0$. We discussed polynomials of degrees 1 and 2 in Chapter 1. For $|x|$ large

$$P(x) = x^n \left(\frac{a_0}{x^n} + \frac{a_1}{x^{n-1}} + \frac{a_2}{x^{n-2}} + \cdots + \frac{a_n}{1} \right) \approx a_n x^n.$$

If n is *odd* and $a_n > 0$, then $P(x) \longrightarrow +\infty$ as $x \longrightarrow +\infty$ and $P(x) \longrightarrow -\infty$ as $x \longrightarrow -\infty$. An analogous statement holds when $a_n < 0$. By the Intermediate Value Theorem, we now have the following result of Euler: if P is a polynomial of degree $2n + 1$ then the equation

$$P(x) = 0$$

has a least *one real solution*.

The Intermediate Value Theorem does not characterize continuous functions and we give later a class of functions that are not necessarily continuous but that satisfy the conclusion of the Intermediate Value Theorem.

9.3   Fundamental Existence Theorem for Maxima and Minima

In this section we obtain the fundamental result, first rigorously proved by Weierstrass, that a continuous real-valued function defined on a closed bounded interval admits a maximum and minimum (see Figure 9.2).[2]

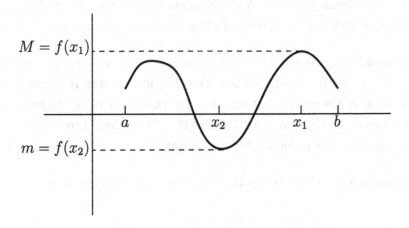

Fig. 9.2

Theorem 9.10. (*Fundamental Existence Theorem for Maxima and Minima*) *If* $f : [a, b] \mapsto \mathbf{R}$ *is continuous, then*

$$im(f) = \{f(x) : x \in [a, b]\} = [m, M]$$

where

$$m = \min\{f(x) : x \in [a, b]\} = glb\{f(x) : x \in [a, b]\}$$

and

$$M = \max\{f(x) : x \in [a, b]\} = lub\{f(x) : x \in [a, b]\}.$$

---

[2]If $f : A \longrightarrow B$ and $im(f)$ is bounded above (respectively below) then the expression *supremum* (respectively *infimum*) is also used for the least upper bound (respectively greatest lower bound) of the image of f and we write $sup(f) = lub(f(A))$ and $inf(f) = glb(f(A))$. When the supremum (respectively infimum) is *achieved* at a point in A, we call it the *maximum* (respectively the *minimum*) of f over A.

Proof.   Suppose the set $\{f(x) : x \in [a, b]\}$ is not bounded above. Then, for each integer n, there exists $x_n \in [a, b]$ such that $f(x_n) > n$. By Theorem 8.6, the sequence $(x_n)_{n=1}^{\infty}$ contains a subsequence $(x_{n_j})_{j=1}^{\infty}$ which converges to $x \in [a, b]$. By continuity $\lim_{j \to \infty} f(x_{n_j}) = f(x)$ and this contradicts the fact that $f(x_{n_j}) > n_j$ for all n_j. Hence, $\{f(x) : x \in [a, b]\}$ is bounded above. Let $M = lub\{f(x) : x \in [a, b]\}$. For each positive integer n there exists $y_n \in [a, b]$ such that $f(y_n) > M - \frac{1}{n}$. Again, by Theorem 8.6, we can choose a subsequence of $(y_n)_{n=1}^{\infty}$ that converges to $y \in [a, b]$. By continuity, $f(y) = M$. Similarly, we can show that $\{f(x) : x \in [a, b]\}$ is bounded below and there exists $z \in [a, b]$ such that $f(z) = m := glb\{f(x) : x \in [a, b]\}$. By Corollary 9.7, $\{f(x) : x \in [a, b]\} = [m, M]$. This completes the proof. $\square$

Combining this result, with the obvious fact that strictly increasing and decreasing functions are injective, we obtain a criterion for detecting bijective functions.

Corollary 9.11. *If* $f : [a, b] \longrightarrow [m, M]$, *where* m *is the minimum and* M *is the maximum of* f *on* $[a, b]$, *is continuous and either strictly increasing or strictly decreasing, then* f *is bijective.*

In the next chapter we will see, using the differential calculus, how to find m and M and how to show that functions are either strictly increasing or strictly decreasing.

Corollary 9.12. *If* $f : [a, b] \longrightarrow [m, M]$ *is a continuous bijective function, then* $f^{-1} : [m, M] \longrightarrow [a, b]$ *is continuous.*

Proof.   Let $y_n \in [m, M] \longrightarrow y \in [m, M]$ as $n \longrightarrow \infty$. We need to show that $f^{-1}(y_n) =: x_n \longrightarrow x := f^{-1}(y)$ as $n \longrightarrow \infty$. Suppose this is not true. Then, by Theorem 8.4, there exists an open interval containing x, (r, s), and an infinite set of positive integers $(n_j)_{j=1}^{\infty}$ such that $x_{n_j} \notin (r, s)$ for all j. By Theorem 8.6, $(x_{n_j})_{j=1}^{\infty}$ contains a subsequence which converges to some point $w \in [a, b]$. By Theorem 8.4, $w \notin (r, s)$ and, in particular, $w \neq x$. By continuity, $f(x_{n_j}) \longrightarrow f(w)$ as $j \longrightarrow \infty$. But, $f(x_{n_j}) = y_{n_j} \longrightarrow y$ as $j \longrightarrow \infty$ and hence $f(w) = y = f(x)$. Since f is bijective this implies

$w = x$. This is impossible and f^{-1} is continuous.   □

A similar result holds for continuous bijective functions on *open intervals* and, combining this with Example 9.9, we see that log : $\mathbf{R}^+ \longrightarrow \mathbf{R}$ is continuous.

Example 9.13. The function $f : (0, 1) \longrightarrow \mathbf{R}$,

$$f(x) = \frac{1}{x} + \frac{1}{x-1}$$

is continuous and, by sketching f (see Figure 9.3), it is easily seen that f has neither a minimum nor a maximum on $(0, 1)$. This shows that a closed interval is necessary in Theorem 9.10.

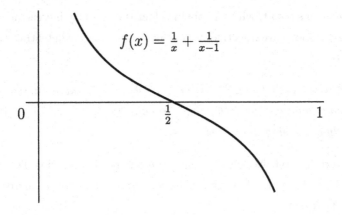

Fig. 9.3

Example 9.14. Suppose the power series $\sum_{n=0}^{\infty} a_n x^n$ converges absolutely when $|x| = l > 0$. Let $f(x) = \sum_{n=0}^{\infty} a_n x^n$ for $|x| \leq l$. Then $f(0) = a_0$ and, by Theorem 8.15, there is a positive real number M such that $|a_n l^n| \leq M$ for all n. For $|\Delta x| < l$ and all n we have $|a_n (\Delta x)^n| = |a_n l^n (\Delta x/l)^n| \leq M |(\Delta x/l)^n|$ and, hence,

$$|f(\Delta x) - f(0)| \leq M \sum_{n=1}^{\infty} (|\Delta x|/l)^n = M \frac{|\Delta x|/l}{1 - (|\Delta x|/l)} = \frac{M|\Delta x|}{l - |\Delta x|}.$$

This implies, as in (9.7), that f is continuous at the origin.

Suppose the power series $\sum_{n=0}^{\infty} a_n x^n$ and $\sum_{n=0}^{\infty} b_n x^n$ converge to $f(x)$ and $g(x)$, respectively, when $|x| < r$. If $a_n = b_n$ for all n then $f(x) = g(x)$ for $|x| < r$. Suppose, conversely, that $f(x) = g(x)$ for $|x| < \delta$ where δ is positive but perhaps much smaller than r. Then $a_0 = f(0) = g(0) = b_0$. Suppose $a_n = b_n$ for all $n < m$ and $a_m \neq b_m$. If $|x| < \delta$, then

$$x^m \cdot \sum_{n=m}^{\infty} a_n x^{n-m} = f(x) - \sum_{n=0}^{m-1} a_n x^n = g(x) - \sum_{n=0}^{m-1} b_n x^n = x^m \cdot \sum_{n=0}^{\infty} b_n x^{n-m}$$

and hence $\sum_{n=m}^{\infty} a_n x^{n-m} = \sum_{n=m}^{\infty} b_n x^{n-m}$ for $0 < |x| < \delta$. Continuity at the origin implies $a_m = b_m$, a contradiction unless $a_n = b_n$ for all n and $f(x) = g(x)$ for $|x| < r$. We have shown that *two power series coincide everywhere they converge provided they coincide on an arbitrarily small interval of positive length.*[3]

Example 9.15. Let $\phi : (a, b) \longrightarrow \mathbf{R}$ be convex. If $a < x < y < z < b$, then

$$0 < \frac{z-y}{z-x} < 1, \qquad 0 < \frac{y-x}{z-x} < 1, \qquad \frac{z-y}{z-x} + \frac{y-x}{z-x} = 1,$$

and

$$y = \frac{z-y}{z-x} \cdot x + \frac{y-x}{z-x} \cdot z.$$

By convexity

$$\phi(y) \leq \frac{z-y}{z-x} \cdot \phi(x) + \frac{y-x}{z-x} \cdot \phi(z).$$

This implies, since $x < z$, that

$$(z - y + y - x)\phi(y) \leq (z-y)\phi(x) + (y-x)\phi(z)$$

and

$$(y-x)(\phi(y) - \phi(z)) \leq (z-y)(\phi(x) - \phi(y)).$$

Hence

$$\frac{\phi(x) - \phi(y)}{x - y} \leq \frac{\phi(y) - \phi(z)}{y - z}. \tag{9.9}$$

---

[3]We considered an interval about the origin but an arbitrary interval suffices. Rationalism, an important philosophical system during the 18th century, maintained that everything was predictable once sufficiently many scientific facts were known. The fact that power series were fully determined by the information contained in a small interval, impressed those who subscribed to this philosophy. See also Exercise 11.22.

Fix $a < u < v < z < w < b$. If $x, y \in [v, z], x \neq y$ then, by (9.9),

$$A := \frac{\phi(u) - \phi(v)}{u - v} \leq \frac{\phi(x) - \phi(y)}{x - y} \leq \frac{\phi(z) - \phi(w)}{z - w} =: B.$$

As x approaches y, $x - y$ is close to 0 and $1/(x - y)$ is either very large and positive or very large and negative. Either way, we have a contradiction unless $\phi(y) \longrightarrow \phi(x)$ as $y \longrightarrow x$. We have shown that *convex functions are continuous*.

9.4   Exercises

(9.1) Let

$$f : [0, 8] \mapsto \mathbf{R}, f = 21_{[1,2]} + 31_{[4,5]} + 61_{[1,7]}.$$

Sketch the graph of f and find the set of points where f is continuous. Justify your remarks. Show that the Intermediate Value Theorem *does not apply* to f on some interval $[a, b] \subset [0, 8]$.

(9.2) If $f : [a, b] \longrightarrow \mathbf{R}$ is a strictly increasing function, show that $\{x \in [a, b] : f$ is not continuous at $x\}$ is a countable set.

(9.3) Let $f : [a, b] \longrightarrow \mathbf{R}$ be continuous. Show that $im(f)$ is either finite or uncountable. Give examples of both possibilities.

(9.4) Show, using Exercise 8.6 and Theorem 9.5 or otherwise, that $\lim_{n \to \infty} n^{1/n} = 1$.

(9.5) Show that

$$P(x) = -x^5 - 3x^2 + 4x + 1 = 0$$

has at least two solutions in $[-2, 0]$.

(9.6) Show that

$$P : \mathbf{R} \longrightarrow \mathbf{R}, \quad P(x) = 2x^4 - 9x^3 + 7x^2 + x = 0,$$

has four real zeros.

(9.7) If $f : [a, b] \longrightarrow [c, d]$ is continuous and injective show that f is either strictly increasing or strictly decreasing.

(9.8) Let f and g denote continuous functions on $[a, b]$. If $f(a) > 0$, $g(a) < 0$, $f(b) < 0$ and $g(b) > 0$ show, using the Intermediate Value Theorem, that there exists c, $a < c < b$, such that $f(c) = g(c)$. Sketch a diagram illustrating this result.

(9.9) Show that 1_Q, the indicator function of the set of rational numbers Q, is nowhere continuous.

(9.10) If $f : (a, b) \longrightarrow R$ is continuous at c, where $a < c < b$, and $f(c) \neq 0$, show that there exists $\delta > 0$, $a < c - \delta < c + \delta < b$, such that $f(y) \neq 0$ for all $y \in (c - \delta, c + \delta)$.

(9.11) Let

$$f : R \to R, \quad f(x) = \frac{x^2 + 1}{x^4 + x^2 + 1}, \quad A = im(f),$$

and

$$g : [0, 1] \to A, \quad g(x) = f(x).$$

Which of the following functions are defined $f \circ f, g \circ f, f \circ g, g \circ g$? Justify your answers. Find $im(g)$.

(9.12) Find

$$\lim_{n \to \infty} \log \left(\frac{n^2 - 1}{n^2 + 1} \right).$$

(9.13) Let $(f_m)_{m=1}^{\infty}$ denote a sequence of continuous real-valued functions defined on $[a, b]$. If, for each m, there exists a positive number M_m such that $|f_m(x)| < M_m$ for all $x \in [a, b]$ and all m and $\sum_{n=1}^{\infty} M_m$ is finite, show that $f := \sum_{m=1}^{\infty} f_m$ is a continuous function on $[a, b]$. If

$$g(x) = \sum_{n=1}^{\infty} \frac{1}{n^2 + x^2}$$

for all $x \in R$, show that g is continuous.

(9.14) If $f : [a, b] \longrightarrow R$ is increasing and continuous and $A \subset [a, b]$, show that $f(lub(A)) = lub(f(A))$.

(9.15) If $A := \{\frac{1}{n}\}_{n=1}^{\infty} \cup \{0\}$, show that $f : A \longrightarrow R$ is continuous if and only if $\lim_{n \to \infty} f(\frac{1}{n}) = f(0)$. If $f : A \longrightarrow R$ is continuous, show that there exists a continuous function $g : R \longrightarrow R$ such that $g(x) = f(x)$ for all $x \in A$.

(9.16) Show that $f : \mathbf{R} \longrightarrow \mathbf{R}$ is continuous if and only if, for all intervals (a, b), the set $\{x \in \mathbf{R} : a < f(x) < b\}$ is a countable union of open intervals.

(9.17) For $f : (a, b) \to \mathbf{R}$, show that the following conditions are equivalent: (a) f is continuous, (b) for *any* sequence $(y_n)_{n=1}^{\infty}$, $y_n \neq 0$ for all n and $\lim_{n \to \infty} y_n = 0$ and any $x \in (a, b)$,

$$\lim_{n \to \infty} \max\{|f(x + y) - f(x)| : |y| \leq |y_n|\} = 0,$$

(c) for *some* sequence $(y_n)_{n=1}^{\infty}$, $y_n \neq 0$ for all n and $\lim_{n \to \infty} y_n = 0$ and any $x \in (a, b)$,

$$\lim_{n \to \infty} \max\{|f(x + y) - f(x)| : |y| \leq |y_n|\} = 0.$$

(9.18) Let $f : (0, 1) \longrightarrow \mathbf{R}$ be defined as follows: $f(x) = 0$ if x is irrational and $f(\frac{p}{q}) = \frac{1}{q}$ whenever p, q are positive integers with $p < q$, p and q having no common factor greater than 1. Show that f is continuous at x if and only if x is irrational. (A common factor of p and q is an integer $r > 1$ such that r divides both p and q.)

References [6; 9; 19; 29; 30]

Chapter 10

DIFFERENTIABLE FUNCTIONS

In signs one observes an advantage in discovery
which is greatest when they express the exact
nature of a thing briefly and, as it were,
picture it; then indeed the labor of thought
is wonderfully diminshed.

Gottfried Wilhelm Leibniz

Summary

We define and prove basic results about differentiable functions such as the Product Rule, the Chain Rule and the Mean Value Theorem. We prove that the exponential and log functions are differentiable, relate the sign of the derivative to the direction of increase and show how to locate local maxima and minima.

10.1 Differentiable Functions

Two mathematical problems of perennial interest are the calculation of lengths of curves and the determination of tangents to curves. The first requires an examination of the complete curve while the second problem is local. It took the genius of the two independent founders of the calculus,

Isaac Newton and Gottfried Wilhelm Leibniz,[1] to recognize that the local problem, involving the differential calculus, was inverse to the global problem, involving the integral calculus. They both discovered, essentially, the same results and between them they gave us many of the techniques, notation and results that one finds in any modern calculus[2] textbook. Neither, however, supplied proofs that would have satisfied Archimedes and their approach and notation were very different. Because of delays in publishing a controversy regarding priorities arose. In 1715, an apparently objective anonymous historical review, now believed to have been written by Newton, appeared in which Leibniz is accused of having stolen Newton's ideas. As a result Continental and English mathematicians ceased to exchange ideas and this hindered the development of English mathematics for most of the 18th century.

The simplest non-constant functions are the *linear* functions. These have the form $f(x) = mx + c$ and are fully determined by two real numbers one of which, m, is the slope. Their graphs are straight lines. In this chapter we consider real–valued functions which can be *locally approximated by linear functions*. To approximate $f : (a, b) \longrightarrow \mathbf{R}$ near $x \in (a, b)$, we

---

[1]Newton, 1642-1716, was born in Woolsthorpe, 100 miles north of London, on Christmas day, and studied in Cambridge. He made fundamental contributions to optics, astronomy, mechanics and mathematics. While the great plague of 1665-1666 raged Newton retreated to Woolsthorpe and made his major discoveries during twenty months of intense creativity in the peace and quiet of the countryside. These include the calculus. He wrote three versions of his results for private circulation in 1669, 1671 and 1676 but, because of his notorious aversion to publication, these only appeared publicly in 1704, 1711 and 1736. In fact, some of Newton's papers were only published in the 20th century. Leibniz, 1646-1716, was born in Leipzig and studied philosophy before turning to mathematics. He obtained his ideas on the calculus between 1673 and 1675 while on an extended stay in Paris and published his first results on the differential calculus in 1684 and on the integral calculus in 1686. He was very conscious of the importance of notation and corresponded with other leading mathematicians on their preferences, often delaying publishing until he felt certain that the notation was suitable. He gave us the notation $\frac{d}{dx}$ and $\int$. Newton and Leibniz both filled positions for the state; Newton became *Master of the Mint* and Leibniz was *librarian and genealogist* to the Duke of Brunswick-Lüneberg.

[2]From now on we use the term *calculus* in place of *integral and differential calculus*.

must consider *all points* near x. We follow traditional notation here, and not our standard convention, by writing $x + \Delta x$, where x is a fixed point or *constant* in (a, b), while Δx, the *variable*, is close to 0. Soon Δx will disappear from our investigations and, at that point, we let x vary, that is become a variable. The straight line that approximates f near x depends on x and, to recognize this relationship, we denote its slope by m_x. We want $f(x + \Delta x) \approx c_x + m_x \Delta x$ for some real number c_x, and as the difference should be 0 when $\Delta x = 0$, we let $c_x = f(x)$. This means that the straight line

$$\Delta x \longrightarrow f(x) + m_x \Delta x$$

approximates f at $x + \Delta x$ when Δx is close to 0. The error in the approximation will depend on Δx but, as x changes, it too will change and we write it as $g_x(\Delta x)$. We now have

$$f(x + \Delta x) = f(x) + m_x \Delta x + g_x(\Delta x) \tag{10.1}$$

where $g_x(\Delta x)$ measures the *error* in our approximation. Clearly, we would like a good approximation or, equivalently, a small error. We do, however, have to be careful. If we are too ambitious and demand too much, we end up with a small collection of functions. On the other hand if we try to include too large a set of functions, the approximation will be weak. We consider approximations with error $g_x(\Delta x) = h_x(\Delta x) \cdot (\Delta x)^a$ where $a \geq 0$ and $h_x(\Delta x) \longrightarrow 0$ as $\Delta x \longrightarrow 0$. If $a = 0$, we obtain, by (9.5), the continuous functions. If $a > 1$, we are reduced to the constant functions.[3] For these reasons, we let $g_x(\Delta x) = h_x(\Delta x)\Delta x$ and suppose $h_x(\Delta x) \longrightarrow 0$ as $\Delta x \longrightarrow 0$. Rewriting (10.1), we obtain

$$f(x + \Delta x) = f(x) + m_x \Delta x + h_x(\Delta x)\Delta x \tag{10.2}$$

for Δx close to 0. Hence, $\lim_{\Delta x \to 0} f(x + \Delta x) = f(x)$ and f is *continuous* at x and (10.2) implies that the slope of the best linear approximation is given by

$$\lim_{\Delta x \neq 0, \Delta x \to 0} \frac{f(x + \Delta x) - f(x)}{\Delta x} = m_x. \tag{10.3}$$

---

[3]This can be deduced from the Mean Value Theorem. Functions satisfying the inequality $|f(x + \Delta x) - f(x)| \leq C|\Delta x|^a, 0 < a \leq 1$, are important in studying differential equations and are named after Rudolf Lipschitz, 1832-1903, a student of Dirichlet.

At this stage, we revert to traditional notation and denote the limit in (10.3) by $f'(x)$, or $\frac{df}{dx}$, or $\frac{df}{dx}(x)$ and call it the *derivative* of f at x. Putting these ingredients together, we arrive at the following definition.

Definition 10.1. A function $f : (a, b) \longrightarrow \mathbf{R}$ is differentiable at $x \in (a, b)$ if there exists a function h_x, defined on an open interval about the origin, I, and a real number, $f'(x)$, such that $\lim_{\Delta x \to 0} h_x(\Delta x) = 0$ and

$$f(x + \Delta x) = f(x) + f'(x)\Delta x + h_x(\Delta x)\Delta x \qquad (10.4)$$

for all $\Delta x \in I$. We say that f is differentiable on (a, b) if it is differentiable at all points in (a, b).

In Exercise 10.12 we give an example of a continuous function that is nowhere differentiable[4] and, in Figure 10.1, we illustrate the local approximation in (10.4) of differentiable functions by linear functions. The line that lies closest to the graph of f near $(x, f(x))$ is called the *tangent line* to f at x. Thus, the *derivative* of f at x is the *slope of the tangent line* to f at x.

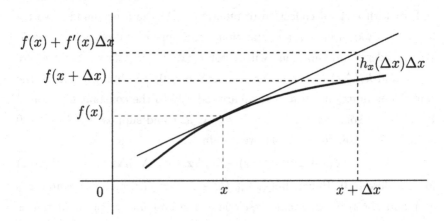

Fig. 10.1

---

[4]The fact that $f : (a, b) \longrightarrow \mathbf{R}$ is differentiable does not imply that f' is continuous. However, by a result of René Baire, 1874-1932, every open subinterval in (a, b) contains points were f' is continuous and f' has the Intermediate Value Property (see the proof of Corollary 10.9 and Exercises 10.8).

It is sometimes convenient to show that f is differentiable and to find the derivative at the *same time* and, moreover, it is always useful to have different approaches to the same problem.

If $l \in \mathbf{R}$, then f is differentiable at x and $f'(x) = l$ if and only if

$$\lim_{\Delta x \neq 0, \Delta x \to 0} \frac{f(x + \Delta x) - f(x) - l\Delta x}{\Delta x} = 0. \tag{10.5}$$

Another easily derived useful characterization is the following: $f : (a, b) \longrightarrow \mathbf{R}$ is differentiable at $x \in (a, b)$ if and only if

$$f(x + \Delta x) = f(x) + k_x(\Delta x)\Delta x \tag{10.6}$$

where k_x is defined on an open interval containing the origin and is continuous at the origin. When (10.6) holds, $\lim_{\Delta x \to 0} k_x(\Delta x) = f'(x)$.

Our next theorem consists of a few very simple but useful observations about differentiable functions. The proofs are straightforward and are omitted. However, the results are used so often that we wish to highlight them. Later in this chapter we will use the Mean Value Theorem to obtain converses to many of these.

We recall that $f : (a, b) \longrightarrow \mathbf{R}$ has a *local maximum* (*minimum*) at x if the function f, restricted to some *open interval I* about x, has a maximum (minimum) over I at x (see Figure 10.2).

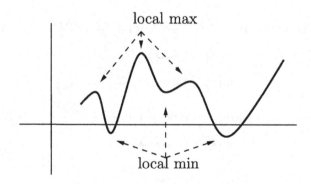

local max

local min

Fig. 10.2

Theorem 10.2. *Let* $f, g : (a, b) \longrightarrow \mathbf{R}$ *be differentiable and let* $c \in \mathbf{R}$.

(a) *If $f(x) = mx + c$ then $f'(x) = m$ for all $x \in (a, b)$.*

(b) *If $f(x) = c$ for all $x \in (a, b)$ then $f'(x) = 0$ for all $x \in (a, b)$.*

(c) *If f is an increasing function on (a, b) then $f'(x) \geq 0$ and if f is decreasing then $f'(x) \leq 0$.*

(d) *If f has a local maximum or minimum at $x \in (a, b)$ then $f'(x) = 0$.*

(e) *$f \pm g$ and cf are differentiable and $(f \pm g)' = f' \pm g'$ and $(cf)' = cf'$.*

In view of (d) we introduce the following definition.

Definition 10.3. If $f : (a, b) \longrightarrow \mathbf{R}$ is differentiable and $f'(x) = 0$, we call x a *critical point* of f.

Item (d) says that local maxima and minima are critical points.

10.2   Basic Functions and Rules for Differentiation

To proceed we establish some basic rules, the product, quotient and chain rules, and produce new examples. We first consider the exponential function and simply extend the technique developed in Lemma 9.4 and Theorem 9.5.

Theorem 10.4. *The exponential function is differentiable and*

$$\frac{d}{dx}(\exp x) = \exp(x).$$

Proof.   Fix $x \in \mathbf{R}$. If $f_x(\Delta x) = \exp(x)\left(1 + \frac{\Delta x}{2!} + \frac{(\Delta x)^2}{3!} + \cdots\right)$ then

$$\exp(x + \Delta x) = \exp(x) \cdot \exp(\Delta x)$$
$$= \exp(x) \cdot \left(1 + \Delta x + \frac{(\Delta x)^2}{2!} + \cdots\right)$$
$$= \exp(x) + \exp(x)\left(1 + \frac{\Delta x}{2!} + \frac{(\Delta x)^2}{3!} + \cdots\right)\Delta x$$
$$= \exp(x) + f_x(\Delta x)\Delta x.$$

Since

$$\left|\frac{\Delta x}{2!} + \frac{(\Delta x)^2}{3!} + \cdots\right| \leq \sum_{n=1}^{\infty} |\Delta x|^n = \frac{|\Delta x|}{1 - |\Delta x|} \longrightarrow 0$$

as $|\Delta x| \longrightarrow 0$ we have $f_x(\Delta x) \longrightarrow \exp(x)$ as $|\Delta x| \longrightarrow 0$. By (10.6), exp is differentiable and $\frac{d}{dx}(\exp x) = \exp(x)$. $\square$

The following theorem contains the principal rules for differentiating functions.

Theorem 10.5. (a) (*The Product Rule*)[5] *If* $f, g : (a, b) \longrightarrow \mathbf{R}$ *are differentiable, then* $f \cdot g$ *is differentiable and*

$$(f \cdot g)' = f' \cdot g + f \cdot g'. \tag{10.7}$$

(b) (*The Chain Rule*) *If* $f : (a, b) \longrightarrow A \subset \mathbf{R}$ *and* $g : (c, d) \longrightarrow \mathbf{R}$ *are differentiable and, if* $A \subset (c, d)$, *then* $g \circ f$ *is differentiable for all* $x \in (a, b)$ *and*

$$(g \circ f)'(x) - g'(f(x)) \cdot f'(x). \tag{10.8}$$

(c) (*The Quotient Rule*) *If* $f, g : (a, b) \longrightarrow \mathbf{R}$ *are differentiable and* $g(x) \neq 0$ *for all* $x \in (a, b)$, *then* f/g *is differentiable and*

$$\left(\frac{f}{g}\right)' = \frac{f' \cdot g - f \cdot g'}{g^2}. \tag{10.9}$$

Proof. For all parts of the proof we suppose

$$f(x + \Delta x) = f(x) + f_x(\Delta x) \cdot \Delta x$$

and

$$g(y + \Delta y) = g(y) + g_y(\Delta y) \cdot \Delta y$$

where $f_x(\Delta x) \longrightarrow f'(x)$ as $\Delta x \longrightarrow 0$ and $g_y(\Delta y) \longrightarrow g'(y)$ as $\Delta y \longrightarrow 0$.
When (a) holds let $x = y$. Then

$$f(x + \Delta x) \cdot g(x + \Delta x) = (f(x) + f_x(\Delta x)\Delta x) \cdot (g(x) + g_x(\Delta x)\Delta x)$$
$$= f(x) \cdot g(x) + h_x(\Delta x)\Delta x$$

where

$$h_x(\Delta x) = f_x(\Delta x) \cdot g(x) + f(x) \cdot g_x(\Delta x) + f_x(\Delta x) \cdot g_x(\Delta x) \cdot \Delta x.$$

Since

$$\lim_{\Delta x \to 0} h_x(\Delta x) = f'(x) \cdot g(x) + f(x) \cdot g'(x) + f'(x) \cdot g'(x) \cdot 0$$

an application of (10.6) proves (a).

---

[5] Also known as Leibniz's rule.

If (b) holds, let $y = f(x)$ and $\Delta y = f_x(\Delta x)\Delta x$. Then $\Delta y \longrightarrow 0$ as $\Delta x \longrightarrow 0$ and

$$g_y(\Delta y) = g_{f(x)}(f_x(\Delta x)\Delta x) \longrightarrow g'(f(x)) \text{ as } \Delta x \longrightarrow 0.$$

Hence

$$g(f(x + \Delta x)) = g(f(x) + f_x(\Delta x)\Delta x)$$
$$= g(y + \Delta y)$$
$$= g(y) + g_y(\Delta y)\Delta y$$
$$= g(f(x)) + g_y(\Delta y)f_x(\Delta x)\Delta x.$$

By (10.6), $g \circ f$ is differentiable and

$$(g \circ f)'(x) = g'(f(x)) \cdot f'(x).$$

This proves (b).

We derive (c) using (a) and (b). Suppose the hypothesis in (c) holds. If $h(x) = 1/x$ for all $x \in \mathbf{R}\backslash\{0\}$, then

$$\frac{1}{x + \Delta x} = \frac{1}{x} + \frac{(-1)}{x(x + \Delta x)} \cdot \Delta x.$$

By (10.6), h is differentiable when $x \neq 0$ and $h'(x) = -1/x^2$.

Hence, $(h \circ g)(x) = h(g(x)) = 1/g(x)$. By (b),

$$\frac{d}{dx}\left(\frac{1}{g(x)}\right) = (h \circ g)'(x) = h'(g(x)) \cdot g'(x) = \frac{-g'(x)}{g(x)^2}$$

and

$$\left(\frac{f}{g}\right)'(x) = (f \cdot (h \circ g))(x) = f'(x) \cdot (h \circ g)(x) + f(x) \cdot (h \circ g)'(x)$$
$$= f'(x) \cdot \frac{1}{g(x)} + f(x) \cdot \frac{-g'(x)}{g^2(x)}$$
$$= \frac{f'(x) \cdot g(x) - f(x) \cdot g'(x)}{g^2(x)}.$$

This completes the proof. $\qquad\qquad\qquad\qquad\qquad\qquad\qquad\qquad\square$

10.3 Mean Value Theorem

The main theoretical result that we prove about differentiable functions is given in the following theorem. It allows us to refine the results in Theorem 10.2.

Theorem 10.6. *(Mean Value Theorem)*[6] *If* $f : [a, b] \longrightarrow \mathbf{R}$, *is continuous on* $[a, b]$ *and differentiable on* (a, b), *then there exists* $c \in (a, b)$ *such that*

$$\frac{f(b) - f(a)}{b - a} = f'(c). \tag{10.10}$$

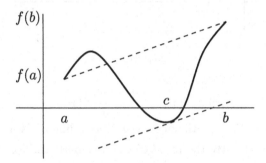

Fig. 10.3

The line joining the points $(a, f(a))$ and $(b, f(b))$ has slope

$$\frac{f(b) - f(a)}{b - a}$$

while the tangent line to the graph of f at $c, a < c < b$, has slope $f'(c)$. The Mean Value Theorem tells us that these lines are *parallel*, as two lines have the same slope if and only if they are *parallel* (see Figure 10.3).

---

[6]Often attributed to Lagrange, who gave an analytic proof in 1791, although the geometric relationship had been known previously. The case $f(a) = f(b) = 0$ is called Rolle's Theorem after Michel Rolle who proved that between any two solutions of the polynomial equation $P(x) = 0$ there is a solution of the equation $P'(x) = 0$. Rolle, 1652-1719, was a self-educated French mathematician from Ambert who worked as a scribe and arithmetical expert in Paris and published works on algebra, geometry and the theory of equations. His ability was recognized and rewarded by election to the French Royal Academy of Sciences and a state pension.

Proof.   First suppose $f(a) = f(b)$. If $f(x) = f(a)$ for all x then any c will do as $f'(c) = 0 = f(b) - f(a)$. If f is not constant, then it has a maximum or minimum inside (a, b) (see Figure 10.4) and, hence, there exists a point $c \in (a, b)$ such that $f'(c) = 0 = f(b) - f(a)$. Now suppose f is arbitrary. If

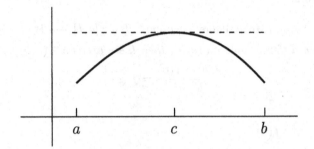

Fig. 10.4

$$g(x) = f(x) - \frac{f(b) - f(a)}{b - a}(x - a)$$

then $g : [a, b] \longrightarrow \mathbf{R}$ is continuous on $[a, b]$ and differentiable on (a, b) and $g(a) = f(a) = g(b)$. By the proof of the first part, there exists $c \in (a, b)$ such that

$$g'(c) = 0 = f'(c) - \frac{f(b) - f(a)}{b - a}.$$

This completes the proof.   □

Corollary 10.7. *Suppose $f : (a, b) \longrightarrow \mathbf{R}$ is differentiable.*

(a) *If $f'(x) = 0$ for all $x \in (a, b)$, then f is a constant function.*

(b) *If $f'(x) > 0$ (respectively $f'(x) < 0$) for all $x \in (a, b)$, then f is strictly increasing (respectively decreasing) and $im(f)$ is an open interval.*[7]

Proof.   If $x \neq y$, then, by the Mean Value Theorem,

$$\frac{f(y) - f(x)}{y - x} = f'(z)$$

[7]We take it as understood from now on, that for every result involving *increasing* functions, we have an analogous result for *decreasing* functions, although sometimes it is necessary to be careful (see Exercises 4.21 and 11.4).

for some z, $x < z < y$. If $f'(z) = 0$ for all $z \in (a, b)$, then $f(x) = f(y)$ and f is a constant function. This proves (a).

If $f'(z) > 0$ for all $z \in (a, b)$ and $a < x < y < b$ then

$$\frac{f(y) - f(x)}{y - x} > 0.$$

This implies $f(y) > f(x)$ and f is strictly increasing. For each $x \in (a, b)$ there exists $c_x > 0$ such that $(x - c_x, x + c_x) \subset (a, b)$ and $f(x - c_x) < f(x) < f(x + c_x)$ and, as f is continuous, the Intermediate Value Theorem implies $(f(x - c_x), f(x + c_x))$ is an open interval in the image of f which contains $f(x)$. This shows the $im(f)$ is an open interval and completes the proof when $f' > 0$. A similar proof applies when $f' < 0$. $\qquad\square$

Theorem 10.8. *If* [8] $f : (a, b) \longrightarrow (c, d)$ *is a bijective differentiable function, then* $f^{-1} : (c, d) \longrightarrow (a, b)$ *is differentiable at* $y = f(x) \in (c, d)$ *if and only if* $f'(x) \neq 0$. *When* $f'(x) \neq 0$ *and* $y = f(x)$

$$(f^{-1})'(y) = \frac{1}{f'(f^{-1}(y))} = \frac{1}{f'(x)}. \tag{10.11}$$

Proof. By Corollary 9.12, f^{-1} is continuous. Fix $y \in (c, d)$, let $x = f^{-1}(y)$ and suppose f^{-1} is differentiable at y. Applying the chain rule to $f \circ f^{-1}$ and noting that $f \circ f^{-1}(y) = y$ for all $y \in (c, d)$, we obtain[9]

$$f'(f^{-1}(y)) \cdot (f^{-1})'(y) = f'(x) \cdot (f^{-1})'(y) = 1$$

and hence $f'(x) \neq 0$.

We now prove the converse. Suppose $y = f(x)$ and $f'(x) \neq 0$. We wish to show that f^{-1} is differentiable at y. Let

$$y + \Delta y = f(x + \Delta x) = f(x) + f_x(\Delta x)\Delta x$$

where $f_x(\Delta x) \longrightarrow f'(x)$ as $\Delta x \longrightarrow 0$. Since $f'(x) \neq 0$, we have $f_x(\Delta x) \neq 0$ for all Δx in some open interval about the origin (see Theorem 8.4 and Exercise 9.10). Hence

$$\Delta x = \frac{\Delta y}{f_x(\Delta x)}$$

---

[8] Due to Brook Taylor.
[9] Using $ab \neq 0$ if and only if $a \neq 0$ and $b \neq 0$. See Chapters 1 and 5.

for all Δx close to 0. Since f is bijective, this implies

$$f^{-1}(y + \Delta y) = x + \Delta x = f^{-1}(y) + \frac{1}{f_x(\Delta x)}\Delta y$$

and, as f^{-1} is continuous, $\Delta y \longrightarrow 0$ when $\Delta x \longrightarrow 0$. By (10.6), f^{-1} is differentiable at y and

$$(f^{-1})'(y) = \lim_{\Delta x \to 0} \frac{1}{f_x(\Delta x)} = \frac{1}{f'(x)} = \frac{1}{f'(f^{-1}(y))}.$$

Figure 10.5 illustrates the result in this theorem.

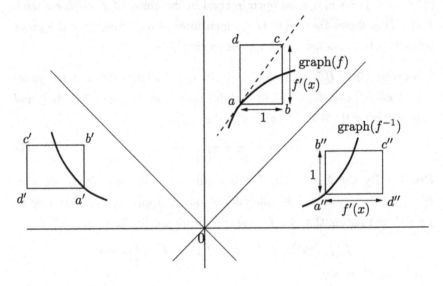

Fig. 10.5

□

Corollary 10.9. *If $f : [a, b] \longrightarrow im(f)$ is continuous on $[a, b]$, differentiable on (a, b) and $f'(x) \neq 0$ for all $x \in (a, b)$, then $im(f) = [m, M]$ where m is the minimum and M is the maximum of f over $[a, b]$, f is bijective and f^{-1} is continuous on $[m, M]$ and differentiable on (m, M).*

Proof.   We first show that f' is either always positive or always negative.[10] Suppose we can find $u, v \in (a, b)$ such that $f'(u) < 0 < f'(v)$.

[10]This intermediate value result for differentiable functions is due to the French mathematician Gaston Darboux, 1842-1917 (see Exercise 10.8).

We may suppose $u < v$. Let $g : [u, v] \longrightarrow \mathbf{R}, g(x) = f(x)$. By Corollary 10.7(b), g is strictly decreasing near u and strictly increasing near v (see Figure 10.6) and g does not achieve its minimum over $[u, v]$ at either u

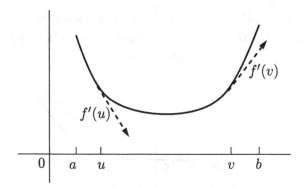

Fig. 10.6

or v. If the minimum occurs at w, then $u < w < v$ and, by Theorem 10.2(d), $f'(w) = g'(w) = 0$, and this is impossible. A further application of Corollary 10.7(b) shows that f is either strictly increasing or strictly decreasing, and in both cases it is injective. By Corollary 9.11, f is bijective and, by Theorem 9.10, $im(f) = [m, M]$. An application of Theorem 10.8 completes the proof. $\qquad\qquad\square$

Example 10.10. (a) Let $f : \mathbf{R}^+ \longrightarrow \mathbf{R}^+, f(x) = x^2$. By the product rule, $f'(x) = 2x$. We have $f^{-1} : \mathbf{R}^+ \longrightarrow \mathbf{R}^+, f^{-1}(x) = +\sqrt{x}$, and hence, $f'(f^{-1}(x)) = 2f^{-1}(x) = 2\sqrt{x}$. Since $f' \neq 0$, Theorem 10.8 implies

$$\frac{d}{dx}(x^{1/2}) = \frac{d}{dx}(\sqrt{x}) = \frac{1}{f'(f^{-1}(x))} = \frac{1}{2\sqrt{x}} = \frac{1}{2}x^{-1/2}.$$

(b) Let $f : \mathbf{R} \longrightarrow \mathbf{R}^+, f(x) = \exp(x)$, then f is bijective and differentiable. Moreover, $f'(x) = \exp(x) > 0$ and $f^{-1} : \mathbf{R}^+ \longrightarrow \mathbf{R}, f^{-1}(x) = \log(x)$. By Theorem 10.8,

$$\frac{d}{dx}(\log x) = \frac{1}{f'(f^{-1}(x))} = \frac{1}{\exp(f^{-1}(x))} = \frac{1}{\exp(\log x)} = \frac{1}{x}.$$

(c) Let $a \in \mathbf{R}$ and let $f : \mathbf{R}^+ \longrightarrow \mathbf{R}, f(x) = x^a$. Since $x^a = \exp{(a \log x)}$, the chain rule implies

$$\frac{d}{dx}(x^a) = \frac{d}{dx}(\exp{(a \log x)}) = \exp{(a \log x)} \cdot \frac{a}{x} = x^a \cdot \frac{a}{x} = a \cdot x^{a-1}.$$

(d) The function $f : \mathbf{R} \longrightarrow \mathbf{R}, f(x) = x^3$ is bijective and differentiable with inverse $f^{-1} : \mathbf{R} \longrightarrow \mathbf{R}, f^{-1}(x) = x^{1/3}$ (see Figure 10.7). Since $f'(0) = 0$, the inverse function f^{-1} is not differentiable at 0.

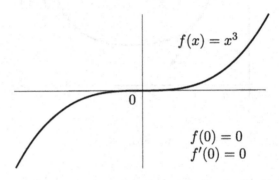

Fig. 10.7

Example 10.11. In this example we show, in two different ways and without using a calculator[11], that

$$P(x) = x^6 - 6x^3 + 12x^2 + 6x - 2$$

is an injective function on $\mathbf{R}_0^+$. The first approach uses only the definition of a critical point whereas the second makes extensive use of the differential calculus via Corollary 10.7(b). It clearly suffices to show that P is either strictly increasing or strictly decreasing and, as P is differentiable and P'

---

[11]The philosopher, mathematician and mystic, Blaise Pascal, 1623-1662, designed and manufactured an adding machine in 1641 to help his father, an accountant. During the 1670's Leibniz designed a machine that also performed multiplication. In both cases, the designs were remarkably practical and some of Leibniz's innovations were still in place in the 1940's when the last mechanical calculators were built. Calculators and computers have an important place in modern science and technology because of their ability to handle vast amounts of data. However, in seeking to appreciate mathematics they should be used sparingly and, at times, a deeper understanding of concepts may result by avoiding the use of calculators.

is continuous, the Intermediate Value Theorem implies that this will be the case if $P'(x) \neq 0$ for $x \geq 0$, that is if P has no positive critical points. We have $P'(x) = 6(x^5 - 3x^2 + 4x + 1)$ and we must show $P'(x) \neq 0$ for all $x \geq 0$. Since P' is continuous, the Intermediate Value Theorem implies that this will be the case if and only if P' is either always positive or always negative on $[0, \infty)$. Since $P'(0) = 6 > 0$, we must show that $P'(x) > 0$ for $x > 0$.

Method 1. The only negativity in $P'(x)$ arises from the $-3x^2$ term and our strategy is to combine this term with a positive term so that the combination is still positive. We have

$$P'(x) = 6(x^2(x^3 - 3) + 4x + 1) > 0$$

if $x^3 > 3$, that is $x > 3^{1/3}$. We also have

$$P'(x) = 6(x^5 + x(4 - 3x) + 1) > 0$$

if $4 > 3x$, that is $x < 4/3$. Since $(4/3) < 3^{1/3} \Longleftrightarrow 4^3 < 3^4 \Longleftrightarrow 64 < 81$, the two cases considered do not cover $\mathbf{R}_0^+$. We proceed by splitting the negative term and combining different parts with positive terms. We obtain

$$P'(x) = 6(x^2(x^3 - 1) + 2x(2 - x) + 1) > 0$$

when $x^3 - 1 > 0$ and $2 - x > 0$, that is for $1 < x < 2$. Since $1 < (4/3) < 3^{1/3} < 2$ (see Figure 10.8), this completes the first proof.

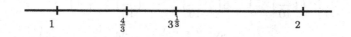

$$\begin{array}{c|c|c|c} 1 & \frac{4}{3} & 3^{\frac{1}{3}} & 2 \end{array}$$

Fig. 10.8

Method 2. For this method we introduce the notation f^n for the nth derivative[12]

$$f^{n+1}(x) := (f^n)'(x) = \left(\frac{d}{dx} f^n\right)(x) =: \frac{d^{n+1} f}{dx^{n+1}}(x).$$

Since $P'(0) = 6 > 0$ it suffices, to show that P' is strictly increasing on $[0, \infty)$ and for this, it is sufficient to prove that $P^2(x) = 6(5x^4 - 6x + 4) > 0$ for $x > 0$. Examining the third derivative, we obtain $P^3(x) := 6(20x^3 - 6)$

---

[12]Some authors use f^n to denote the n-fold composition of f with itself, $f \circ f \circ \cdots \circ f$, while others use $f^{(n)}$ to denote the nth derivative of f.

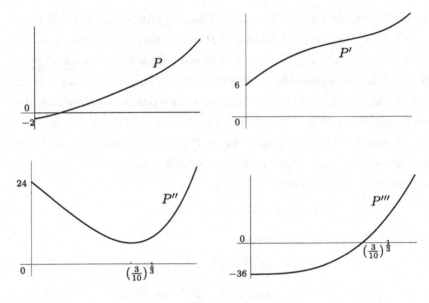

Fig. 10.9

and, since $20x^3 > 6 \iff x > (3/10)^{1/3}$ we see that $P^3(x) < 0$ for $x < (3/10)^{1/3}$ and $P^3(x) > 0$ for $x > (3/10)^{1/3}$. Hence P^2 is decreasing for $x < (3/10)^{1/3}$ and increasing for $x > (3/10)^{1/3}$ and the minimum of P^2 is achieved at $(3/10)^{1/3}$. We have

$$P^2\left(\left(\frac{3}{10}\right)^{1/3}\right) = 6\left(5\left(\frac{3}{10}\right)^{4/3} - 6\left(\frac{3}{10}\right)^{1/3} + 4\right)$$
$$= 30 \cdot \frac{3}{10}\left(\frac{3}{10}\right)^{1/3} - 36\left(\frac{3}{10}\right)^{1/3} + 24$$
$$= -27\left(\frac{3}{10}\right)^{1/3} + 24$$

and

$$-27\left(\frac{3}{10}\right)^{1/3} + 24 > 0 \iff 27\left(\frac{3}{10}\right)^{1/3} < 24$$
$$\iff \left(\frac{3}{10}\right)^{1/3} < \frac{8}{9}$$
$$\iff \frac{3}{10} < \left(\frac{8}{9}\right)^3 = \frac{8^3}{9^3}$$
$$\iff 3 \cdot 9^3 < 10 \cdot 8^3$$
$$\iff 2187 < 5120.$$

Since the final equality is true, all the equivalent conditions are also true and $P^2(x) > 0$ for all $x > 0$. Hence, P' is strictly increasing and thus always positive. In Figure 10.9 we have illustrated the crucial observations connected with the second method.

There are many interesting results about differentiable functions that we are now competent to investigate. A good review of convergent series and the fundamentals of differentiation can be obtained by examining the following result in the manner and with the attitude that we approached Theorem 1.1 on quadratic equations in Chapter 1.

If $f(x) = \sum_{n=0}^{\infty} a_n x^n$ has radius of convergence $r > 0$, then f is differentiable on the interval $(-r, +r)$, the power series $\sum_{n=1}^{\infty} n a_n x^{n-1}$ has radius of convergence r, and

$$f'(x) = \frac{d}{dx}\left(\sum_{n=0}^{\infty} a_n x^n\right) = \sum_{n=0}^{\infty} \frac{d}{dx}\left(a_n x^n\right) = \sum_{n=1}^{\infty} n a_n x^{n-1}. \quad (10.12)$$

10.4 Exercises

(10.1) Give the natural domain of definition of each of the following functions and find their derivatives

$$\exp\left(x^2\right) \cdot \log(x^2 + 1), \quad \frac{2^x}{3^x + 1}, \quad \frac{e^{x^2}}{(\log x)^2 + 1}, \quad \frac{\log(x^2) \cdot (\log x)^2}{\log(\log x)}.$$

(10.2) Show, for $x > 0$, that

$$\lim_{n \to \infty} n(x^{1/n} - 1) = \log x.$$

(10.3) Verify the Mean Value Theorem for the function $f(x) = x^3 - x$ on $[0, 2]$ and the function $g(x) = x^5 + 5x^3 + 40x$ on $[1, 2]$.

(10.4) Find where the following functions are increasing and decreasing

$$\frac{3}{x^4 - 2x^2 + 1}, \quad \frac{x^2 + 1}{x}, \quad \frac{x}{x^2 + 1}.$$

Hence find where each function has local maxima and minima. Draw a rough sketch of each function.

(10.5) Solve for x

$$1 - e^{-2x} = 0.$$

Find the interval where the function $f : \mathbf{R} \longrightarrow \mathbf{R}$

$$f(x) = \frac{e^x}{1 - e^{-2x}}$$

is increasing. Find

$$\lim_{x \longrightarrow +\infty} \frac{e^{-2x}}{1 - e^{-x}}, \quad \lim_{x \longrightarrow -\infty} \frac{e^{-2x}}{1 - e^{-x}}.$$

(10.6) Find appropriate intervals on which $f(x) = \log(\frac{1}{1+x^2})$ and $g(x) = (\log x)^{\log x}$ are defined and differentiable. Find the derivatives.

(10.7) Find $(f^{-1})'$ where $f(x) = \exp(x^2 + 1)$ (a) by first finding f^{-1}, (b) by using the identity $f \circ f^{-1}(x) = x$ for all x, and (c) by writing f^{-1} as a composition and using the chain rule.

(10.8) (Darboux's Theorem) Let $f : \mathbf{R} \longrightarrow \mathbf{R}$ denote a differentiable function. If $a < b$ are real numbers and $f'(a) < r < f'(b)$ show that there exists c, $a < c < b$, such that $f'(c) = r$.

(10.9) Using only f', find the maximum and minimum of $f : \mathbf{R} \longrightarrow \mathbf{R}$ where

$$f(x) = \frac{8x}{3x^2 + 4}.$$

(10.10) Show that the function

$$f : \mathbf{R} \longrightarrow \mathbf{R}, \quad f(x) = 3x^5 + 10x^3 - 45x^2 + 15x + 7$$

has at least two critical points in $[0, 2]$.

(10.11) If $f : \mathbf{R} \longrightarrow \mathbf{R}$ is differentiable and there exists $c > 0$ such that $f'(x) \geq c$ for all $x \in \mathbf{R}$ show that f is bijective.

(10.12) Let $f : \mathbf{R} \longrightarrow [0, 1/2]$, $f(x) =$ distance of x to the nearest integer, and let

$$g(x) = \sum_{n=0}^{\infty} 2^{-n} f(2^n x).$$

Show that g is everywhere continuous and nowhere differentiable.

(10.13) Let $f : (a, b) \longrightarrow \mathbf{R}$ denote a differentiable function. Show that $f'(x) \geq 0$ for all $x \in (a, b)$ if and only if $f(x) \geq f(y)$ for all $x, y \in (a, b)$, $x > y$.

(10.14) Show that

$$f : \mathbf{R} \longrightarrow \mathbf{R}, \quad f(x) = e^x - \frac{1}{2}x^2 - 1,$$

is a bijective function.

(10.15) If p and q are real numbers show that

$$f : \mathbf{R} \longrightarrow \mathbf{R}, \quad f(x) = x^3 + px + q,$$

is bijective if and only if $p \geq 0$.

(10.16) Let $f(x) = \log x$. Define natural domains of definition for the functions $(f(x))^2$, $f(x^2)$ and $f(f(x))$. In each case find the derivative of the function.

(10.17) If a and b are real numbers with $a < b$, show that

$$e^a < \frac{e^b - e^a}{b - a} < e^b.$$

(10.18) Find all real numbers m such that $e^x \geq 1 + mx$ for all $x \in \mathbf{R}$.

(10.19) If f and g are defined and have derivatives up to order n on (a, b) show that $f \cdot g$ has derivatives up to order n and

$$(f \cdot g)^n(x) = \sum_{k=0}^{n} \binom{n}{k} f^k(x) \cdot g^{n-k}(x)$$

where $f^0(x) = f(x)$ and $g^0(x) = g(x)$ for all $x \in (a, b)$.

References [3; 6; 9; 25]

Chapter 11

APPLICATIONS OF DIFFERENTIATION

> *The recognition that mathematical concepts are suggested, though not defined, by intuition thus easily accounts for the fact that the results of mathematical deductive reasoning are in apparent agreement with those of inductive reasoning.*
>
> Carl Boyer, 1949

Summary

We continue developing the methods of the differential calculus and obtain new approaches to sequences, series, limits and curve sketching.

11.1   Sequences

Example 11.1. In this example we use different methods to show that the sequence $(2^n n^{-2})_{n=3}^{\infty}$ is increasing. Even though some of the methods overlap, it is still an interesting exercise to fill in the details and to avoid using a calculator.

Method 1. The most direct method of showing that the sequence $(a_n)_{n=3}^{\infty}$ is increasing is to verify that $a_{n+1} \geq a_n$ for all $n \geq 3$. We have

$$\frac{2^{n+1}}{(n+1)^2} \geq \frac{2^n}{n^2} \Longleftrightarrow \frac{2^{n+1}}{2^n} \geq \frac{(n+1)^2}{n^2} \Longleftrightarrow 2 \geq \left(\frac{n+1}{n}\right)^2 \Longleftrightarrow 2 \geq \left(1+\frac{1}{n}\right)^2$$

$$\Longleftrightarrow \sqrt{2} \geq 1+\frac{1}{n} \Longleftrightarrow (\sqrt{2}-1) \geq \frac{1}{n} \Longleftrightarrow n \geq \frac{1}{\sqrt{2}-1}.$$

To show that the sequence is increasing, it suffices to show that $\frac{1}{\sqrt{2}-1} < 3$. This can easily be verified using a calculator but it is interesting to verify it directly. We have

$$\frac{1}{\sqrt{2}-1} < 3 \Longleftrightarrow 1 < 3(\sqrt{2}-1) \Longleftrightarrow 4 < 3\sqrt{2} \Longleftrightarrow 4^2 = 16 < 18 = (3\sqrt{2})^2$$

and this completes the proof.

Method 2. In this case we show that $a_{n+1} - a_n \geq 0$ for $n \geq 3$. We have

$$\frac{2^{n+1}}{(n+1)^2} - \frac{2^n}{n^2} \geq 0 \Longleftrightarrow \frac{2^{n+1}n^2 - 2^n(n+1)^2}{(n+1)^2 n^2} \geq 0$$

$$\Longleftrightarrow \frac{2n^2 - (n+1)^2}{(n+1)^2 n^2} \geq 0$$

$$\Longleftrightarrow 2n^2 - (n^2 + 2n + 1) \geq 0$$

$$\Longleftrightarrow n^2 \geq 2n + 1$$

$$\Longleftrightarrow n \geq 2 + \frac{1}{n}$$

and this last inequality is clearly satisfied when $n \geq 3$.

Method 3. Since the sequence has all positive terms, it suffices to show

$$\frac{a_{n+1}}{a_n} \geq 1$$

for $n \geq 3$. We have

$$\frac{\frac{2^{n+1}}{(n+1)^2}}{\frac{2^n}{n^2}} \geq 1 \Longleftrightarrow \frac{2^{n+1}}{2^n} \cdot \frac{n^2}{(n+1)^2} \geq 1$$

$$\Longleftrightarrow \frac{2n^2}{(n+1)^2} \geq 1$$

$$\Longleftrightarrow 2n^2 \geq (n+1)^2$$

$$\Longleftrightarrow n^2 \geq 2n + 1$$

and we have already shown in the previous method that this last inequality is satisfied whenever $n \geq 3$.

Method 4. Here we consider the function $f : [3, \infty) \longrightarrow \mathbf{R}$ where $f(x) = 2^x/x^2$ and show that it is increasing. Since

$$f(x) = \frac{2^x}{x^2} = \exp(x \log 2) \cdot \exp(-2 \log x) = \exp(x \log 2 - 2 \log x)$$

it suffices, by Corollary 10.7(b), to show $f'(x) > 0$ for $x \geq 3$. We have

$$f'(x) = x^{-2} \exp(x \log 2 - 2 \log x) \cdot (x^2 (\log 2) - 2x).$$

Now, $x^{-2} \exp(x \log 2 - 2 \log x) > 0$ for $x > 0$ and $x^2 (\log 2) - 2x$ is a quadratic function (Figure 11.1) with zeros at 0 and $2/\log 2$. We need to verify that

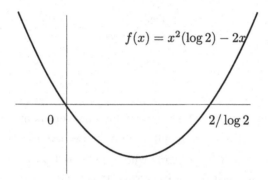

$$f(x) = x^2 (\log 2) - 2x$$

0                                      $2/\log 2$

Fig. 11.1

$2/\log 2 < 3$. Since the exponential function is strictly increasing,

$$\frac{2}{\log 2} < 3 \Longleftrightarrow 2 < 3 \log 2 = \log 8 \Longleftrightarrow e^2 < 8$$

and, using the exponential and geometric series, we see that

$$e^2 < 1 + 2 + \frac{2^2}{2!} + \frac{2^3}{3!}\left(1 + \frac{1}{2} + \frac{1}{2^2} + \cdots\right) = 5 + \frac{8}{6}\frac{1}{1/2} = \frac{23}{3} < 8$$

and this completes the proof. In Figure 11.2 we include all the above estimates but we need to check the relationship between $1/(\sqrt{2} - 1)$ and $2/\log 2$. We have

$$\frac{1}{\sqrt{2} - 1} < \frac{2}{\log 2} \Longleftrightarrow \log 2 < 2(\sqrt{2} - 1) \Longleftrightarrow 2 + \log 2 < 2\sqrt{2}$$

$$\Longleftrightarrow 4 + 4 \log 2 + (\log 2)^2 < 8$$

$$\Longleftrightarrow \log 2 + \frac{1}{4}(\log 2)^2 < 1.$$

Since

$$\log 2 < 1 \iff 2 < e^1 = e = 1 + 1 + \frac{1}{2!} + \cdots,$$

$(\log 2)^2 < 1$ and it suffices to show $\log 2 < 3/4$. This means we need to show $4 \log 2 = \log 2^4 = \log 16 < 3$ or that $16 < e^3$. We have

$$e^3 > 1 + 3 + \frac{3^2}{2!} + \frac{3^3}{3!} + \frac{3^4}{4!} = 4 + 9 + \frac{27}{8} = 16\frac{3}{8} > 16$$

as required.

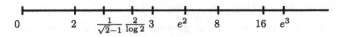

Fig. 11.2

11.2   Limits

For different reasons, both Newton and Leibniz were poor communicators of the calculus to the general public and it was left to others to spread the message. Two mathematicians who appreciated, at an early stage, its potential were the Bernoulli brothers from Basel, Jacob and Johann.[1] They studied, digested and shaped much of the basic calculus into the form we know it today. The Marquis de L'Hôpital hired Johann Bernoulli, during

---

[1] Jacob, 1654-1705, and Johann, 1667-1748, were the first two mathematicians in a family that was to produce a number of outstanding mathematicians over three generations. The family had migrated to Switzerland from Belgium to escape religious persecution. Jacob, a professor of mathematics at Basel from 1687 until he died in 1705, was noted for his work on probability theory, which included the *Law of Large Numbers*, and his contributions to the calculus of variations, mechanics and the theory of infinite series. Johann was professor of mathematics at Groningen for ten years and succeeded his brother as professor at Basel. Three of his children, Daniel, who collaborated with Euler, Johann II, who succeeded his father as professor at Basel, and Nicolaus became mathematicians. Johann travelled widely and had a very extensive correspondence with Leibniz. The Bernoulli brothers started out as collaborators but this soon degenerated into hostility and a most unseemly public row with very questionable comments being made by each about the other. Johann appeared to have been a particularly belligerent individual and, as the self-appointed chief supporter of Leibniz, he stoked the fires of controversy with the followers of Newton over the discovery of the calculus.

a visit to Paris by Johann, as a tutor in calculus during the first half of 1692 and the lessons continued by correspondence after Johann returned to Basel. In 1696 the first textbook on the differential calculus, *Analyse des infiniment petits pour l'intelligence des lignes courbes*, was published by L'Hôpital[2] and in it appeared, for the first time, our next theorem. We now know that this book was just a corrected version of the notes of Johann Bernoulli written during 1691-1692 and *L'Hôpital's Rule* is due to Johann Bernoulli. However, such was Johann's reputation that when he claimed the result, nobody believed him. We present one, from among the many, versions of L'Hôpital's Rule now available (see Exercise 11.26).

Theorem 11.2. (*L'Hôpital's Rule*) *If f is defined and differentiable on an open interval about a, $f(a) = g(a) = 0$ and $g'(x) \neq 0$ for all x near a then*

$$\lim_{x \to a} \frac{f(x)}{g(x)} = \lim_{x \to a} \frac{f'(x)}{g'(x)} \tag{11.1}$$

whenever the limit on the right-hand side exists.

Proof. We consider an interval about a, $[a, x]$, where $g' \neq 0$ and g is injective (see Corollary 10.7(b)). Let $H(y) = f(y)g(x) - g(y)f(x)$ for all $y \in [a, x]$.[3] Then H is continuous on $[a, x]$, differentiable on (a, x), and $H(a) = H(x) = 0$. By the Mean Value Theorem, there exists $\theta_x, a < \theta_x < x$, such that $H'(\theta_x) = 0$. Then $f'(\theta_x)g(x) = g'(\theta_x)f(x)$ and, since $f(a) = g(a) = 0$,

$$\frac{f(x) - f(a)}{g(x) - g(a)} = \frac{f(x)}{g(x)} = \frac{f'(\theta_x)}{g'(\theta_x)}$$

and, as $\lim_{x \to a} \theta_x = a$,

$$\lim_{x \to a} \frac{f(x)}{g(x)} = \lim_{x \to a} \frac{f'(\theta_x)}{g'(\theta_x)} = \lim_{x \to a} \frac{f'(x)}{g'(x)}.$$

This completes the proof. □

---

[2]Guillaume François Antoine Marquis de L'Hôpital, 1661-1704, was from Paris.

[3]We follow a convention previously used and suppose that x may be on either side of a. During the initial stage of the proof x is fixed and y is the variable and, in the final stage, we allow x to vary.

Example 11.3. Let $f(x) = \log x$ and $g(x) = x^2 - 1$ for $x > 1$. Then $f(1) = g(1) = 0$. By L'Hopitâl Rule and continuity

$$\lim_{x \to 1} \left(\frac{\log x}{x^2 - 1} \right)^5 = \left(\lim_{x \to 1} \frac{\log x}{x^2 - 1} \right)^5 = \left(\lim_{x \to 1} \frac{(1/x)}{2x} \right)^5 = \frac{1}{2^5} = \frac{1}{32}.$$

Example 11.4. We have shown in Example 7.10 that

$$\sum_{j=0}^{n} r^j = 1 + r + r^2 + \cdots + r^n = \frac{r^{n+1} - 1}{r - 1} \tag{11.2}$$

when $r \neq 1$. Note that the left and center parts of this equality are both equal to $n + 1$ when $r = 1$ while the right is not defined. However, by L'Hôpital's Rule,

$$\lim_{r \to 1} \frac{r^{n+1} - 1}{r - 1} = \lim_{r \to 1} \frac{(n+1)r^n}{1} = n + 1.$$

If we differentiate (11.2), we obtain

$$\sum_{j=1}^{n} j r^{j-1} = 1 + 2r + 3r^2 + \cdots + nr^{n-1} = \frac{(n+1)r^n(r-1) - (r^{n+1} - 1)}{(r-1)^2}$$

$$= \frac{nr^{n+1} - (n+1)r^n + 1}{(r-1)^2}.$$

Applying L'Hopitâl's Rule twice we obtain

$$\sum_{j=1}^{n} j = \lim_{r \to 1} \frac{n(n+1)r^n - n(n+1)r^{n-1}}{2(r-1)} = \lim_{r \to 1} \frac{n(n+1)r^{n-1}(r-1)}{2(r-1)}$$

$$= \lim_{r \to 1} \frac{n(n+1)r^{n-1}}{2}$$

$$= \frac{n(n+1)}{2}.$$

This formula and similar formula given in Exercise 11.1 are usually proved by mathematical induction, a method that requires prior knowledge of the formula. This approach required no such knowledge.

If an amount A is earning interest at a rate r, compounded n times per year, then after t years it will be worth $A(1 + \frac{r}{n})^{nt}$. To calculate the continuously compounded interest one needs to let n tend to infinity. We calculated this limit previously when $r > 0$, see Theorem 7.23, and we derive it again in the following theorem.

Theorem 11.5. *For any real number* r

$$\lim_{n \longrightarrow \infty} \left(1 + \frac{r}{n}\right)^n = e^r.$$

Proof. We have

$$\frac{d}{dx} \log(x) = \lim_{\Delta x \longrightarrow 0} \frac{\log(x + \Delta x) - \log x}{\Delta x} = \frac{1}{x}.$$

If we let $x = 1$ and $\Delta x = r/n$, then $\Delta x \longrightarrow 0$ as $n \longrightarrow \infty$. Since $\log 1 = 0$, this implies

$$\lim_{n \longrightarrow \infty} \frac{\log(1 + \frac{r}{n})}{\frac{r}{n}} = \lim_{n \longrightarrow \infty} \frac{n}{r} \log\left(1 + \frac{r}{n}\right) = \frac{1}{r} \lim_{n \longrightarrow \infty} \log\left(1 + \frac{r}{n}\right)^n = 1.$$

Hence, $\lim_{n \to \infty} \log(1 + \frac{r}{n})^n = r$ and, as exp and log are both continuous, this implies

$$\lim_{n \longrightarrow \infty} \left(1 + \frac{r}{n}\right)^n = \exp\left\{\lim_{n \longrightarrow \infty} \log\left(1 + \frac{r}{n}\right)^n\right\} = \exp(r) = e^r.$$

This completes the proof. $\qquad\square$

Theorem 11.5 allows one to compare, assuming a given interest rate, different amounts of money at specified future times. This is an essential feature of financial markets. The present worth of an amount A at a future time t is Ae^{-rt} if the interest rate is r continuously compounded.

11.3 Curve Sketching

Example 11.6. We revisit quadratic functions and use them to study cubics. Let $f : \mathbf{R} \longrightarrow \mathbf{R}, f(x) = ax^2 + bx + c$ where $a > 0$. We have $f'(x) = 2ax + b$. This implies $f'(x) < 0$ and f is decreasing for $x < -b/2a$ and $f'(x) > 0$ and f is increasing for $x > -b/2a$. At $x = -b/2a$, f has a minimum (see Figure 11.3). We now consider a *cubic*, $P(x) = ax^3 + bx^2 + cx + d$ for all $x \in \mathbf{R}$, where $a \neq 0$. We have already noted that $P(x) \longrightarrow \pm\infty$ as $x \longrightarrow \pm\infty$ and that the equation $P(x) = 0$ has at least one and at most three real solutions. The derivative $P'(x) = 3ax^2 + 2bx + c$ is a quadratic function. As $P'(x) = 0$ has at most two solutions, P has at most two critical points. From the graph of P', we can get some idea of the shape of P. For example, suppose

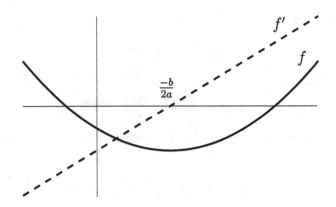

Fig. 11.3

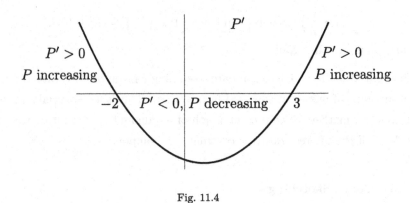

Fig. 11.4

$P(x) = 2x^3 - 3x^2 - 36x + 12$. Then $P'(x) = 6x^2 - 6x - 36 = 6(x-3)(x+2)$ and we display the graph of P' in Figure 11.4. Since P' is continuous, the Intermediate Value Theorem implies that it will have the same sign between critical points and hence P will either be strictly increasing or strictly decreasing between these points. From the diagram, we see that P is increasing up to -2, has a local maximum at -2, is decreasing between -2 and $+3$, has a local minimum at $+3$ and is increasing when $x > 3$. We also note that $P(-2) = 56$ and $P(3) = -69$ (see Figure 11.5). This allows us to sketch the graph of P in Figure 11.6.

In Theorem 1.3 we saw that the sign of $b^2 - 4ac$ enabled us to count the

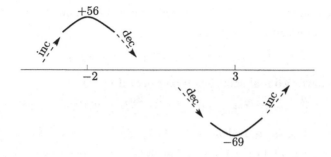

Fig. 11.5

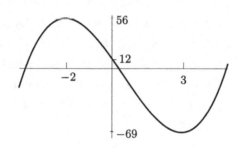

Fig. 11.6

number of real solutions of the quadratic equation $ax^2 + bx + c = 0, a \neq 0$.
We now show that the expression $4p^3 + 27q^2$ plays a similar role for the
cubic equation (∗) $f(x) := x^3 + px + q = 0$ with real coefficients p and q.
Exercise 1.3 shows how an arbitrary cubic equation can be converted into
one of this type. We already know that (∗) has at least one (Example 9.9)
and at most three (Exercise 1.2) real solutions. The graph of f is similar
to either Figure 10.7 or Figure 11.6 and from these we see that (∗) has one
real solution if f has either no local maxima or minima or if it has one of
each, then both are above or both are below the horizontal axis. We have
two real solutions if f has either a local maximum or a local minimum on
the horizontal axis and we have three real solutions if the local maximum is
positive and the local minimum negative. Since $f'(x) = 3x^2 + p$, f has no
critical point and (∗) has a unique real solution when $p > 0$. When $p = 0$,
$f(x) = 0 \iff x = (-q)^{1/3}$ and again (∗) has only one real solution.

We now suppose $p < 0$. Then f has a local maximum at $-(-p/3)^{1/2}$, a local minimum at $(-p/3)^{1/2}$ and

$$f(\pm(-p/3)^{1/2}) = \pm\frac{2p}{3}(-p/3)^{1/2} + q.$$

Thus, we have two real solutions if and only if either

$$\frac{2p}{3}(-p/3)^{1/2} + q = 0 \quad \text{or} \quad -\frac{2p}{3}(-p/3)^{1/2} + q = 0.$$

Since $p < 0$, this is equivalent to: $-\frac{2p}{3}(-p/3)^{1/2} = |q|$ and we may square both sides without losing information to obtain two real solutions if and only if $4p^3 + 27q^2 = 0$. If α and β are the two real solutions then $(x^3 + px + q) - (\alpha^3 + p\alpha + q) = (x - \alpha)(x^2 + \alpha x + \alpha^2 + p)$ and the equation $x^2 + \alpha x + \alpha^2 + p = 0$ has at least one solution β. By Theorem 1.3 we either have $x^2 + \alpha x + \alpha^2 + p = (x - \beta)(x - \alpha)$ or $x^2 + \alpha x + \alpha^2 + p = (x - \beta)^2$. Suppose the first case holds (the other is handled similarly) then $x^3 + px + q = (x-\alpha)^2(x-\beta)$. Comparing coefficients we see that $2\alpha + \beta = 0$ and $-\alpha^2\beta = 2\alpha^3 = q$. Hence $\alpha = (q/2)^{1/3}$ and $\beta = (-4q)^{1/3} = -2(q/2)^{1/3}$ (see Exercises 4.25 and 11.28).

We have three real solutions if and only if

$$\frac{2p}{3}(-p/3)^{1/2} + q < 0 < \frac{-2p}{3}(-p/3)^{1/2} + q$$

that is if and only if

$$q < \frac{-2p}{3}(-p/3)^{1/2} \quad \text{and} \quad -q < \frac{-2p}{3}(-p/3)^{1/2},$$

or, equivalently, $|q| < \frac{-2p}{3}(-p/3)^{1/2}$. By positivity we may again take squares. Hence, always assuming $p < 0$, we obtain three distinct real solutions if and only if $q^2 < (\frac{-2p}{3}(-p/3)^{1/2})^2 = \frac{-4p^3}{27}$, that is if and only if $27q^2 + 4p^3 < 0$. Since we have now eliminated all other possibilities, this means that when $p < 0$ we have a unique real solution when $27q^2 + 4p^3 > 0$. As $p > 0 \implies 27q^2 + 4p^3 > 0$, we may combine the two cases where we have a unique real solution and see that $(*)$ has one real solution if and only if either $p = 0$ or $27q^2 + 4p^3 > 0$.

Example 11.7. We have already drawn graphs using a direct approach in Chapter 2. For example, if $f : \mathbf{R}\backslash\{\pm 1\} \longrightarrow \mathbf{R}$,

$$f(x) = \frac{3}{x^4 - 2x^2 + 1} = \frac{3}{(x^2 - 1)^2}$$

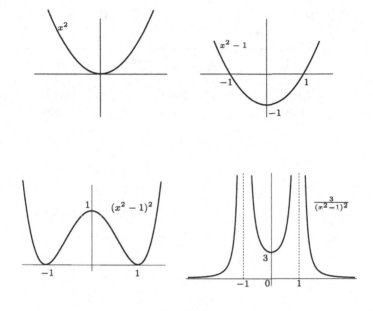

Fig. 11.7

then we obtain the following sequence (Figure 11.7). We now wish to sketch the same graph using the differential calculus. As preliminary observations, we note that $f \geq 0$, and hence the graph will always lie above the x-axis, and as we approach ± 1 the function gets very large. We have

$$f'(x) = \frac{-12x}{(x^2 - 1)^3} = \frac{-12x(x - 1) \cdot (x + 1)}{(x^2 - 1)^4} = \frac{-12(x^3 - x)}{(x^2 - 1)^4}.$$

Since $(x^2 - 1)^2 \geq 0$ and the cubic $x \longrightarrow -12x^3 + 12x$ is positive for $x < -1$ and $0 < x < 1$, f is strictly increasing on these intervals and, similarly, it is strictly decreasing on the intervals $(-1, 0)$ and $(1, +\infty)$. Since f is not defined at ± 1, its only critical point, a local minimum, is at 0, see Figure 11.8. It is also clear that $\lim_{x \to \pm\infty} f(x) = 0$. Putting all this together, we obtain the following rough diagram (Figure 11.9). Note that as x approaches ± 1 and $\pm\infty$, we have two choices, one increasing and one decreasing. The sign of the derivative gives the correct one and we have marked it in a more obvious way in Figure 11.9. It is now a simple matter to sketch f (see Figure 11.7). Note that f behaves as if it has local maxima

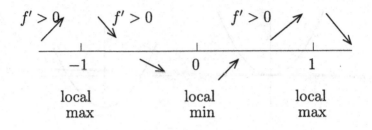

Fig. 11.8

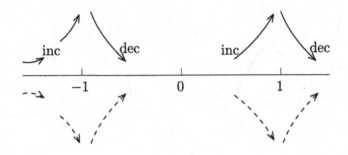

Fig. 11.9

at ±1.

Example 11.8. In this example we use the differential calculus to sketch

$$f : \mathbf{R}\backslash\{1\} \longrightarrow \mathbf{R}, f(x) = \frac{x}{x^3 - 1}.$$

We first note that

$$f(x) = \frac{1}{x^2 - (1/x)} \longrightarrow 0 \text{ as } x \longrightarrow \pm\infty.$$

We have

$$f'(x) = \frac{x^3 - 1 - x(3x^2)}{(x^3 - 1)^2} = \frac{-1 - 2x^3}{(x^3 - 1)^2}.$$

Since $-1 - 2x^3 > 0 \iff x^3 < -1/2 \iff x < -1/2^{1/3} \approx -0.8$, f is increasing when $x < -0.8$ and decreasing when $x > -0.8$. Hence, f has a local maximum at -0.8 and $f(-0.8) \approx 0.53$. Since $f(x) \longrightarrow \pm\infty$ as $x \longrightarrow 1$, we now have good outline information on the shape of the graph and sketch it in Figure 11.10. In Figure 11.11 we sketch the same graph without using the differential calculus.

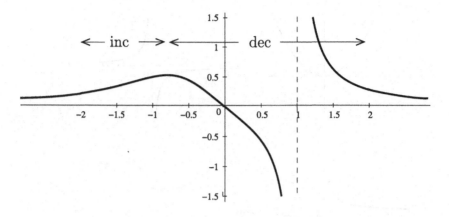

Fig. 11.10

11.4 Applications of the Mean Value Theorem

Theorem 11.9. *If $f; (a,b) \longrightarrow \mathbf{R}$ has first and second derivatives at all points and $f''(x) \geq 0$ for all $x \in (a,b)$, then f is convex.*

Proof. Since $f'' \geq 0$, the function f' is increasing. Suppose $a < x < y < b$ and $0 < t < 1$. Inequality (4.6), which defines convexity, is easily seen after cross multiplication of the two middle terms, to be equivalent to the following inequality

$$\frac{f(tx + (1-t)y) - f(x)}{(tx + (1-t)y) - x} = \frac{f(tx + (1-t)y) - f(x)}{(1-t)(y-x)}$$

$$\leq \frac{f(y) - f(tx + (1-t)y)}{t(y-x)}$$

$$= \frac{f(y) - f(tx + (1-t)y)}{y - (tx + (1-t)y)}.$$

By Theorem 10.6, there exist z_1 and z_2, $x \leq z_1 \leq tx + (1-t)y \leq z_2 \leq y$, such that

$$f'(z_1) = \frac{f(tx + (1-t)y) - f(x)}{(tx + (1-t)y) - x}$$

$$\leq \frac{f(y) - f(tx + (1-t)y)}{y - (tx + (1-t)y)}$$

$$= f'(z_2).$$

This completes the proof.   □

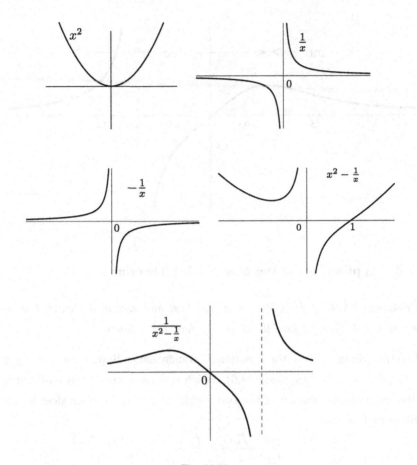

Fig. 11.11

Theorem 11.9 is useful in verifying easily that certain functions are convex, e.g. f and $g : \mathbf{R} \longrightarrow \mathbf{R}, f(x) = x^2, g(x) = e^{ax}$ and also $f + g$.

Definition 11.10. If $f : (a, b) \longrightarrow \mathbf{R}$ is differentiable and non-zero we call the mapping

$$x \longrightarrow \frac{f'(x)}{f(x)}$$

the *proportional rate of change* of f and the mapping

$$x \longrightarrow \frac{x f'(x)}{f(x)} =: El(f)(x)$$

the *elasticity* of f.

Both of these functions are important in economics. When $f(x) > 0$ we have

$$\frac{f'(x)}{f(x)} = \frac{d}{dx}(\log f(x))$$

and, if we let $y = f(x)$, then when Δx is *small*

$$\frac{\frac{\Delta y}{y} \times 100}{\frac{\Delta x}{x} \times 100} \approx \lim_{\Delta x \longrightarrow 0} \frac{\Delta y}{\Delta x} \frac{x}{y} = \frac{x \frac{dy}{dx}}{y} = El(f)(x). \tag{11.3}$$

This allows us to interpret particular examples. For example, suppose p is the price of a certain good and that q is the amount sold. We denote the relationship between p and q by $q = f(p)$. Then p and q are both positive and, clearly, as p increases q decreases and, if f is differentiable, $f' \leq 0$ and $El(f) \leq 0$. If $El(f) = -1.7$ when $p = 134$ then a 2% increase in price, when the price is 134, will result in a $2 \times 1.7 = 3.4\%$ decrease in sales.

Example 11.11. We wish to find f given that $f : \mathbf{R}^+ \longrightarrow \mathbf{R}^+$ has constant elasticity 3 and $f(2) = 4$. We have

$$(\log f)'(x) = \frac{f'(x)}{f(x)} = \frac{El(f)(x)}{x} = \frac{3}{x}.$$

To proceed we need to find a function whose derivative is $3/x$. This amounts, as we will see in Chapter 13, to finding the *integral* of $3/x$. In the meantime we rely on sensible guesses. Since $\frac{d}{dx}(\log x) = \frac{1}{x}$ we have $\frac{d}{dx}(3\log x) = \frac{d}{dx}\log(x^3) = \frac{3}{x}$. This implies

$$\frac{d}{dx}(\log f(x) - \log(x^3)) = \frac{d}{dx}\left(\log \frac{f(x)}{x^3}\right) = 0.$$

By Corollary 10.7(a), there is a constant C such that

$$\log\left(\frac{f(x)}{x^3}\right) = C.$$

Hence, $f(x) = x^3 e^C$ and $f(2) = 4 = 8e^C$ implies $f(x) = x^3/2$ for all $x \in \mathbf{R}$.

Example 11.12. Many different natural phenomena change with time and are characterized by the fact that the rate of change is proportional to the amount present. If $f(t)$ denotes the value of such a function at time t, then over a small period of time Δt, we have

$$f(t + \Delta t) - f(t) \approx c\Delta t f(t) \tag{11.4}$$

and, assuming that f is differentiable, we obtain

$$f'(t) = cf(t). \tag{11.5}$$

If we suppose $f(t) > 0$ for all t, then

$$\frac{d}{dt}\big(\log f(t)\big) = \frac{f'(t)}{f(t)} = c = \frac{d}{dt}(ct) \tag{11.6}$$

and $\log(f(t)) = a + ct$ and $f(t) = e^{a+ct}$ where a is some constant. Note that $c > 0$ means that f is increasing while $c < 0$ implies that f is decreasing.

Although such functions model many different natural phenomena, they are often only a first approximation and we now develop this model by supposing that the system has a *carrying capacity*, K. We are thus supposing $f : [0, \infty) \longrightarrow (0, K)$ and that, in particular, $0 < f(t) < K$ for all $t \geq 0$. It is natural to attempt to modify (11.5). We assume smooth growth, that is $f'(t) > 0$, but that this growth tapers off to 0 as $f(t) \longrightarrow K$. We have many different choices, for instance

$$f'(t) = rf(t)\Big(1 - \frac{f(t)}{K}\Big)$$

or

$$f'(t) = rf(t)\Big(1 - \frac{f(t)}{K}\Big)^3.$$

Each of these would model the physical situation and experimental data could later be used to provide the criterion to choose between the models. We stay with the first of these and obtain

$$\frac{f'(t)}{f(t)} = \frac{r}{K}(K - f(t))$$

and hence

$$\frac{f'(t)}{f(t)(K - f(t))} = \frac{f'(t)}{Kf(t)} + \frac{f'(t)}{K(K - f(t))} = \frac{r}{K}.$$

This implies

$$\frac{d}{dt}\big(\log f(t)\big) - \frac{d}{dt}\big(\log(K - f(t))\big) = \frac{d}{dt}(rt)$$

and, by the Mean Value Theorem, there exists a constant a such that

$$\log\Big(\frac{f(t)}{K - f(t)}\Big) = a + rt.$$

Hence

$$\frac{f(t)}{K - f(t)} = e^{a+rt}, \quad \frac{f(0)}{K - f(0)} = e^a, \quad f(t) = \frac{Ke^{a+rt}}{1 + e^{a+rt}}$$

and

$$f(t) = \frac{K}{1 + e^{-a}e^{-rt}} = \frac{K}{1 + \left(\frac{K - f(0)}{f(0)}\right)e^{-rt}}.$$

Note that $f(t) \longrightarrow K$ as $t \longrightarrow +\infty$.

Example 11.13. Suppose $\sum_{n=0}^{\infty} a_n x^n$ is a power series which converges absolutely when $|x| < r$. Fix $x, |x| < r$, and $k \in \mathbf{N}$ and choose $\delta > 0$ such that $|x| + 2\delta < r$. By Theorem 8.15, $\sum_{n=0}^{\infty} |a_n|(|x| + 2\delta)^n < \infty$. The ratio test, Theorem 8.12(b), implies that the series $\sum_{n=1}^{\infty} n^k r^n$ is absolutely convergent for any positive integer k when $|r| < 1$ and, by Theorem 8.8, there exists, for each k, a positive real number M such that

$$n^k \left(\frac{|x| + \delta}{|x| + 2\delta}\right)^n \leq M$$

for all n. Hence

$$\sum_{n=1}^{\infty} n^k |a_n|(|x| + \delta)^n = \sum_{n=1}^{\infty} n^k \left(\frac{|x| + \delta}{|x| + 2\delta}\right)^n |a_n|(|x| + 2\delta)^n$$

$$\leq M \sum_{n=1}^{\infty} |a_n|(|x| + 2\delta)^n$$

$$< \infty$$

and $\sum_{n=1}^{\infty} n^k a_n x^n$ converges absolutely for $|x| < r$.

Example 11.14. Let $f(x) = \sum_{n=0}^{\infty} a_n x^n$ for $|x| < r$. By Example 11.13, the series $\sum_{n=1}^{\infty} n a_n x^{n-1}$ and $\sum_{n=2}^{\infty} n(n-1)a_n x^{n-2}$ converge absolutely when $|x| < r$. Fix $x, |x| < r$, and choose $\delta > 0$ such that $\delta_x := |x| + \delta < r$. For each n we apply the Mean Value Theorem to $x \longrightarrow x^n$ over $[x, x + \Delta x]$, where $|\Delta x| < \delta$, and obtain, $\theta_n, |\theta_n| < 1$, such that

$$f(x + \Delta x) - f(x) = \Delta x \sum_{n=1}^{\infty} n a_n (x + \theta_n \Delta x)^{n-1} =: \Delta x f_x(\Delta x).$$

A further application of the Mean Value Theorem implies

$$\sum_{n=1}^{\infty} n a_n (x + \theta_n \Delta x)^{n-1} - \sum_{n=1}^{\infty} n a_n x^{n-1} = \Delta x \sum_{n=2}^{\infty} n(n-1)a_n \theta_n (x + \theta_n' \Delta x)^{n-2}$$

where $|\theta_n'| < |\theta_n| < 1$ for all n. Hence

$$\left| f_x(\Delta x) - \sum_{n=1}^{\infty} n a_n x^{n-1} \right| \leq |\Delta x| \sum_{n=2}^{\infty} n(n-1)|a_n|\delta_x^{n-2} \longrightarrow 0$$

as $\Delta x \longrightarrow 0$. This implies $\lim_{\Delta x \to 0} f_x(\Delta x) = \sum_{n=1}^{\infty} n a_n x^{n-1}$ and $f'(x) = \sum_{n=1}^{\infty} n a_n x^{n-1}$ by (10.6). Proceeding in this way, we see that f has derivatives of all orders at all points x, $|x| < r$, and $f^n(x)/n! = \sum_{m=n}^{\infty} \binom{m}{n} a_m x^{m-n}$ where $\binom{m}{n} = \frac{m!}{n!(m-n)!}$. When $x = 0$, $f^n(0) = n!a_n$.

As a particular example we obtain, (see Exercise 8.12 when $x = 1$),

$$\frac{d}{dx}\big(\log(1+x)\big) = \frac{1}{1+x} = \sum_{n=0}^{\infty}(-1)^n x^n = \frac{d}{dx}\Big(\sum_{n=0}^{\infty}\frac{(-1)^n x^{n+1}}{n+1}\Big)$$

when $|x| < 1$. Since $\log 1 = 0$, Corollary 10.7(a) implies

$$\log(1+x) = \sum_{n=0}^{\infty}\frac{(-1)^n x^{n+1}}{n+1} = x - \frac{x^2}{2} + \frac{x^3}{3} - \cdots . \qquad (11.7)$$

Example 11.15. The Riemann Zeta function $\zeta(s) = \sum_{n=1}^{\infty} n^{-s}$, introduced in Chapter 7, converges for all $s > 1$ by the condensation test (Exercise 8.9) or the integral test (Example 14.12(a)). Moreover, by comparing entries in the appropriate series, we see that $(\zeta(k))_{k=2}^{\infty}$ is a decreasing sequence of positive numbers. Hence

$$\sum_{k=2}^{\infty}\frac{\zeta(k)}{k(k+1)} = \sum_{k=2}^{\infty}\frac{1}{k(k+1)}\Big(\sum_{n=1}^{\infty}\frac{1}{n^k}\Big) =: M < \infty. \qquad (11.8)$$

We now prove a result which is, apart from an unspecified constant, a very useful classical limit of Stirling from 1730. If $n \in \mathbf{N}$, let

$$R(n) = \log n! - n\log n = \log\Big(\frac{n!}{n^n}\Big).$$

If $m > 1$ then, applying the power series (11.7) and rearranging terms, we obtain,

$$R(m) = \log\Big(\frac{(m-1)!}{m^{m-1}}\Big) = \log\Big(\frac{(m-1)!}{(m-1)^{m-1}}\Big) + \log\Big(\frac{(m-1)^{m-1}}{m^{m-1}}\Big)$$

$$= R(m-1) + (m-1)\log\Big(1 - \frac{1}{m}\Big)$$

$$= R(m-1) - (m-1)\Big(\frac{1}{m} + \frac{1}{2m^2} + \frac{1}{3m^3} + \cdots\Big)$$

$$= R(m-1) - 1 + \Big(\frac{1}{m} - \frac{1}{2m}\Big) + \Big(\frac{1}{2m^2} - \frac{1}{3m^2}\Big)$$

$$+ \Big(\frac{1}{3m^3} - \frac{1}{4m^3}\Big) + \cdots$$

and, proceeding in this way and noting that $R(1) = 0$, we obtain

$$R(n) = -n + \frac{1}{1 \cdot 2} \sum_{m=1}^{n} \frac{1}{m} + \sum_{k=2}^{\infty} \frac{1}{k(k+1)} \Big(\sum_{m=1}^{n} \frac{1}{m^k} \Big). \qquad (11.9)$$

By Exercise 8.12, $(\sum_{m=1}^{n} \frac{1}{m} - \log n) := \gamma_n \longrightarrow \gamma$ (Euler's Constant) as $n \longrightarrow \infty$. Substituting this result into (11.9) we obtain

$$R(n) = \log n! - n \log n = -n + \frac{1}{2} \log n + \frac{\gamma_n}{2} + \sum_{k=2}^{\infty} \frac{1}{k(k+1)} \Big(\sum_{m=1}^{n} \frac{1}{m^k} \Big)$$

and, on applying (11.8) and the exponential function, we obtain

$$\frac{n! e^n}{n^{(2n+1)/2}} \longrightarrow e^{\frac{\gamma}{2} + M} \qquad \text{as} \qquad n \longrightarrow \infty. \qquad (11.10)$$

A version of *Stirling's Formula* states that

$$\lim_{n \longrightarrow \infty} \frac{\sqrt{2\pi} e^{-n} n^{(2n+1)/2}}{n!} = 1$$

and shows that $e^{\gamma + 2M} = 2\pi$. By (11.8), this implies

$$2\pi = e^{\gamma} \prod_{n=2}^{\infty} \exp \Big(\frac{\zeta(n)}{\binom{n+1}{2}} \Big).$$

Our approach in this example follows Luis J. Boya.

Our two final examples concern equations involving an unknown function and its derivatives, that is a differential equation. Such equations appear frequently when constructing mathematical models of real life situations (see Example 11.12). There are many different types of differential equations and a vast literature on the subject exists. A differential equation involving only functions of a single real variable is called an *ordinary differential equation* and the highest order derivative that appears is called the *order* of the equation. The principal questions that arise naturally are existence, uniqueness and the form of solutions (see Exercise 13.12). We confine ourselves to a few selective first and second order ordinary differential equations merely to illustrate concepts we have already introduced.

In contrast to the theory of quadratic equations, additional complications may arise because of varied notation. This is not due to disagreement over the choice of notation but because the same equation may have different applications and there are several ways of approaching the same material. We briefly consider notation.

Consider, as an example, the following equation. Given $g : \mathbf{R} \longrightarrow \mathbf{R}$, a continuous function, we wish to find $f : \mathbf{R} \longrightarrow \mathbf{R}$ a twice continuously differentiable function such that

$$\frac{d^2 f}{dx^2} + 5\frac{df}{dx} + 6f = g. \tag{11.11}$$

Motivated by the notation $y = f(x)$ in Chapter 4 we write y in place of f and rephrase (11.11) as

$$y'' + 5y' + 6y = g. \tag{11.12}$$

Again if we let $D = \frac{d}{dx}$ and $D^2 = \frac{d^2}{dx^2}$ then $Df = \frac{df}{dx}$, $D^2 f = D(Df) = \frac{d^2 f}{dx^2}$ and (11.12) becomes

$$D^2 y + 5Dy + 6y = (D^2 + 5D + 6)y = g. \tag{11.13}$$

If $\mathcal{D}^2(\mathbf{R})$ denotes the set of all real-valued functions defined on $\mathbf{R}$ which have continuous first and second order derivatives and $\mathcal{C}(\mathbf{R})$ denotes the continuous real-valued functions of $\mathbf{R}$ then we may rephrase the above by asking if the mapping

$$T : \mathcal{D}^2(\mathbf{R}) \longrightarrow \mathcal{C}(\mathbf{R}),$$

where $T(f) := (D^2 + 5D + 6)f$ for all $f \in \mathcal{D}^2(\mathbf{R})$, is *surjective*.

Example 11.16. Consider the second order ordinary differential equation

$$(x^2 + 1)y'' + 2xy' = 0. \tag{11.14}$$

We suppose initially that there exists a solution y which admits a power series expansion, $y(x) = \sum_{n=0}^{\infty} a_n x^n$, with positive radius of convergence. By Example 11.14, we can differentiate the power series term by term to obtain

$$(x^2 + 1) \sum_{n=2}^{\infty} n(n-1)a_n x^{n-2} + 2x \sum_{n=1}^{\infty} na_n x^{n-1} = 0,$$

that is

$$(x^2 + 1) \cdot (2a_2 + 6a_3 x + 12a_4 x^2 \cdots) + 2x(a_1 + 2a_2 x + 3a_3 x^2 \cdots) = 0.$$

When $x = 0$, we have $a_2 = 0$ and, from the coefficient of x, we see that $6a_3 + 2a_1 = 0$, that is $a_3 = -\frac{1}{3}a_1$. For $n \geq 2$, we obtain $n(n-1)a_n + (n+2)(n+1)a_{n+2} + 2na_n = 0$, that is $(n^2+n)a_n + (n+2)(n+1)a_{n+2} = 0$ and, hence,

$$a_{n+2} = -\frac{n(n+1)}{(n+2)(n+1)}a_n = -\frac{n}{(n+2)}a_n$$

for all $n > 0$. Since $a_2 = 0$, this implies $a_{2n} = 0$ for all $n > 0$. For $n = 2m + 1, m > 0$ we have

$$a_{2m+1} = -\frac{2m-1}{2m+1}a_{2m-1} = (-1)^2\frac{2m-1}{2m+1} \cdot \frac{2m-3}{2m-1}a_{2m-3}$$
$$= (-1)^2\frac{2m-3}{2m+1}a_{2m-3}$$
$$= (-1)^m\frac{1}{2m+1}a_1.$$

Hence

$$y(x) = a_0 + a_1\left[\sum_{m=0}^{\infty}(-1)^m\frac{x^{2m+1}}{2m+1}\right]. \tag{11.15}$$

The ratio test shows that the series in (11.15) has radius of convergence 1 and, by Example 11.14, y has derivatives of all orders. One may now show, by differentiating term by term and varying a_0 and a_1, that all power series solutions to (11.14) are given by (11.15).

Our final example connects the theme of the first chapter, quadratic equations, with differential equations.

Example 11.17. Consider the equation

$$a\frac{d^2y}{dx^2} + b\frac{dy}{dx} + cy = e^{rx}. \tag{11.16}$$

If y_1 and y_2 are solutions to (11.16), then $y_1 - y_2$ is a solution to the homogeneous differential equation

$$a\frac{d^2y}{dx^2} + b\frac{dy}{dx} + cy = 0 \tag{11.17}$$

and it suffices to find one solution for (11.16) and to add to it all solutions to (11.17). From (11.17) we obtain the quadratic equation

$$ax^2 + bx + c = 0$$

and we suppose that this has two distinct real solutions, r_1 and r_2, neither of which equals r. From Chapter 1 we know that this will be the case if and only if $b^2 - 4ac > 0$ and $ar^2 + br + c \neq 0$. Using the D notation in (11.13), we see that (11.17) is equivalent to

$$(D - r_1)(D - r_2)y = 0.$$

By Example 11.12, we see that $(D - r_2)y(x) = Ae^{r_1 x}$. Hence

$$\frac{d}{dx}(y(x)e^{-r_2 x}) = \frac{dy}{dx}e^{-r_2 x} - r_2 y e^{-r_2 x} = e^{-r_2 x}\left(\frac{d}{dx} - r_2\right)y(x) = Ae^{(r_1 - r_2)x}$$

and

$$\frac{d}{dx}(y(x)e^{-r_2 x}) = \frac{d}{dx}\left(\frac{Ae^{(r_1 - r_2)x}}{r_1 - r_2}\right). \tag{11.18}$$

If we let $B = A/(r_1 - r_2)$, then we see that there exists a constant C such that

$$y(x) = e^{r_2 x}(Be^{(r_1 - r_2)x} + C) = Be^{r_1 x} + Ce^{r_2 x}$$

and, by choosing different constants B and C, we obtain all solutions to (11.17). This means that if we find one particular solution to (11.16), y_0, then all solutions will be given by

$$f(x) = y_0(x) + Be^{r_1 x} + Ce^{r_2 x}$$

where B and C are arbitrary real numbers. Since the right-hand side of (11.16) equals e^{rx} we try, on the basis that $\frac{d}{dx}(e^{rx}) = re^{rx}$, to see if Ae^{rx} is a solution. We have

$$A(ar^2 + br + c)e^{rx} = e^{rx}$$

and

$$A = \frac{1}{ar^2 + br + c}. \tag{11.19}$$

This implies that the general solution to (11.16) is given by

$$y(x) = \frac{e^{rx}}{ar^2 + br + c} + Be^{r_1 x} + Ce^{r_2 x}$$

where B and C are arbitrary real numbers. By (11.18) and (11.19), we see that our approach required the hypotheses $r_1 \neq r_2$ and $ar^2 + br + c \neq 0$. Nevertheless, solutions do exist when these conditions are not satisfied and an investigation, similar to that undertaken in Chapter 1, may be followed to unearth them.[4] For example the power series approach in Example 11.16 leads, with moderate effort, to the required solutions.

---

[4]See for instance Exercise 11.33(v).

11.5 In Retrospect

The controversy over the discovery of the calculus combined with the superior notation adopted by continental mathematicians led to almost a century of mathematical isolation for Britain. During the century 1730-1830, the main mathematical developments came from French and German mathematicians. Mathematics in England[5] had little prestige except perhaps at Cambridge where students were required to pass the *Mathematical Tripos* examination if they wished to graduate with honors, no matter what their specialty. The name of the examination dates from the 15th century when the examiner sat on a three legged stool. The examination itself was the intellectual equivalent of a sequence of olympic sprints which rewarded cleverness and quick thinking over creativity. Serious candidates were often trained for up to three years by a professional coach. To *wrangle* means to argue and the examination results were published in order of merit with the best candidate getting the title of *Senior Wrangler* and the lowest ranking graduate being presented unofficially with a large *wooden spoon*.

An important factor in the modernization of the mathematics curriculum at Cambridge and, by extension, at other English universities, was a revolt initiated by a group of Cambridge undergraduates, led by Charles Babbage, known later as the father and grandfather of *computer science* and *operations research*, John Herschel, 1792-1871, the future astronomer, chemist, botanist, experimental photographer and senior wrangler from 1813, and George Peacock, 1791-1858. They formed *The Analytical Society* for the cultivation of mathematics. This evolved after a few years into the *Cambridge Philosophical Society*. Convinced that the notation of Newton was *a strong impediment to the progress of English science* and aware that they, young and unknown, were taking on a powerful educational establishment they decided to publish their deliberations. In *Memoirs of the Analytic Society* they, diplomatically and with some justification, promoted Fermat as the discoverer of the calculus with supporting roles for Newton

---

[5]The situation was better in Scotland.

and Leibniz and, in 1816, three years later, they published a translation of the best known continental textbook on the calculus; Lacroix's *Sur le Calcul Différential et Intégral.* A few years later, Peacock, now a junior lecturer, was given the task of setting and correcting undergraduate examination papers and introduced, in spite of opposition and controversy, Leibniz's notation. Their progress was slow but effective and paved the way for British mathematics to rejoin the international mathematical research community later in the 19th century.

The academic interests of Babbage[6] were varied, he was Lucasian Professor of Mathematics at Cambridge for eleven years and an actuary for a year. He resigned from both positions to devote himself to his life's main work, the development of his calculating machines. He called the simplest one a Difference Engine and the more advanced one an Analytic Engine.[7] His work on computers, as we now call them, achieved limited success during his lifetime and did not achieve the physical completion he wanted for them. His youngest son, Henry Prevost Babbage, 1824-1918, created six working engines based on his father's design and, in 1991, a perfectly func-

---

[6]Charles Babbage, 1791-1871, from London considered himself a philosopher, and indeed, he was a lover of all knowledge. His guiding principle was to always endeavor to *discover those laws of mind by which man's intellect passes from the known to the discovery of the unknown.* As a youth, he was not strong and his parents instructed his teacher not to press too much knowledge on him: a mission, he reported, was faithfully accomplished. His innate curiosity, appetite for understanding and discipline compensated for any deficiencies. He made, like Descartes and Leibniz, original contributions to a wide range of subjects and his writings made him well known during his lifetime. His list of active interests included; astronomy, engineering, volcanoes, codes, the art of picking locks, humor, railway safety, miracles, mechanical games, taxation, detecting forgeries and confidence tricksters, nautical communication, political economy, etc. He was, by all accounts, an honest eccentric who took people as they were and, as a result, he had a wide circle of friends ranging from European royalty to academic professors to ordinary tradesmen to street-beggars. He hated hypocrisy, nepotism, and swindling from any source. In one incident he tracked down, by spending three hours questioning all passersby on Monmouth Street, a vagrant down on his luck who he discovered had taken advantage of his generosity, got him arrested and jailed for a week.

[7]One of the many design innovations in the Analytic Machine was the use of programs with punched cards, similar to those used in computers in the middle of the twentieth century.

tioning Difference Engine was constructed from Babbage's original plans by the London Science Museum. Menabrea, who had attended the lectures of Cauchy in Turin, wrote up in French, a seminar that Babbage gave in Turin on his Analytic Engine. At Babbage's request it was translated into English by Ada Lovelace[8] and supplemented with her own extensive notes, including what is now considered the first computer program.

Babbage, no doubt influenced by his experiences with the Analytical Society at Cambridge, was interested all his life in the process by which human activities are *organized*. He was actively involved in the formation of the (Royal) Astronomical Society, the British Association for the Advancement of Science, the (Royal) Statistical Society, and a Society of Authors to Protect Copyright, and his coaxing and enthusiasm led several of his continental friends to organize informal scientific seminars. He was twice the active chairman of a committee that successfully saw William Cavendish, a second wrangler, elected as a Liberal MP to the House of Commons from Cambridge. His brief analysis, based on this experience, of committee dynamics is accurate, perceptive and as relevant today as the day it was written.

11.6 Exercises

(11.1) By developing the method used in Example 11.4, find

$$\sum_{j=1}^{n} j^2 \text{ and } \sum_{j=1}^{n} j^3.$$

(11.2) Let $f(x) = 3x^9 + x^3 - 6x^2 + 12x - 8$. Show that f is an increasing function and find the image of f. Show that f has an inverse f^{-1} and that f^{-1} is differentiable. Find $(f^{-1})'(-8)$.

---

[8]Augusta Ada King, Countess Lovelace, 1815-1852, was born in London, the daughter of the poet Lord Byron. She was privately educated and Augustus de Morgan, one of her later tutors, considered her capable of becoming an original mathematical investigator. She met and corresponded with Babbage on numerous occasions and foresaw the great potential in his machines. She is credited as the first to write an algorithm to be read by a machine.

(11.3) Find the intervals where the functions $x^2 e^{-x}$ and $x^4 e^{-x^2}$ are increasing, decreasing and convex. Find the maximum and minimum of the functions on $[0, 4]$.

(11.4) If $g : \mathbf{R} \longrightarrow \mathbf{R}$ is a twice differentiable increasing convex function and $U : \mathbf{R} \longrightarrow \mathbf{R}$ is a twice differentiable convex function, show that $g \circ U$ is convex. By using the function $g(x) = e^{-x}$, show that the composition of convex functions is not always convex.

(11.5) Find the derivative, the proportional rate of change and the elasticity of the following functions

$$\exp\left(\frac{x}{x+1}\right), \quad \exp\left(x^2\right) \cdot \log(x^2), \quad \exp\left(2\log(x^2)\right), \quad x \exp\left(x^2 + 1\right).$$

(11.6) Use differentiation to show that the sequence $\left(\frac{n^2}{n^2+1}\right)_{n=1}^{\infty}$ is increasing.

(11.7) How much money should be deposited at 5.85% interest compounded continuously, in order to accumulate \$2000 at the end of 3 years? How much longer would the same amount take if interest was compounded monthly?

(11.8) A mortgage of \$250,000 is to be repayed over 20 years in equal monthly instalments. Find a lower bound for the repayments. If the interest rate is 5.2% per annum, continuously compounded, and the interest is added to the mortgage at the beginning of each year, find the total monthly repayments. Find the total amount repayed.

(11.9) Use L'Hôpital's rule to find the following limits:

$$(a)\ \lim_{x \to 1} \frac{x-1}{x^2-1}, \quad (b)\ \lim_{x \to 1} \left(\frac{x-1}{x^2-1}\right)^5, \quad (b)\ \lim_{x \to 0} \left\{ \exp\left(\frac{e^x - 1}{\log(1+x)}\right) \right\}.$$

(11.10) Sketch the graph of $f(x) = 2x^2/(x^4 + 1)$ and find the maximum of f over $[0, \infty)$ and $\mathbf{R}$. Does the function have a minimum over these intervals?

(11.11) Use the Intermediate Value Theorem to verify the Mean Value Theorem for the function

$$f : \mathbf{R} \longrightarrow \mathbf{R},\ f(x) = x^4 - x^3 + 2x^2 - x$$

on $[1, 2]$.

(11.12) If $g : \mathbf{R} \longrightarrow \mathbf{R}$ is differentiable and $El(g) = x^2 e^{-x}$ find g.

(11.13) Let $f(x) = x \exp(x^2 + 1)$ for all $x \in \mathbf{R}$. Find f', the proportional rate of change of f and the elasticity of f.

(11.14) Evaluate

$$\sum_{n=1}^{\infty} \frac{n^2}{2^n} \quad \text{and} \quad \sum_{n=1}^{\infty} \frac{n^3 x^{3n}}{4^n}.$$

(11.15) Find the intervals where

$$f : \mathbf{R} \backslash \{\pm 2\} \longrightarrow \mathbf{R}, \quad f(x) = \frac{2x^2}{4 - x^2}$$

and

$$g : \mathbf{R} \backslash \{0\} \longrightarrow \mathbf{R}, \quad g(x) = \frac{e^x}{1 - e^{-2x}}$$

are convex. Sketch the graphs of both functions.

(11.16) Give the natural domains for the following functions. Sketch their graphs (a) without using the differential calculus and (b) using the differential calculus,

$$\frac{1}{1 + x^3}, \quad \frac{x}{8 + x^3}, \quad 5x^5 - 20x^3 + 6x, \quad \frac{e^x}{x^2 - 3}, \quad \frac{2x^3 - x^2 + x + 1}{x^2}.$$

(11.17) The cost of producing x items of a certain good is $a + bx$ and the selling price is $c - dx$. Find the level of production that maximizes profits. How much of a tax t should the manufacturer absorb in order to maximize profits. (The constants a, b, c, d and t are all positive and $b > d$.)

(11.18) A truck has a top speed of 75 mph and, when traveling at x mph, consumes gasoline at the rate of

$$\frac{1}{100} \left(\frac{3200}{x} + x \right)$$

miles per gallon. If gasoline costs \$3 per gallon find the cost of a 120 mile journey undertaken at a constant speed of x mph. Find the most economical speed. If the driver is paid \$15 per hour, find the most economical speed.

(11.19) Let a denote a strictly positive real number. By considering the derivative of the function

$$f : \mathbf{R}^+ \longrightarrow \mathbf{R}, \quad f(x) = \log(ax) - \log(x)$$

establish the identity $\log(ab) = \log a + \log b$ for all $a, b \in \mathbf{R}^+$.

(11.20) Show that $f : \mathbf{R} \longrightarrow \mathbf{R}$, $f(x) = 2x^6 - 5x^4 - 20x^3 + 90x^2$ is convex.

(11.21) Let x and y denote the number of items of two different products produced in a factory. If the cost of producing these items is given by $C(x, y) = 10 + 18x^2 + 6xy + 9y^2$ and the total to be produced is 56, how many of each should be produced in order to minimize cost?

(11.22) If the functions f and g are represented by power series $\sum_{n=0}^{\infty} a_n x^n$ and $\sum_{n=0}^{\infty} b_n x^n$, respectively, in the interval $(-r, +r)$, $0 < r \leq \infty$, and if $f(x_n) = g(x_n)$ for all n where $(x_n)_{n=1}^{\infty}$ is a sequence of distinct real numbers converging to 0, show that $f(x) = g(x)$ when $|x| < r$.

(11.23) A car rental agency has 24 identical cars. When cars are rented out at \$100 per day, all cars are rented out. However, each time the price is increased by \$10 one of the cars is not rented out. If the agency loses \$20 for every car that is not rented out find the rental price that would maximize profits.

(11.24) Show that
$$f : \mathbf{R}_0^+ \longrightarrow \mathbf{R}_0^+, \quad f(x) = \log(1 + x) - x$$
is strictly decreasing and, hence, deduce that $\log(1 + \frac{1}{k}) < \frac{1}{k}$ for every positive integer k. Hence show that
$$\sum_{k=1}^{n} \log\left(1 + \frac{1}{k}\right) = \log(1 + n) < \sum_{k=1}^{n} \frac{1}{k}$$
and prove that the series $\sum_{n=1}^{\infty} \frac{1}{n}$ diverges.

(11.25) Let $f : (a, b) \longrightarrow (c, d)$ denote a twice differentiable surjective convex function and suppose $f'(x) \neq 0$ for all $x \in (a, b)$. Show that $f^{-1} : (c, d) \longrightarrow (a, b)$ is convex if and only if f is a decreasing function. Sketch the graphs of $f : \mathbf{R} \longrightarrow \mathbf{R}^+, f(x) = e^{-x}$, and its inverse, displaying the fact that both are decreasing and convex.

(11.26) Let $f, g : [a, b] \longrightarrow \mathbf{R}$ denote continuous functions which are differentiable on (a, b). If $f(a) = g(a) = 0$ and $f(x)g(x) \neq 0$ for all $x \in (a, b)$, show that
$$\lim_{x > a, x \to a} \frac{f(x)}{g(x)} = \lim_{x > a, x \to a} \frac{f'(x)}{g'(x)}$$

whenever the right-hand limit exists. Let $f, g : \mathbf{R}^+ \longrightarrow \mathbf{R}$ denote differentiable functions and suppose $\lim_{x\to+\infty} f(x) = \lim_{x\to+\infty} g(x) = 0$. Using the first part of the exercise and the change of variable $y = 1/x$, show, whenever the right-hand limit exists, that

$$\lim_{x\to+\infty} \frac{f(x)}{g(x)} = \lim_{x\to+\infty} \frac{f'(x)}{g'(x)}.$$

Hence, or otherwise, find

$$\lim_{x\to+\infty} \frac{\log x}{x} \quad \text{and} \quad \lim_{x\to+\infty} x^{1/x}.$$

(11.27) Show that $e^x(1 + sx) \le 1$ for all $x \in \mathbf{R}$ if and only if $s = -1$.

(11.28) Let $P(x) = x^3 + px + q = 0$. If $x = z - \frac{p}{3z}$, show that $z^6 + qz^3 - \frac{p^3}{27} = 0$ and, using quadratic equations, deduce that[9]

$$z = \sqrt[3]{\frac{-q}{2} \pm \sqrt{\frac{p^3}{27} + \frac{q^2}{4}}}.$$

Show that

$$x = \sqrt[3]{\frac{-q}{2} + \sqrt{\frac{p^3}{27} + \frac{q^2}{4}}} + \sqrt[3]{\frac{-q}{2} - \sqrt{\frac{p^3}{27} + \frac{q^2}{4}}}.$$

Find all solutions of the cubic equations $x^3 - x = 0$ and $32x^3 + 64x^2 - 78x - 135 = 0$.

(11.29) If $f : \mathbf{R}\backslash\{4\} \longrightarrow \mathbf{R}$, $f(x) = x(x - 3)(x - 4)^{-1}$, show that the image of f does not intersect the open interval $(1, 9)$. Sketch the graphs of f and $g : \mathbf{R}\backslash\{0\} \longrightarrow \mathbf{R}$, $f(x) = 5x^{-1}$ on the same diagram and hence show there are no positive solutions of the equation

$$\frac{5}{x} = \frac{x(x-3)}{x-4}.$$

(11.30) If $y(x) = (Ax + B)e^{2x}$, $A, B \in \mathbf{R}$, solves the differential equation $y'' - y = 4xe^{2x}$, find A and B. Find all solutions of the equation.

(11.31) Let

$$f(x) = \sum_{n=0}^{\infty} (-1)^n \frac{x^{2n+1}}{(2n+1)!} \quad \text{and} \quad g(x) = \sum_{n=0}^{\infty} (-1)^n \frac{x^{2n}}{(2n)!}.$$

Show that f and g are defined for all $x \in \mathbf{R}$ and that $f'(x) = g(x)$ and $g'(x) = -f(x)$. Show, using the Mean Value Theorem or otherwise, that $(f(x))^2 + (g(x))^2 = 1$ for all $x \in \mathbf{R}$.

---

[9]Paradoxically this method of solving cubic equations, when all solutions are real, requires the square root of negative real numbers and thus complex numbers.

(11.32) Find the power series solutions for the differential equations $y'' - xy' + 2y = 0$ and $(1 - x^2)y'' - xy' + 4y = 0$. Find the radius of convergence of each power series.

(11.33) Find all solutions to the following differential equations:

$$(i) \quad y'' + 4y' - 12y = e^{-2x}, \qquad (ii) \quad y'' - 5y' + 6y = 3e^{2x},$$

$$(iii) \quad y'' - 4y' + y = 4e^{3x}, \qquad (iv) \quad y'' - 4y' + y = 12e^{2x} + x,$$

$$(v) \quad y'' - 4y' + 4y = x^2 - 2x, \qquad (vi) \quad y''' - 2y'' - y' + 2y = 4e^{5x}.$$

Find the particular solutions which satisfy $y'(0) = y(0) = 0$.

References [2; 5; 6; 9; 10; 15; 34; 38]

Chapter 12

THE REAL NUMBERS

Every measurement of a quantity is a
real number.

René Descartes

A true understanding of mathematics must involve
an explanation of which set-theory notions have
'mathematical content', and this question is
manifestly not reducible to a problem of logic,
however broadly conceived.

Leon A. Henkin, 1965

Summary

We construct the real numbers and discuss recent notions of rigor and logic.

12.1   Upper Bounds

In Chapter 5, using equivalence relationships, we enlarged the number system N by defining the rational numbers Q and showed, geometrically, that the number line contains points which do not correspond to any rational number. Subsequently, we developed, assuming only an intuitive notion of real number and the Upper Bound Principle, the differential calculus. To place this on a firm logical footing we need to define the real numbers as an extension of the rational numbers and show that they satisfy an Upper Bound Principle. We first show that $\mathbf{Q}$ does not satisfy an Upper Bound Principle.

Example 12.1. Let $A = \{x \in \mathbf{Q} : x^2 < 2\}$. If $y \in \mathbf{Q}$ and $y > 2$ then $y^2 > 4$. Hence $x < 2$ for all $x \in A$. This shows that A is bounded above. Suppose $U \in \mathbf{Q}$ is a least upper bound for A. Clearly, $U > 0$ and, by Theorem 5.9, $U^2 \neq 2$. If $U^2 < 2$ then there exists, by Theorem 5.12, a positive integer n such that $n(2 - U^2) > 2U + 1$. Hence

$$\left(U + \frac{1}{n}\right)^2 = U^2 + \frac{2U}{n} + \frac{1}{n^2} < U^2 + \frac{2U}{n} + \frac{1}{n} < 2$$

and $U + \frac{1}{n} \in A$. This contradicts the fact that U is an upper bound for A.

If $U^2 > 2$ then $U^2 - 2 > 0$ and, by Theorem 5.12, we can find a positive integer m such that $m(U^2 - 2) > 2U$. If $x \in A$, then

$$\left(U - \frac{1}{m}\right)^2 = U^2 - \frac{2U}{m} + \frac{1}{m^2} > 2 > x^2$$

and $U - \frac{1}{m}$ is an upper bound for A which is smaller than the least upper bound. This shows that no least upper bound exists and $\mathbf{Q}$ does not satisfy an Upper Bound Principle.

We could join $\pm\sqrt{2}$ to the rationals to obtain

$$\mathbf{Q}(\sqrt{2}) := \{r + s\sqrt{2} : r, s \in \mathbf{Q}\}$$

and it is quite easy to define addition, multiplication and an order on $\mathbf{Q}(\sqrt{2})$ and to embed $\mathbf{Q}$ in $\mathbf{Q}(\sqrt{2})$ so that these operations extend those already defined on $\mathbf{Q}$. However, afterwards, we would find that the equation $x^2 = 3$ does not have a solution in $\mathbf{Q}(\sqrt{2})$ and, as above, that $\mathbf{Q}(\sqrt{2})$ does not satisfy an Upper Bound Principle. Indeed, it can be shown that even if all algebraic numbers[1] were added to $\mathbf{Q}$, we would still miss real numbers like π and e and still not obtain the desired principle. Consequences, that we have already seen, indicate how we might proceed. By the *Intermediate Value Theorem* one cannot go from above to below the real number line in a smooth fashion without hitting a real number. Similarly, the fundamental

---

[1] Real numbers which satisfy a polynomial equation with integer coefficients are called *algebraic numbers*, all other real numbers are called *transcendental*.

existence theorem for maxima and minima says that there is no gap or hole in the real number line. Thus, any time one cuts the real number line the cut goes through a unique real number.

Richard Dedekind realized on November 24, 1858, while preparing a class on the differential calculus, that this process could be used to *define* the real numbers and thus put on a logical basis the concept of irrational number, a difficulty that had caused problems for close to two thousand years. He used what we now call *Dedekind cuts or sections* of rational numbers. Cantor used equivalence classes of Cauchy sequences of rational numbers to construct the real numbers.

The fact that the solution to a major foundational problem within mathematics was discovered, within the confines of an elementary course, supports an opinion, widely held by mathematicians though not generally shared by administrators, that teaching and research complement one another. Mathematical truths are communicated to new generations, while being tested and distilled into their essential parts through teaching, as researchers are inspired and challenged by questions that arise in the classroom.

For many years Guido Zapata used Dedekind's approach in his lectures at Universidade Federal do Rio de Janeiro before developing the variation that we follow in our presentation. It is based on an unpublished article of Zapata on the *Bisection Principle*. The fundamental idea underlying this approach has already appeared in Theorems 8.6 and 9.6 and will appear again in the next chapter when we construct the Riemann integral. Dedekind's approach is existential, uses the rational numbers, and two sections. Zapata's is constructive. It uses the dyadic numbers, a single section, increasing sequences, and relies on the intuitive notion that real numbers are precisely those numbers that admit decimal and, of course, dyadic expansions. All these constructions lead to essentially the same set of real numbers but we do not prove this assertion.

12.2   Dyadic Sections and the Bisection Principle

Definition 12.2.

(a) Let

$$\mathbf{D} := \left\{ a \in \mathbf{Q} : \quad a = \frac{p}{2^n} = p2^{-n}, \quad p \in \mathbf{Z}, \quad n \in \mathbf{N}_0 \right\}.$$

The elements in $\mathbf{D}$ are called dyadic numbers.

(b) A dyadic segment is a subset S of $\mathbf{D}$ with the following properties:

    1. it is non-empty,

    2. it is bounded above,

    3. it does not contain a largest element,

    4. if $x \in S, y \in \mathbf{D}$ and $y \le x$ then $y \in S$.

We let $\mathbb{S}(\mathbf{D})$ denote the set of all dyadic segments.

The following lemma collects simple facts about the dyadic numbers.[2] We leave the proofs as exercises. Note that Theorem 5.12 is required for parts (d) and (e).

Lemma 12.3. *If $a, b, c \in \mathbf{D}$ then:*

    (a) $\mathbf{Z} \subset \mathbf{D} \subset \mathbf{Q}$,

    (b) $a + b$, $\frac{a}{2}$ and $ab \in \mathbf{D}$,

    (c) if $a < b$ then $a < \frac{a+b}{2} < b$,

    (d) if $0 < a$ then there is an $n \in \mathbf{N}_0$ such that $2^{-n} < a$,

    (e) if $0 < a$ and $b \le a2^{-n}$ for all $n \in \mathbf{N}_0$ then $b \le 0$.

In the first two parts of the following example we construct dyadic segments. Parts (ii) and (iii) show how increasing sequences can be used to give concrete representations of any dyadic segment. Part (iii) is a preview of bisectional sequences. For the sake of clarity we use, in the remainder of this section, Roman letters to denote dyadic numbers and Greek letters for dyadic segments.

---

[2]Because of (a) the commutative, associative, order relations, and the identity laws for addition and multiplication are inherited by the dyadic numbers from the rational numbers.

Example 12.4.

(i) If $a \in \mathbf{D}$ then $S(a) := \{b \in \mathbf{D}, b < a\} \in \mathbb{S}(\mathbf{D})$.

(ii) If $(a_n)_{n=0}^{\infty}$ is an *increasing* sequence in $\mathbf{D}$ which is bounded above then

$$S((a_n)_{n=0}^{\infty}) := \bigcup_{n=0}^{\infty} S(a_n) \in \mathbb{S}(\mathbf{D}).$$

(iii) Let $\alpha \in \mathbb{S}(\mathbf{D})$. Let $a_0 = lub\{p \in \mathbf{Z} : p \in \alpha\}$ and let $b_0 = a_0 + 1$. Suppose $a_n \in \alpha$ and $b_n \in \alpha^c := \{x \in \mathbf{D} : x \notin \alpha\}$, the complement of α in $\mathbf{D}$, have been chosen. If $2^{-1}(a_n + b_n) \in \alpha$ let $a_{n+1} = 2^{-1}(a_n + b_n)$, otherwise let $a_{n+1} = a_n$. In either case let $b_{n+1} = a_{n+1} + 2^{-n-1}$. Then $b_{n+1} \notin \alpha$ and, since $a_n \in \alpha$ for all n, we have $S((a_n)_{n=0}^{\infty}) \subset \alpha$.

Let $c \in \alpha$ be arbitrary. Since α does not have a largest element we can find $d \in \alpha$ such that $c < d$. By Lemma 12.3(d), $2^{-p} < d - c$ for some positive integer p. Hence $c + 2^{-p} \in \{a \in \mathbf{D} : a \leq d\} \subset \alpha$. If $c \geq a_p$ then $c + 2^{-p} \geq a_p + 2^{-p} = b_p$. This is impossible since b_p is an upper bound for α and α has no largest element. Hence $c < a_p$ and $c \in S(a_p) \subset S((a_n)_{n=0}^{\infty})$. This shows that $\alpha = S((a_n)_{n=0}^{\infty})$.

Thus, any $\alpha \in \mathbb{S}(\mathbf{D})$ can be written as $S((a_n)_{n=0}^{\infty})$ where $(a_n)_{n=0}^{\infty} \subset \alpha$ is increasing and bounded above, and for which there exists a decreasing sequence of dyadic rationals $(b_n)_{n=0}^{\infty}$ in α^c such that the set $\{(b_n - a_n)^{-1}\}_{n=0}^{\infty}$ is not bounded above. When we write $\alpha = S((a_n)_{n=0}^{\infty})$ we are assuming that these conditions are satisfied. Generally, any such representation can be used in proofs but, for simplicity, we usually take $b_n = a_n + 2^{-n}$. It is, however, necessary to check that the proofs are independent of the sequences used but verification is usually routine and left to the reader (see Exercise 12.10).

Our model of the real numbers is $\mathbb{S}(\mathbf{D})$. Until our construction is fully in place we use the notation $\widehat{0}$ and $\widehat{1}$ in place of $S(0)$ and $S(1)$ respectively.

We now endow $\mathbb{S}(\mathbf{D})$ with an order structure. If $\alpha, \beta \in \mathbb{S}(\mathbf{D})$ we let $\alpha \leq \beta$ if $\alpha \subset \beta$ (as sets). Clearly, if $\alpha = S((a_n)_{n=0}^{\infty})$ and $\beta = S((c_n)_{n=0}^{\infty})$ then $\alpha \leq \beta$ if and only if for each positive integer n there exists a positive integer m such that $a_n \leq c_m$. Hence, if $\alpha, \beta \in \mathbb{S}(\mathbf{D})$, then either $\alpha = \beta$ or $\alpha < \beta$ or $\alpha > \beta$. We have

$$\mathbf{D}^- := \mathbf{D} \cap \mathbf{Q}^- = S(0) = \{d \in \mathbb{D} : d < 0\} = \bigcup_{n=0}^{\infty} S(-2^{-n}) =: \widehat{0} \in \mathbb{S}(\mathbf{D}).$$

If $\alpha \in \mathbb{S}(\mathbf{D})$ then $\alpha \leq \widehat{0}$ if and only if $\alpha \subset \mathbf{D}^-$ and $\alpha > \widehat{0}$ if and only if $\alpha \cap \mathbf{D}^+ \neq \emptyset$.

If $\alpha = S((a_n)_{n=0}^{\infty}) \in \mathbb{S}(\mathbf{D})$, $b_n := a_n + 2^{-n} \notin \alpha$ for all n and $(b_n)_{n=0}^{\infty}$ is a decreasing sequence we let $-\alpha := S((-b_n)_{n=0}^{\infty})$. It is easily verified that this is a well defined dyadic section. For example, if $a_n = -2^{-n-1}$ then $b_n = -2^{-n-1} + 2^{-n} = 2^{-n-1}$ and we see that $-\widehat{0} = S((-2^{-n-1})_{n=0}^{\infty}) = \widehat{0}$.

Next we define addition on $\mathbb{S}(\mathbf{D})$ by letting

$$\alpha + \beta := \{a + b : a \in \alpha, b \in \beta\}$$

for all $\alpha, \beta \in \mathbb{S}(\mathbf{D})$. For example, if $\alpha = S((a_n)_{n=0}^{\infty})$ and $\beta = S((b_n)_{n=0}^{\infty})$, then $\alpha + \beta = S((a_n + b_n)_{n=0}^{\infty}) = S((b_n + a_n)_{n=0}^{\infty}) = \beta + \alpha$. If $\alpha \in \mathbb{S}(\mathbf{D})$, then there is a unique dyadic segment β satisfying $\beta + \beta = \alpha$. In fact, it is not difficult to show that $\beta = \{2^{-1}x : x \in \alpha\}$ and we write $2^{-1}\alpha$ in place of β. Proceeding in this way we define $2^{-n}\alpha$ for any positive integer n and any $\alpha \in \mathbb{S}(\mathbf{D})$. For any $\alpha \in \mathbb{S}(\mathbf{D})$, any $x \in \mathbf{D}$, and any positive integer n, we have $x \in \alpha \iff 2^{-n}x \in 2^{-n}\alpha$.

Operating with dyadic sections may initially appear strange and to show it is really not difficult we have included the details in the following lemma.

Lemma 12.5. *If $\alpha, \beta, \gamma \in \mathbb{S}(\mathbf{D})$ then*

   (i) $\alpha > \widehat{0} \iff 0 \in \alpha \iff \alpha$ contains a positive dyadic number,
   (ii) If $\alpha \in \mathbb{S}(\mathbf{D})$ then either $\alpha > \widehat{0}$, $\alpha < \widehat{0}$ or $\alpha = \widehat{0}$.
   (iii) $\widehat{0} \in \mathbb{S}(\mathbf{D})$ is the identity for addition,
   (iv) $\alpha - \alpha := \alpha + (-\alpha) = \widehat{0}$,
   (v) $\alpha + (\beta - \alpha) = \beta$,
   (vi) If, for every positive integer n,

$$\alpha \leq \beta + 2^{-n}\gamma,$$

   then $\alpha \leq \beta$.

Proof. (i) If $\alpha > \widehat{0}$ then α contains all the negative dyadic numbers and some non-negative dyadic number a. Since $a \geq 0$ and $\{b \in \mathbf{D} : b \leq a\} \subset \alpha$ this implies $0 \in \alpha$.

If $0 \in \alpha$, then, since a dyadic segment has no largest element, α contains a positive dyadic number.

If α contains a positive dyadic number a then $\mathbf{D}^- \subset \{d \in \mathbf{D} : d \leq a\} \subset \alpha$. Hence $\alpha \geq \widehat{0}$ and $\alpha \neq \widehat{0}$, that is $\alpha > 0$. This proves (i).

(ii) Suppose $\alpha \neq \widehat{0}$. If $0 \in \alpha$ then (i) implies $\alpha > 0$. If $0 \notin \alpha$ and $a \in \mathbf{D}^+ := \mathbf{D} \cap \mathbf{Q}$ then $a \notin \alpha$. Hence $\alpha \subset \mathbf{D}^-$, that is $\alpha \leq 0$. Since $\alpha \neq 0$ this implies $\alpha < 0$.

(iii) If $a \in \widehat{0}$, then $a < 0$. Hence, $a + b < b$ for any $b \in \alpha$ and $\alpha + \widehat{0} \leq \alpha$. If $c \in \alpha$, then there exists a positive integer n such that $c + 2^{-n} \in \alpha$. Since $-2^{-n} \in \widehat{0}$, $c = c + 2^{-n} - 2^{-n} \in \alpha + \widehat{0}$. Hence $\alpha \leq \alpha + \widehat{0}$ and $\alpha \subset \alpha + \widehat{0} \subset \alpha$. This implies $\alpha = \alpha + \widehat{0}$.

(iv) If $\alpha = S\big((a_n)_{n=0}^{\infty}\big) \in \mathbb{S}(\mathbf{D})$, $b_n := a_n + 2^{-n} \notin \alpha$, and $(b_n)_{n=0}^{\infty}$ is a decreasing sequence, then $-\alpha := S\big((-b_n)_{n=0}^{\infty}\big)$ and

$$\alpha + (-\alpha) = S((a_n - b_n)_{n=0}^{\infty}) = S((-2^{-n})_{n=0}^{\infty}) = \widehat{0}.$$

(v) By parts (iii) and (iv) and the associative and commutative laws:

$$\alpha + (\beta - \alpha) = \alpha + (\beta + (-\alpha)) = (\alpha + (-\alpha)) + \beta = \widehat{0} + \beta = \beta.$$

(vi) Let $a \in \alpha$ be arbitrary. We are required to show that $a \in \beta$. Choose, using Lemma 12.3(d), a positive integer p such that $a + 2^{-p} \in \alpha$ and let $d \in \mathbf{D}$ denote an upper bound for the set γ. Using Theorem 5.12, choose $q \in \mathbf{N}$ such that $2^{-q}d \leq 2^{-p-1}$. For some $b \in \beta$ and some $c \in \gamma$ we have

$$a + 2^{-p} = b + 2^{-q}c \leq b + 2^{-q}d \leq b + 2^{-p-1}.$$

This implies

$$a < a + 2^{-p} - 2^{-p-1} \leq b.$$

Hence $a < b$ and $a \in \beta$. This completes the proof. $\qquad \square$

For $\alpha, \beta \in \mathbb{S}(\mathbf{D})$, $\alpha \leq \beta$, let $[\alpha, \beta] := \{\gamma \in \mathbb{S}(\mathbf{D}) : \alpha \leq \gamma \leq \beta\}$.

Definition 12.6. A bisectional sequence in $\mathbb{S}(\mathbf{D})$ is a sequence of intervals $([\alpha_n, \beta_n])_{n=0}^{\infty}$, where $\alpha_n, \beta_n \in \mathbb{S}(\mathbf{D})$ and $\alpha_n \leq \beta_n$ such that, for all $n \in \mathbf{N}_0$,
$$\alpha_{n+1} = \alpha_n \text{ or } \frac{\alpha_n + \beta_n}{2} \quad \text{and} \quad \beta_{n+1} = \alpha_{n+1} + \frac{\beta_n - \alpha_n}{2}.$$
This implies $\beta_{n+1} = \frac{\alpha_n + \beta_n}{2}$ or β_n and, in both cases, $\beta_n - \alpha_n = \frac{\beta_0 - \alpha_0}{2^n}$ for all $n \in \mathbf{N}_0$ (see Figure 12.1).

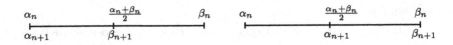

Fig. 12.1

Theorem 12.7. *(The Bisection Principle) If $([\alpha_n, \beta_n])_{n=0}^{\infty}$ is a bisectional sequence in $\mathbb{S}(\mathbf{D})$, then there exists a unique $\gamma \in \mathbb{S}(\mathbf{D})$ such that $\alpha_n \leq \gamma \leq \beta_n$ for all n.*

Proof.   Let $\gamma := \bigcup_{n=0}^{\infty} \alpha_n$. Clearly, γ is non-empty. For all n, $\alpha_n \subset \beta_n \subset \beta_1$, and the set γ is bounded above. If $d \in \gamma$ then $d \in \alpha_n$ for some n and we can find $d' \in \alpha_n$ such that $d' > d$. Hence $d' \in \gamma$ and γ does not contain a largest element. Moreover, if $d'' \leq d$, then $d'' \in \alpha_n \subset \gamma$. Hence $\gamma \in \mathbb{S}(\mathbf{D})$. For n and m in $\mathbf{N}_0$, we have $\alpha_n \leq \alpha_{n+m} \leq \beta_{n+m} \leq \beta_m$. This implies $\alpha_n \leq \bigcup_{n=0}^{\infty} \alpha_n = \gamma \leq \beta_m$ for all n and m and, in particular, $\alpha_n \leq \gamma \leq \beta_n$ for all n. It remains to show that this determines γ.

Suppose $\gamma_1 \in \mathbb{S}(\mathbf{D})$ and $\alpha_n \leq \gamma_1 \leq \beta_n$ for all n. This implies $\alpha_n \subset \gamma_1$ for all n and hence $\gamma = \bigcup_{n=0}^{\infty} \alpha_n \subset \gamma_1$, that is $\gamma \leq \gamma_1$. For any n
$$\gamma_1 \leq \beta_n = \alpha_n + (\beta_n - \alpha_n) = \alpha_n + 2^{-n}(\beta_0 - \alpha_0) \leq \gamma + 2^{-n}(\beta_0 - \alpha_0)$$
and Lemma 12.5(vi) implies $\gamma_1 \leq \gamma$. Hence, $\gamma_1 = \gamma$ and γ is the unique element in $\mathbb{S}(\mathbf{D})$ satisfying $\alpha_n \leq \gamma \leq \beta_n$ for all n. $\square$

Theorem 12.8. *(Upper Bound Principle) If A is a non-empty subset of $\mathbb{S}(\mathbf{D})$ which is bounded above then A has a least upper bound.*

Proof. The set U of all upper bounds for A is non-empty by hypothesis. If $\alpha \in A \cap U$ then, clearly, $\alpha = lub(A)$.

If $\alpha \cap U = \emptyset$, we define a bisectional sequence in $\mathbb{S}(\mathbf{D})$ as follows. Choose any $\alpha_0 \in A$ and any $\beta_0 \in U$. Then $\alpha_0 < \beta_0$. Suppose $\alpha_k \notin U$ and $\beta_k \in U$ have been defined for all $k \leq n$. Let $\beta_{n+1} = 2^{-1}(\alpha_n + \beta_n)$ if $2^{-1}(\alpha_n + \beta_n) \in U$, otherwise let $\beta_{n+1} = \beta_n$. In either case let $\alpha_{n+1} = \beta_{n+1} - 2^{-1}(\beta_n - \alpha_n)$. By the Bisection Principle, there exists a unique $\gamma \in \mathbb{S}(\mathbf{D})$ such that $\alpha_n \leq \gamma \leq \beta_n$ for all n. Let $\alpha \in A$. Since $\beta_n \in U$ for all $n \in \mathbf{N}_0$ we have

$$\alpha \leq \beta_n = \alpha_n + (\beta_n - \alpha_n) \leq \gamma + 2^{-n}(\beta_0 - \alpha_0).$$

By Lemma 12.5(vi), $\alpha \leq \gamma$ and γ is an upper bound for A. If $n \in \mathbf{N}_0$, then, since $\alpha_n \notin U$, we can choose $\alpha \in A$ such that $\alpha_n \leq \alpha$. If $\beta \in U$, then

$$\gamma \leq \beta_n = \alpha_n + 2^{-n}(\beta_0 - \alpha_0) \leq \alpha + 2^{-n}(\beta_0 - \alpha_0) \leq \beta + 2^{-n}(\beta_0 - \alpha_0).$$

By Lemma 12.5(vi), $\gamma \leq \beta$ and hence γ is the least upper bound for A. $\square$

Finally we define the operation of multiplication on $\mathbb{S}(\mathbf{D})$. If $\alpha, \beta \in \mathbb{S}(\mathbf{D})^+$, let

$$\alpha \cdot \beta := \{a \in \mathbf{D} : a < b \cdot c \text{ for some } b \in \mathbf{D}^+ \cap \alpha \text{ and some } c \in \mathbf{D}^+ \cap \beta\}.$$

If $\alpha > \widehat{0}$ and $\beta < \widehat{0}$, let $\alpha \cdot \beta := -(\alpha \cdot (-\beta))$. If $\alpha < \widehat{0}$ and $\beta > \widehat{0}$, let $\alpha \cdot \beta := -((-\alpha) \cdot \beta)$. If $\alpha < \widehat{0}$ and $\beta < \widehat{0}$, let $\alpha \cdot \beta := (-\alpha) \cdot (-\beta)$. If $\alpha = \widehat{0}$ or $\beta = \widehat{0}$, let $\alpha \cdot \beta := \widehat{0}$.

It is easily seen that $\widehat{1} := S(1) = \{a \in \mathbf{D} : a < 1\}$ is the identity for multiplication. It is tedious but routine to show that all the standard properties, such as associativity, commutativity, etc. hold for the real numbers and that every real number has an additive inverse and every non-zero real number has a multiplicative inverse.

To embed the rational numbers $\mathbf{Q}$ in $\mathbb{S}(\mathbf{D})$, we require the following lemma.

Lemma 12.9. *If $r, s \in \mathbf{Q}$ and $r < s$, then we can find $d \in \mathbf{D}$ such that $r < d < s$.*

Proof.   First suppose $r \in \mathbf{Q}^+$. By Theorem 5.12, we can find $n \in \mathbf{N}$ such that $2^n(s-r) > 1$ and $2^n r > 1$. Fix such an n. If $A := \{m \in \mathbf{N} : m \leq 2^n r\}$, then $1 \in A$ and clearly A is bounded above. Let m' denote the least upper bound of A in $\mathbf{N}$. Then $m' + 1 > 2^n r$ and

$$r < \frac{m'+1}{2^n} = \frac{m'}{2^n} + \frac{1}{2^n} < r + (s - r) = s.$$

Since $2^{-n}(m' + 1) \in \mathbf{D}$ this completes the proof when r is positive.

If $r \leq 0$ let $k = -r + 2$. Then $1 < r + k$ and, by the above, we can find $d' \in \mathbf{D}$ such that $r + k < d' < s + k$. This completes the proof since $r < d' - k < s$ and $d' - k \in \mathbf{D}$.   □

The proof of the following theorem is similar to the proofs of Theorems 6.11 and 6.12. We supply the details in the case of addition, but leave the multiplication part, in which the different cases have to be considered separately as an exercise.

Theorem 12.10. *The mapping*

$$q \in \mathbf{Q} \longrightarrow S(q) \in \mathbb{S}(\mathbf{D})$$

is injective and has the following properties:

1. *$q_1 < q_2 \iff S(q_1) < S(q_2)$,*
2. *$S(q_1) + S(q_2) = S(q_1 + q_2)$,*
3. *$S(q_1) \cdot S(q_2) = S(q_1 \cdot q_2)$*

where $q_1, q_2 \in \mathbf{Q}$.

Proof.   If $q_1 < q_2$ then, by Lemma 12.9, we can choose $d \in \mathbf{D}$, such that $q_1 < d < q_2$. If $r \in S(q_1)$, then $r < q_1 < d < q_2$. This implies $S(q_1) \leq S(q_2)$, and since $d \notin S(q_1)$ and $d \in S(q_2)$, $S(q_1) \neq S(q_2)$. Hence (1) holds and, moreover, S is injective.

If $d_1 \in S(q_1)$ and $d_2 \in S(q_2)$, then $d_1 + d_2 < q_1 + q_2$. Hence $d_1 + d_2 \in S(q_1 + q_2)$ and $S(q_1) + S(q_2) \subset S(q_1 + q_2)$. If $d \in S(q_1 + q_2)$ then, by Lemma 12.9, we can choose $d' \in \mathbf{D}$ such that $d - q_2 < d' < q_1$. Then $d' \in S(q_1)$ and $d - d' \in S(q_2)$. Hence $d = d' + (d - d') \in S(q_1) + S(q_2)$ and $S(q_1 + q_2) \subset S(q_1) + S(q_2)$. These two inequalities establish (2).   □

From now on we write **R** in place of $\mathbb{S}(\mathbf{D})$, 0 in place of $\widehat{0}$, 1 in place of $\widehat{1}$, and call **R** the set of real numbers.

12.3 Foundational Developments

It appears we have finally achieved the clarity, rigor and certainty that we sought in Chapter 1. But can we be certain? Confidence, certainty and rigor are interrelated. Classical Greek mathematics laid down a standard of rigor and achieved a level of certainty that was lost for centuries and only regained during the 19th century. When mathematicians in the 16th and 17th centuries such as Viéte, Fermat and Descartes sought to find the secrets of the classical Greek mathematicians, they found, instead, different treasures; analytic geometry and results that would lead Newton and Leibniz to the calculus. The excitement generated by the calculus and its success in application made it easy to overlook the apparent lack of rigor. The precision, found in the writings of Archimedes, gradually lapsed. Many felt, during the 18th century, that the ancient Greeks had been over careful in their analysis and that more progress would have resulted from greater reliance on intuitive notions. At the beginning of the 19th century, mathematicians knew much more mathematics than ever before but their understanding had become more geometric and intuitive and less rigorous and analytic. We have mentioned the moves taken in the 19th century to remedy the situation by Bolzano, Cauchy, Gauss, Weierstrass, and others, which culminated in the construction of the real numbers by Dedekind, in the development of set theory by Cantor, and in the reduction of logic to algebra by Boole.

Mathematics is a human endeavor and human beings make mistakes. Every year, new mathematical results are discovered and published in learned journals. Before publication, these results are examined by experts and corrected and yet errors still go undetected. Most of these are of no consequence but some are important and may take time to identify and, if possible, rectify. The 20th century saw the appearance of some major mathematical results, for example *Fermat's Last Theorem,* which took years to complete, and few mathematicians have either the time or the ability to check the details. Moreover, the solution to the *Four Color Problem*[3] includes computer programs that would, literally, take a lifetime to read and, as computers are constructed and programs are written by human beings, both may contain errors. In a few cases proofs are so long and involved that no one has checked the complete proof, leaving instead the task to a team of experts.

Is the situation hopeless? Is it so complex that mathematical knowledge will become the preserve of the few? Does certainty exist in mathematics? No, the situation is not hopeless and the questions we have explicitly mentioned are important and do have partial answers. Mathematicians are constantly building and rebuilding the foundations and strengthening the structures already in place and, where errors exist, they will eventually be discovered and removed. Absolute certainty does not exist except as a limiting ideal and mathematicians are constantly looking at approaches that will increase our confidence in mathematical results.

Consider the following three mathematical theorems:

$$2 + 2 = 1 + 3 \tag{12.1}$$

$$5^2 + 12^2 = 13^2 \tag{12.2}$$

and

$$1 + \frac{1}{2} + \frac{1}{2^2} + \frac{1}{2^3} + \frac{1}{2^4} + \cdots = 2. \tag{12.3}$$

It is safe to say that everyone feels confident about (12.1) and would like to feel equally certain about (12.2) and (12.3). Clearly, an ever so slight understanding of addition, multiplication, and notation leads to a completely satisfactory proof of (12.2). With the construction of the real numbers in the previous section we have proved (12.3) rigorously but it took much more work. We may even have relied on an informal understanding with the reader and we had more opportunities to make errors. Could we have

---

[3]That any map only requires four colors.

done it for the reader who questioned every single step of the proof and allowed no appeal to intuition? Perhaps, but then every logical implication would have to be traced back to the fundamental assumptions, to the axioms or postulates and, to avoid even the unintentional use of intuition, it would be necessary to use the symbolic language developed by Peano. To check that each step was based on strict logic, it would be necessary, as suggested many years ago by Charles Babbage, to use a computer programme to check the proof. This type of extremely rigorous proof is known as a *formal proof*. It is truly amazing that the abstract concepts necessary for formal proofs were conceived in the 19th century long before the physical machines required were constructed. In this area of mathematics, now called *Proof Theory*, standard proofs are translated into formal proofs[4] and checked by a computer. Formal proofs are less intuitive, less susceptible to error, and attain a degree of rigor that even Archimedes did not achieve. Transforming a standard proof into a formal proof is a major undertaking, not least because of the notation involved. The computer programs which carry out many of the tedious steps are known as *proof assistants*. Today mathematicians, working individually or in teams, are developing formal proofs for many of the major results in mathematics and the classical Greek standard of rigor has been recovered and surpassed.[5] Such proofs give us confidence in mathematical results, add to our understanding of the nature of proofs, and help us appreciate the capabilities of computer programmes.

On the other hand, this automated approach to mathematics gave rise, especially at the beginning of the 20th century, to the philosophy that all

---

[4]Babbage originally designed his machines, the various *Difference Machines* and the *Analytic Machine* to perform the arithmetic operations of addition and, later multiplication, so that he could construct accurate mathematical tables. His appreciation, as a student, of the difficulties caused by the different notation of Leibniz and Newton made him aware throughout his life of the importance of notation both for mathematics and engineering. This proved important when incorporating mathematical procedures into his machines. Babbage believed his Difference Engine Number 1 was *the first great step towards reducing the whole science of number to the absolute control of mechanism.*

[5]There are now at least 17 different formal proofs showing that $\sqrt{2}$ is irrational. There are also formal proofs of the Prime Number Theorem and of Dedekind's construction of the real numbers.

mathematics could be reduced to a set of *tautologies*; that is obvious statements such as $2 + 2 = 4$. This suggests that mathematics is devoid of content and merely states the obvious in different ways and that the art of the mathematician is merely to make the obvious more obvious. The working mathematician makes choices and, while there are an infinity of choices available, most lead nowhere. On the other hand proof assistants make no choices, they follow instructions, although they can be programmed to perform randomly and unpredictably. Mathematicians talk in vague generalities about their approach to making choices, usually claiming intuition and the aesthetic beauty of mathematics as there guides. Aesthetic beauty in mathematics, as in other arts, cannot be defined as we define mathematical concepts but it is real, it can be recognized, it can be communicated, and it can be appreciated.

While mathematicians were gaining confidence in known mathematical results they were exposed, completely unexpectedly, to a new kind of uncertainty which has led in recent years to renewed philosophical speculation and quasi-mystical interpretations on the nature and scope of mathematics, human intuition and intelligence. It was only natural that the increasing emphasis on *axioms* or *postulates*[6] and *logic* would lead to a more careful examination of these areas that had basically remained unchanged for over two thousand years. The results were dramatic and led to one of those rare cases where an easily stated mathematical discovery with philosophical nuances enters the public consciousness and almost immediately generates ambiguous interpretations. We refer to the 1931 results of Kurt Gödel[7] who *proved that mathematics contains true but unprovable statements* and

---

[6]To the classical Greek mathematicians, an axiom was a statement so obviously true that it was hardly worth mentioning, while a postulate was an explicit assumption made for the purposes of further reasoning. Nowadays we know that all assumptions, obvious or not, should be mentioned explicitly and, consequently, we do not distinguish between axioms and postulates.

[7]Kurt Gödel, 1906-1978, was born in what is now Brno in the Czech Republic, previously Brunn in the Austrian-Hungarian Empire. He was educated in Vienna and in 1933 joined the faculty in Vienna. He was a guest lecturer at Princeton in 1934 and 1938 and in 1940 joined the permanent staff at Princeton.

that *mathematics cannot prove if it is itself consistent*. Formally his results are the following:

(1) *in any consistent formal axiomatic system that contains* **N**, *there are undecidable propositions*,

(2) *a consistent formal axiomatic system that contains* **N** *cannot prove its own consistency*.

The hypothesis that the axiomatic system contains **N** implies that the system is non-trivial and contains infinite sets while an axiomatic system is consistent if contradictory statements cannot be proved using the axioms. Clearly, inconsistent systems are uninteresting. An undecidable proposition is one that can neither be proved nor disproved. If P is an undecidable proposition and if one is operating in a *two-valued logic*[8], that is one in which every proposition is either true of false, then one of the statements 'P is true' and 'P is false' is true and one is false and thus there are true statements which cannot be proved and false statements which cannot be disproved. All this points to unexpected limitations in the axiomatic approach and suggests that absolute certainty is unattainable. On the other hand this appears, again unexpectedly, to reaffirm the importance of the human intellect and intuition in the discovery or creation of new mathematics.

In 1936 a further result along similar lines was proved independently by Alonzo Church and Alan Turing.[9] They showed that

---

[8]As we have throughout this book. Proof by contradiction is only possible in a two-valued logic environment. Many-valued and even countably infinite valued logics have been constructed.

[9]Alonzo Church, 1903-1995, was born in Washington, D.C., and educated at Princeton. He was a faculty member at Princeton from 1929 until 1967 and between 1967 and 1992 he was Professor of Mathematics and Philosophy at UCLA. Church, who created the λ-calculus, is noted for his contributions to mathematical logic, axiomatic set theory and theoretical computer science. The English mathematician Alan Mathison Turing, 1912-1954, defined abstract machines, now known as *Turing machines*, in 1936 and this led to his acquaintance with Church, culminating in Turing becoming Church's graduate student at Princeton. Turing had very broad interests; probability theory, artificial intelligence, quantum mechanics, abstract algebra, the design of computers,

in any consistent formal axiomatic system that contains **N** *there is no general process which determines if a given proposition is provable.*

The combined results of these three truly original thinkers continues to have profound and, at times, controversial implications for mathematics, philosophy, artificial intelligence, computer science, complexity theory and computability. That axiomatic systems cannot prove their own consistency does not imply that they are inconsistent and history shows that the confusion and doubt generated by such statements will eventually be resolved in ways that will surprise and delight us. In the meantime we will have popular interpretations of various kinds, for instance that it is impossible to *know thyself*, or that all systems are *self referential.*

From our point of view it is interesting to note that crucial roles in their arguments were played by *countability*, Cantor's *diagonal process* (see Theorem 6.6) and the unique factorization of positive integers into a product of *primes* (see Exercise 5.5).

12.4   Infinitesimals

The founders of the calculus, Newton and Leibniz, and their immediate followers, the Bernoulli brothers and Euler, all used *infinitesimals*, that is infinitely small numbers which were not zero but yet smaller than any positive real number, to justify their results. At first glance it would appear that the idea is contrary to all our intuitions. But consider the idea of a line, we accept it and we know that lines exist. A line has length, but no thickness, at least its thickness is not positive and yet it is not zero. The eminent philosopher Berkeley was able to show the weakness in their approach. This was the impetus that led to the development of a precise notion of limit and eventually to the construction of the real numbers. However, infinitesimals, which were used as an intuitive aid by different schools of mathematicians stretching back to ancient Greece, were rehabilitated in

---

and mathematical biology and is well known as the breaker of the *Enigma Code* at Bletchley Park. This saved many lives and hastened the end of World War Two.

1966 with the appearance of *Non-Standard Analysis* by Abraham Robinson.[10] This theory was modelled on Leibniz's use of infinitesimals and, with the aid of modern mathematical logic, these infinitely small objects were precisely and rigorously defined. Nowadays infinitesimals are used throughout analysis, e.g. in Functional Analysis, Probability Theory, and Integration Theory and even in the teaching of elementary calculus.

We briefly mention Robinson's construction. He extended the real numbers $\mathbf{R}$ to a number system $\mathbf{R}^*$ which included both infinitely large and infinitely small numbers. To do so he took as his postulates *all* postulates satisfied by $\mathbf{R}$ and the postulate that there is a number ω such that $\omega > n$ for any $n \in \mathbf{N}$. Since there clearly is a model for any finite subset of this set of postulates, the *Compactness Theorem* from *mathematical logic* says that the full set also has a model. This shows the existence of the infinitely large number and, since any non-zero number in $\mathbf{R}^*$ has a multiplicative inverse, this implies the existence of infinitesimals; that is non-zero numbers which are less than any positive real number.

Other classes of real numbers have also come into prominence in recent years. Turing defined the computable real numbers as *numbers that can be written down by a machine* and showed that the set of all such numbers is countable while Church worked on the definition of a *random sequence* which is related to the notion of *random number*.

12.5 Exercises

(12.1) Find a polynomial equation with integer coefficients which has $\sqrt{2} + \sqrt{3}$ as a solution.

---

[10] Abraham Robinson, 1918-1974, was from Waldenburg in Germany. His family emigrated to Palestine in 1933 and he graduated in 1939 from the Hebrew University in Jerusalem. During World War II he worked as a scientific officer in England and became an expert in *aerodynamics*. In 1946 he became a lecturer at the College of Aeronautics in Cranfield and, in 1949, he obtained a Ph.D. from the University of London in mathematical logic. He held professorships at Toronto, Jerusalem, Los Angeles and Yale. The work of Robinson has permeated, in a very direct and influential way, many branches of mathematics and provided mathematicians with a new and unexpected point of view.

(12.2) Show that the set of algebraic real numbers is countable.

(12.3) If α and β are positive real numbers, show that there exists a positive integer n such that $n\alpha > \beta$.

(12.4) A sequence of real numbers $(x_n)_{n=1}^{\infty}$ is called a Cauchy sequence if, for any $\epsilon > 0$, there exists a positive integer n_ϵ such that $|x_n - x_m| < \epsilon$ for all $n, m > n_\epsilon$. Show that every Cauchy sequence converges.

(12.5) If $(x_n)_{n=1}^{\infty}$ and $(y_n)_{n=1}^{\infty}$ are Cauchy sequences show that $(x_n)_{n=1}^{\infty} \sim (y_n)_{n=1}^{\infty}$ if $(x_n - y_n)_{n=1}^{\infty}$ converges to 0 defines an equivalence relationship on the set of all Cauchy sequences.

(12.6) If $\alpha \in \mathbb{S}(\mathbf{D})$, show that $-\alpha \in \mathbb{S}(\mathbf{D})$.

(12.7) Show that the set of dyadic numbers is countable.

(12.8) If α and β are real numbers, $\alpha < \beta$, show that there exists a $q \in \mathbf{D}$ such that $\alpha < q < \beta$.

(12.9) If $\alpha, \beta \in \mathbb{S}(\mathbf{D})$ show that $\alpha + \beta \in \mathbb{S}(\mathbf{D})$ and $\alpha \cdot \beta \in \mathbb{S}(\mathbf{D})$.

(12.10) Let $(a_n)_{n=0}^{\infty}$ denote an increasing sequence of dyadic numbers that is bounded above and let $\alpha = S\big((a_n)_{n=0}^{\infty}\big) \in \mathbb{S}(\mathbf{D})$. If $(n_j)_{j=0}^{\infty}$ is a strictly increasing sequence of non-negative integers, show that $S\big((a_{n_j})_{j=0}^{\infty}\big) = \alpha$. If $(c_j)_{j=1}^{\infty}$ is any increasing sequence of positive dyadic numbers which is not bounded above, show that we can choose $(n_j)_{j=0}^{\infty}$ so that $a_{n_j} + c_j^{-1} \notin \alpha$ for all j.

(12.11) If $\alpha, \beta \in \mathbb{S}(\mathbf{D})$ show that $\alpha \leq \beta$ if and only if every upper bound for the set β is also an upper bound for the set α.

(12.12) If $q \in \mathbf{Q}$ show, with all details, that $S(q) = \{d \in \mathbf{D} : d < q\}$ is a dyadic section.

(12.13) If A is a subset of $\mathbf{R}$ and $a \in A$ is an upper bound for A show that a is the least upper bound for A.

References [2; 4; 7; 9; 12; 14; 15; 20; 25; 28; 38; 39]

Chapter 13

INTEGRATION THEORY

*I set myself the task of communicating
to you the geometrical result
'Every segment bounded by a parabola and a chord
Qq is equal to four-thirds of the triangle which
has the same base as the segment and equal height'
which I investigated and discovered by means of
mechanics and then proved by means of geometry.*

Archimedes

Summary

We make general remarks about mathematics as a prelude to considering
the background to integration theory. We prove the Fundamental Theorem
of Calculus and discuss methods of integration.

13.1   The Riemann Integral

We have now encountered the basic ingredients that constitute mathematics: logic, number systems, foundational concepts, basic techniques, outstanding results, etc. and, while doing so, have noted, however briefly, the contributions of certain individuals and the extraordinary mathematical

achievements of certain civilizations.[1] The innate regard of the ambient
society may nurture and support the formation of a critical mass of people
with the commitment and resources to pursue fundamental truths, while the
presence of a receptive audience, including some willing to commit themselves to a speculative project, can make a movement out of what otherwise
might have remained an isolated individual effort.[2] Conceptual advances
feed slowly into the general consciousness and, from shared ideas, slightly
different interpretations arise and divergent directions are explored. This
suggests reasons why mathematical discoveries are often made practically
simultaneously but yet independently.

We pause briefly to consider the creative process. There is no reason
to suppose that the process is any different today than previously and it
is relevant in an introductory book: the difference between the creative
and appreciative experiences is merely a matter of degree. One, however,
is clearly more demanding, more intense, and more uncertain. Individuals
often follow parallel creative paths: an initial period of concentrated study
in which a topic is thoroughly examined, next a period of either rest or
confusion or both in which no apparent conscious work is undertaken, followed by sudden insight or inspiration and the final stage, when the insight
is made real, that is realized, in a form that can be shared. Descartes studied for nine years before he found his *Method*, Newton hid himself in the
countryside as the plague closed the universities while contemplating the
calculus and Leibniz spent years visiting Paris and London for information
on all the mathematics that was then known. It may be that, during the
rest period, the unconscious mind is classifying and synthesizing the information and intuitions that were received and analyzed during the conscious
first stage and, when these are properly arranged new patterns emerge. The
final stage turns the dream into a reality.

The growth of mathematics in Babylon and Egypt was due, in no small

---

[1] Babylon, Egypt, classical Greece, the final hundred years of the Renaissance, and the
twentieth century, were periods of exceptional mathematical growth.

[2] A lack of followers meant that the advances due to Bishop Nicolas Oresme in the 14th
century were never fully exploited.

way, to the importance of measuring areas for taxation and inheritance purposes. Many, but not all, such measurements were of regular shapes such as
squares, rectangles, triangles or of simple areas that could be easily subdi-
vided into these regular figures. When circular areas were being measured
approximations to π were required.[3] The classical Greek mathematicians
combined their abstract understanding and computational skills to investigate areas and volumes and took the first steps in what is now called
integration theory. They used the *method of exhaustion* due to Eudoxus
and, to obtain satisfactory proofs, they were led, as we noted in Chapter 6, to consider infinite series. The method of Eudoxus was used almost
exclusively for close to two millennia and all modern integration theories
rely on some form of exhaustion process. For this reason, the theoretical
constructions in this section deserve careful examination.

The *Fundamental Theorem of Calculus*, which shows that differentiation and integration are inverse operation, is often conveniently taken as
the key event in assigning the discovery of the calculus to Newton and Leibniz.[4] This theorem not only led to the computation of many integrals that
were previously inaccessible but also to the possibility of reversing methods
of differentiating to obtain techniques for integrating. Reversing the *product rule* for differentiating yields the technique of *integration by parts* and,
similarly, the *chain rule* reverses to give the *method of substitution*. All this
happened during the final quarter of the 17th century while the concept of

---

[3]The Babylonians and Egyptians used $(16/9)^2$, Archimedes showed that $3\frac{10}{71} < \pi < 3\frac{1}{7}$, while Viète calculated π to ten places of decimals using a polygon with 6×2^{16} sides in 1583. In 1761 the self-taught Swiss mathematician, astronomer, physicist and philosopher, Johann Heinrich Lambert, 1728-1777, proved that π was irrational and an elementary proof was given by Legendre in 1795.

[4]In its modern formulation it is due to Newton and Leibniz but it should be noted that the Scottish mathematician James Gregory, 1638-1675, stated a general version of the Fundamental Theorem of Calculus in 1668, and Isaac Barrow 1630-1677, proved a geometric version of the same result in 1669. Barrow and Gregory died relatively young and history has tended to overlook their contributions. Barrow, Newton's teacher and predecessor at Cambridge, played an influential role in Newton's intellectual development. Gregory discovered many basic results in the differential and integral calculus but had the misfortune to delay publishing until after Newton.

function was slowly emerging. During the 18th century the emphasis was on methods of integration and, as a result, integration was considered anti-differentiation and not given its due credit as a subject in its own right. Cauchy retrieved the ancient Greek exhaustion approach at the beginning of the 19th century and gave the first modern definition of an integral. He showed, as we do, that continuous functions are integrable. Later, Riemann extended the class of integrable functions using Cauchy's definition and today, it is known as *The Riemann Integral*. Since 1900, integration theory has witnessed remarkable advances. Henri Lebesgue[5] introduced an abstract measure and integral that provided *Probability Theory* with an axiomatic foundation and indirectly promoted the growth of *Statistics*, while Norbert Wiener defined another integral which plays a special role in *Financial Mathematics*.[6] To discuss these further would be too great a diversion, we merely mention that refinements of ideas that we have already encountered are important for these theories. For instance, a probability measure on a set Ω is defined as a $[0, 1]$-*valued function* defined on a collection of subsets of Ω which satisfies certain *countability* criteria, and the power of the Lebesgue measure[7] derives from its ability to handle *convergent sequences*

---

[5]Henri Lebesgue, 1875-1941, from Beauvais in northern France published his first results in 1901 on what we now call the Lebesgue integral. In his thesis, which appeared in 1902, and subsequent papers he developed and applied this integral so that today it is one of the most used, useful, and indispensable tools available to mathematicians. Lebesgue's penetrating analysis of major themes from the past led him to his remarkable discoveries. He wrote extensively on mathematical education and promoted an integrated approach to learning, based on motivation, physical interpretation and rigor. His book, *Measure and the Integral* (Holden-Day), is recommended reading for the layman, the student and the professional mathematician.

[6]Norbert Wiener, 1894-1964, was born in Columbia, Missouri. He was a child prodigy, graduating from high school at 11, from university at 14 and obtaining a doctorate in mathematical logic from Harvard at 18. He founded the area of *cybernetics* which he described as *a statistical approach to the theory of communication*. The Wiener measure is defined on measurable subsets of $C[a, b]$, the continuous real-valued functions on $[a, b]$.

[7]It is of interest to note that the theory of absolutely convergent series can be presented as measure theory over a countable set. Moreover, one of the main results in abstract integration theory, *the Monotone Convergence Theorem*, bears comparison with Definition 7.5.

of integrable functions. We are interested in considering integration theory for a reasonably extensive collection of well behaved functions and do not feel it necessary to discuss the Riemann[8] integral in its full generality. For this reason, we confine ourselves to *continuous functions.*

If $f : [a, b] \longrightarrow \mathbf{R}$ is continuous let

$$\overline{\mathcal{R}}(f, n) = \sum_{i=0}^{2^n - 1} f(x_{i,n}^*)(x_{i+1} - x_i) \, , \, \underline{\mathcal{R}}(f, n) = \sum_{i=0}^{2^n - 1} f(x_{i,n}^{**})(x_{i+1} - x_i)$$

where $x_i = a + \frac{i(b-a)}{2^n}$ for $0 \le i \le 2^n$, $f(x_i^*) = \max\{f(x) : x_i \le x \le x_{i+1}\}$ and $f(x_i^{**}) = \min\{f(x) : x_i \le x \le x_{i+1}\}$ for $0 \le i \le 2^n - 1$ (see Figure 13.1). We call $\overline{\mathcal{R}}(f, n)$ and $\underline{\mathcal{R}}(f, n)$ the nth *upper and lower Riemann*

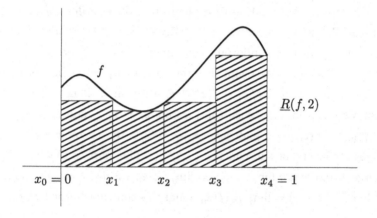

Fig. 13.1

sums for f respectively. If $f \ge 0$, we interpret the Riemann sums as upper

---

[8]Georg Friedrich Bernhard Riemann, 1826-1866, was born in Breselenz, Hanover, and died in Selasca, Italy. The mathematical legacy of his short life is such that he is now regarded as one of the most influential mathematicians of the 19th century. He had contact with a number of those mentioned in this book. Gauss, supervised his thesis and described him as having *a gloriously fertile originality.* His friend and colleague Dirichlet helped him develop an intuitive approach to mathematics. Dedekind attended his lectures and made them available later to the world at large, while Weierstrass's careful reading of his thesis raised questions whose answers led to further mathematical advances. Riemann made contributions to number theory, Fourier series, differential geometry, integration theory and complex analysis.

and lower estimates of the area above the x-axis, between the lines $x = a$ and $x = b$ and under the graph of f. For $n, m \in \mathbb{N}$ we have

$$\underline{\mathcal{R}}(f, n) \leq \underline{\mathcal{R}}(f, n + m) \leq \overline{\mathcal{R}}(f, n + m) \leq \overline{\mathcal{R}}(f, m) \tag{13.1}$$

and, by Figure 13.1,

$$|\overline{\mathcal{R}}(f, n)| \leq M(b - a)$$

where $M = \max\{|f(x)| : x \in [a, b]\}$. By (13.1), the sequences of upper and lower Riemann sums are monotonic and bounded and hence converge and, moreover,

$$\lim_{n \to \infty} \underline{\mathcal{R}}(f, n) \leq \lim_{n \to \infty} \overline{\mathcal{R}}(f, n).$$

At this stage we have a number of choices on how to present the material. One approach would be to formally introduce *uniformly continuous* functions. We do not do so but, essentially show that continuous functions on closed intervals are uniformly continuous. Let δ denote an arbitrary positive real number. We claim that there exists a positive integer n_0 such that if $n \geq n_0$, then $f(x_{i,n}^*) - f(x_{i,n}^{**}) \leq \delta$ for all i. If not, then, for all $n \geq n_0$ we can find an i_n such that, letting $y_n = x_{i_n,n}^*$ and $z_n = x_{i_n,n}^{**}$, we have $f(y_n) - f(z_n) > \delta$ and $|y_n - z_n| < 1/2^n$. By Theorem 8.6, $(y_n)_{n=1}^{\infty}$ contains a convergent subsequence $(y_{n_j})_{j=1}^{\infty}$. This implies that $(z_{n_j})_{j=1}^{\infty}$ is convergent and hence $\lim_{j \to \infty} y_{n_j} = \lim_{j \to \infty} z_{n_j}$. Since f is continuous, we have $\lim_{j \to \infty} f(y_{n_j}) = \lim_{j \to \infty} f(z_{n_j})$ and this contradicts the fact that for all n_j,

$$f(y_{n_j}) - f(z_{n_j}) > \delta.$$

This establishes our claim. Hence, for $n \geq n_0$, we have

$$\overline{\mathcal{R}}(f, n) - \underline{\mathcal{R}}(f, n) = \sum_{i=0}^{2^n - 1} (f(x_{i,n}^*) - f(x_{i,n}^{**})) \cdot (x_{i+1} - x_i)$$

$$\leq \sum_{i=0}^{2^n - 1} \delta(x_{i+1} - x_i)$$

$$= \delta(b - a)$$

and, since $\delta > 0$ was arbitrary, (13.1) implies

$$\lim_{n \to \infty} \overline{\mathcal{R}}(f, n) = \lim_{n \to \infty} \underline{\mathcal{R}}(f, n).$$

We have prepared the ground for the following definition.

Definition 13.1. If f is continuous on $[a, b]$, we let

$$\int_a^b f(x)dx := \int_a^b f := \lim_{n \to \infty} \overline{\mathcal{R}}(f, n) = \lim_{n \to \infty} \underline{\mathcal{R}}(f, n)$$

and call $\int_a^b f$ the Riemann integral of f over $[a, b]$.

In place of taking upper and lower Riemann sums we could have used

$$\mathcal{R}(n, f, w) := \sum_{i=0}^{2^n-1} f(w_{i,n})(x_{i+1,n} - x_{i,n}) =: \sum_{i=0}^{2^n-1} f(w_{i,n})\Delta x_i, \quad (13.2)$$

where, for each i, $w_{i,n}$ is an arbitrary point in $[x_{i,n}, x_{i+1,n}]$ and $w = (w_{i,n})_{i=0}^{2^n-1}$. Clearly,

$$\underline{\mathcal{R}}(f, n) \leq \mathcal{R}(n, f, w) \leq \overline{\mathcal{R}}(f, n)$$

and, for continuous functions, Definition 13.1 implies

$$\lim_{n \to \infty} \mathcal{R}(n, f, w) = \int_a^b f(x)dx. \quad (13.3)$$

The above construction leads naturally to the computation of series for the evaluation of areas and this would have been the approach followed by the ancient Greeks. As the Fundamental Theorem of Calculus leads to a much simpler method of calculation, this approach is now mainly of historical interest and we confine ourselves to just one example.

Example 13.2. A solution to Exercise 11.1 shows that

$$\sum_{j=1}^n j^2 = \frac{n(n+1)(2n+1)}{6}$$

and, hence, if $f(x) = x^2$ on $[0, 1]$ we obtain

$$\underline{\mathcal{R}}(f, n) = \sum_{i=0}^{2^n-1} \left(\frac{i}{2^n}\right)^2 \frac{1}{2^n} = \frac{(2^n - 1)(2^n)(2^{n+1})}{6 \cdot 2^{3n}} = \frac{2}{6} \cdot \frac{2^n - 1}{2^n} \longrightarrow \frac{1}{3}$$

as $n \longrightarrow \infty$. This shows that

$$\int_0^1 x^2 dx = \frac{1}{3}.$$

The quotation by Archimedes, which introduced this chapter, is taken from his article *The Quadrature of the Parabola.*[9] He was interested in finding the area within a parabola (see Figure 13.2) and found the answer using

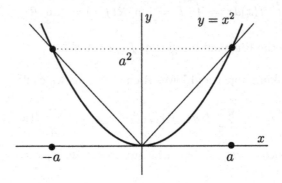

Fig. 13.2

mechanics and geometric series and justified it using proof by contradiction (see (6.1)). He proved that the area equals

$$\frac{4}{3}\left(a^2 \cdot 2a - \frac{1}{2} \cdot 2a \cdot a^2\right) = \frac{4a^3}{3}.$$

A slight variation in our calculation above confirms this result since

$$a^2 \cdot 2a - \int_{-a}^{a} x^2\,dx = 2a^3 - \frac{2a^3}{3} = \frac{4a^3}{3}.$$

The next result follows easily from the definition of Riemann sums and elementary properties of limits (see Theorem 8.2). The proof of (b) requires (13.3), while (d) is obtained by computing the areas of a rectangle and a triangle.

Theorem 13.3. *Let $a < b < c$, $f : [a,c] \longrightarrow \mathbf{R}$ and $g, h : [a,b] \longrightarrow \mathbf{R}$ denote continuous functions. If α is a real number and $g(x) \le f(x) \le h(x)$ for all $x \in [a,b]$, then*

---

[9]The word *quadrature* was quite common within mathematics up to a hundred years ago. It has the same Latin roots as the word *quadratic* and is derived from *quadrus*, a square. In calculating areas the quadrature of a figure meant the square with the same area.

(a)
$$\int_a^b g(x)dx \le \int_a^b f(x)dx \le \int_a^b h(x)dx,$$

(b)
$$\int_a^b (\alpha f(x) + g(x))dx = \alpha \int_a^b f(x)dx + \int_a^b g(x)dx,$$

(c)
$$\int_a^b f(x)dx + \int_b^c f(x)dx = \int_a^c f(x)dx,$$

(d)
$$\int_a^b \alpha dx = \alpha(b-a), \qquad \int_a^b x dx = \frac{b^2 - a^2}{2}.$$

Using (a) and $-|f(x)| \le f(x) \le |f(x)|$ we obtain

$$\left| \int_a^b f(x)dx \right| \le \int_a^b |f(x)|dx \qquad (13.4)$$

and, if $m \le f(x) \le M$ for all $x \in [a, b]$, then

$$m(b-a) \le \int_a^b f(x)dx \le M(b-a). \qquad (13.5)$$

13.2 Fundamental Theorem of Calculus

The Fundamental Theorem of Calculus shows that differentiation and integration may be regarded as inverse processes. We give two versions of this theorem, Theorems 13.4 and 13.5.

Theorem 13.4. *If $f : [a, b] \longrightarrow \mathbf{R}$ is continuous on $[a, b]$, differentiable on (a, b), and f' is continuous,*[10] *on $[a, b]$, then*

$$\int_a^b \frac{d}{dx} f(x)dx = \int_a^b f'(x)dx = f(b) - f(a). \qquad (13.6)$$

[10]That is f' coincides on (a, b) with the restriction to (a, b) of a continuous function on $[a, b]$.

Proof. If $x_i = a + \frac{i(b-a)}{2^n}$ for $0 \leq i \leq 2^n$ then, by the Mean Value theorem, there exists for each i, y_i, $x_i < y_i < x_{i+1}$ such that

$$f(b) - f(a) = \sum_{i=0}^{2^n-1} f(x_{i+1}) - f(x_i) = \sum_{i=0}^{2^n-1} f'(y_i)(x_{i+1} - x_i) = \mathcal{R}(f', n, y)$$

where $y := (y_i)_{i=0}^{2^n-1}$. An application of (13.3) completes the proof.   □

This gives us the following basic integrals.

Derivative	Integral
$\frac{d}{dx}(e^x) = e^x$	$\int_a^b e^x dx = e^x\big]_a^b = e^b - e^a$
$\frac{d}{dx}(\log(x)) = \frac{1}{x}$	$\int_a^b \frac{1}{x}dx = \log(x)\big]_a^b = \log b - \log a, b > a > 0$
$\frac{d}{dx}(\frac{x^{\alpha+1}}{\alpha+1}) = x^\alpha$	$\int_a^b x^\alpha dx = \frac{x^{\alpha+1}}{\alpha+1}\big]_a^b = \frac{b^{\alpha+1}}{\alpha+1} - \frac{a^{\alpha+1}}{\alpha+1}, \alpha \neq -1$

If $f : [a, b] \longrightarrow \mathbb{R}$ then $F : [a, b] \longrightarrow \mathbb{R}$ is called an indefinite integral of f if F is continuous on $[a, b]$, differentiable on (a, b) and $F'(x) = f(x)$ for all $x \in (a, b)$. We write the indefinite integral of f as $\int f(x)dx + c$ and call c the constant of integration. If F and G are indefinite integrals of f, then $(F + G)'(x) = f(x) - f(x) = 0$ and, by Corollary 10.7, there exists a constant c such that $G = F + c$. Conversely, as the derivative of a constant function is 0, $F + c$ is an indefinite integral of f whenever F is one. Thus by varying c, we obtain all indefinite integrals. This explains our terminology. Our next result gives us an indefinite integral.

Theorem 13.5. *If $f : [a, b] \longrightarrow \mathbb{R}$ is continuous and $F(x) = \int_a^x f(t)dt$ for all $x \in [a, b]$ then F is differentiable on (a, b) and $F' = f$. Symbolically, we write*

$$f(x) = \frac{d}{dx}\int f(x)dx, \tag{13.7}$$

Proof. Fix $x \in (a, b)$. If $\Delta x \neq 0$ then, by Theorem 13.3(c),

$$F(x + \Delta x) = \int_a^{x+\Delta x} f(t)dt = \int_a^x f(t)dt + \int_x^{x+\Delta x} f(t)dt$$

and, hence,

$$F(x + \Delta x) = F(x) + \Delta x \cdot \frac{1}{\Delta x}\int_x^{x+\Delta x} f(t)dt.$$

Let $(x_n)_{n=1}^{\infty} \subset \mathbf{R}, x_n \neq 0$ and suppose $\lim_{n\to\infty} x_n = 0$. By continuity and Theorem 9.10, we can choose for each n, y_n, such that $|y_n| \leq |x_n|$ and $\max\{|f(x+t) - f(x)| : |t| \leq |x_n|\} = |f(x+y_n) - f(x)|$. Applying in turn, Theorem 13.3(d), (b), (13.4), (13.5) and (9.6) we obtain

$$\left| \frac{1}{x_n} \int_x^{x+x_n} f(t)dt - f(x) \right| = \left| \frac{1}{x_n} \int_x^{x+x_n} f(t)dt - \frac{1}{x_n} \int_x^{x+x_n} f(x)dt \right|$$

$$= \left| \frac{1}{x_n} \int_x^{x+x_n} (f(t) - f(x))dt \right|$$

$$\leq \frac{1}{|x_n|} \int_x^{x+x_n} |f(t) - f(x)|dt$$

$$\leq \frac{1}{|x_n|} \cdot |x_n| \max\{|f(x+t) - f(x)| : |t| \leq |x_n|\}$$

$$= |f(x+y_n) - f(x)|$$

$$\longrightarrow 0 \text{ as } n \longrightarrow \infty.$$

An application of (10.6) completes the proof. $\square$

Corollary 13.6. *If F is an indefinite integral of the continuous function $f : [a, b] \longrightarrow \mathbf{R}$, then*

$$\int_a^b f(x)dx = F(b) - F(a) =: F(x) \Big]_a^b. \tag{13.8}$$

Proof. By the Mean Value Theorem and Theorem 13.5, $F(x) = \int_a^x f(t)dt + c$ for some constant c. Hence

$$F(b) - F(a) = \int_a^b f(t)dt + c - \int_a^a f(t)dt - c = \int_a^b f(x)dx. \quad \square$$

13.3 Methods of Integration

In this section we give practical methods for evaluating integrals. In view of the Fundamental Theorem of Calculus, this amounts to reversing methods of differentiation. We first consider the product rule for differentiating, Theorem 10.5(a), and begin by rewriting it in a more suitable way:

$$\frac{d(uv)}{dx} = \frac{du}{dx}v + u\frac{dv}{dx}$$

and hence

$$u\frac{dv}{dx} = \frac{d(uv)}{dx} - \frac{du}{dx}v.$$

An application of (13.7) implies

$$\int u\frac{dv}{dx}dx = \int \frac{d(uv)}{dx}dx - \int \frac{du}{dx}vdx = uv - \int \frac{du}{dx}vdx$$

and, with limits of integration,

$$\int_a^b u(x)\frac{dv}{dx}(x)dx = uv\Big]_a^b - \int_a^b \frac{du}{dx}(x)v(x)dx. \qquad (13.9)$$

On the left- and right-hand sides in (13.9) we have two integrals. The strategy is to start with the left-hand side and end up with a simpler integral on the right-hand side. This is called *integrating by parts*.

Example 13.7. (a)

$$\int_1^2 (2x\log x)dx = \int_1^2 \log x\frac{d}{dx}(x^2)dx = x^2\log x\Big]_1^2 - \int_1^2 x^2\frac{d}{dx}(\log x)dx$$

$$= 4\log 2 - \int_1^2 x^2\frac{1}{x}dx = 4\log 2 - \int_1^2 xdx$$

$$= 4\log 2 - \frac{x^2}{2}\Big]_1^2 = 4\log 2 - \frac{3}{2}.$$

(b)

$$\int_1^4 (\log x)^2 dx = \int_1^4 (\log x)^2\frac{d}{dx}(x)dx$$

$$= x(\log x)^2\Big]_1^4 - \int_1^4 x\frac{d}{dx}(\log x)^2 dx$$

$$= 4(\log 4)^2 - \int_1^4 x\cdot 2(\log x)\frac{1}{x}dx$$

$$= 4(\log 4)^2 - 2\int_1^4 \log x dx$$

$$= 4(\log 4)^2 - 2\int_1^4 \log x\frac{d}{dx}(x)dx$$

$$= 4(\log 4)^2 - 2x(\log x)\Big]_1^4 + 2\int_1^4 x\frac{d}{dx}(\log x)dx$$

$$= 4(\log 4)^2 - 8(\log 4) + 2\int_1^4 x\frac{1}{x}dx$$

$$= 4(\log 4)^2 - 8(\log 4) + 6.$$

We now reverse the *Chain Rule* for differentiating to obtain the method of integrating known as *The Method of Substitution*. Two applications of (13.6) imply

$$\int_a^b (g \circ f)'(x)dx = g \circ f(b) - g \circ f(a) = \int_{f(a)}^{f(b)} g'(y)dy.$$

However, by the chain rule, $(g \circ f)'(x) = g'(f(x)) \cdot f'(x)$, and hence

$$\int_{f(a)}^{f(b)} g'(y)dy = \int_a^b g'(f(x)) \cdot f'(x)dx.$$

In practical situations we frequently encounter

$$\int_a^b h(f(x)) \cdot f'(x)dx.$$

If $y = f(x)$ then $\frac{dy}{dx} = f'(x)$ or, symbolically, $dy = f'(x)dx$ and

$$\int_a^b h(f(x)) \cdot f'(x)dx = \int_{f(a)}^{f(b)} h(y)dy.$$

This reduces the original problem to finding an indefinite integral for h.

Example 13.8. (a) Consider the integral

$$\int_1^2 \frac{x^4}{x^5 + 4}dx.$$

If $y := f(x) := x^5 + 4$ then $dy = 5x^4dx$ and

$$\int_1^2 \frac{x^4}{x^5 + 4}dx = \int_{f(1)}^{f(2)} \frac{dy}{5y} = \frac{1}{5}\int_5^{36} \frac{1}{y}dy = \frac{1}{5}\log(y)\Big]_5^{36}$$
$$= \frac{1}{5}\log 36 - \frac{1}{5}\log 5 = \frac{1}{5}\log\left(\frac{36}{5}\right).$$

(b) Consider

$$\int_0^2 2x^3 e^{x^2} dx.$$

If $y = x^2$ then $dy = 2xdx$ and

$$\int_0^1 2x^3 e^{x^2} dx = \int_0^2 x^2 e^{x^2} 2xdx = \int_0^4 ye^y dy.$$

Integration by parts implies

$$\int_0^4 ye^y dy = \int_0^4 y\frac{d}{dy}(e^y)dy = ye^y\Big]_0^4 - \int_0^4 e^y dy = 4e^4 - (e^4 - 1) = 3e^4 + 1.$$

Example 13.9. Suppose $f : \mathbf{R}^+ \longrightarrow \mathbf{R}^+$ is differentiable, $El(f)(x) = x^2 e^{-x}$ and $f(1) = 1$. Then

$$\frac{xf'(x)}{f(x)} = x^2 e^{-x} \text{ and } \frac{f'(x)}{f(x)} = xe^{-x} = \frac{d}{dx}\left(\log f(x)\right).$$

Hence

$$\log f(x) + c = \int \frac{d}{dx}\left(\log f(x)\right)dx = \int xe^{-x}dx = \int x\frac{d}{dx}(-e^{-x})dx$$

$$= -xe^{-x} + \int e^{-x}dx$$

$$= -xe^{-x} - e^{-x}.$$

Since $f(1) = 1$,

$$c = -e^{-1} - e^{-1} = -2e^{-1}$$

and

$$f(x) = \exp(2e^{-1} - xe^{-x} - e^{-x}).$$

Example 13.10. We consider a situation similar to that examined in Example 11.12. Let $f(t)$ denote the value at time t of a function whose rate of change is proportional to its size and where the proportion may vary with time. The approach in Example 11.12 implies that, over the small interval $[t, t + \Delta t]$

$$f(t + \Delta t) - f(t) \approx c(t)f(t)\Delta t. \tag{13.10}$$

In Example 11.12 we assumed that f was differentiable and c was constant but here we suppose only that f and c are both continuous. If we partition the interval $[0, T]$ into 2^n adjacent intervals of equal length, $([t_i, t_{i+1}])_{i=0}^{2^n-1}$, and let $\Delta t_i = t_{i+1} - t_i$ then, applying (13.10) to each interval, we obtain

$$f(T) - f(0) = \sum_{i=0}^{2^n-1}(f(t_{i+1}) - f(t_i)) \approx \sum_{i=0}^{2^n-1} f(t_i)c(t_i)\Delta t_i$$

and, as we take finer and finer partitions, we obtain the *integral equation*

$$f(T) - f(0) = \int_0^T c(t)f(t)dt.$$

By the Fundamental Theorem of Calculus, Theorem 13.6, f is differentiable and the integral equation is equivalent to the differential equation, $f'(t) = c(t)f(t)$. If c is positive valued, then

$$\left(\log f(t)\right)' = \frac{f'(t)}{f(t)} = c(t) = \frac{d}{dt}\left(\int c(t)dt\right)$$

and, by the Mean Value Theorem, there exists a real number a such that

$$\log f(t) = a + \int_0^t c(x)dx.$$

Then $f(0) = e^a$ and

$$f(t) = f(0)\exp\left\{\int_0^t c(x)dx\right\}.$$

For example, if $c(t) = 3t^2$ and $f(0) = 4$ then $f(t) = 4e^{t^3}$.

13.4   Exercises

(13.1) Evaluate

$$\text{(a)}\ \int_4^5 x10^x\,dx, \quad \text{(b)}\ \int_1^{12} \log(x^x)\,dx, \quad \text{(c)}\ \int_1^{12} xe^x\,dx,$$

$$\text{(d)}\ \int_1^4 \frac{dx}{\sqrt{x}(1+\sqrt{x})}, \quad \text{(e)}\ \int_1^9 \frac{1-\sqrt{x}}{\sqrt{x}-x}\,dx.$$

(13.2) Find

$$\frac{d}{dx}\left(\int_2^{x^3} 2^t\,dt\right), \quad \int_5^8 2x(x^2+1)\log(x^2+1)\,dx, \quad \int_{10}^{100} \frac{\log x}{x}\,dx.$$

(13.3) Find the indefinite integrals of

$$\text{(a)}\ \int x^3 e^{x^4}\,dx, \quad \text{(b)}\ \int \frac{\log x}{x^2}\,dx, \quad \text{(c)}\ \int \frac{x}{\sqrt[3]{x^2+1}}\,dx,$$

$$\text{(d)}\ \int x^2 e^{-x^3}\,dx, \quad \text{(e)}\ \int (8x+40)(x^2+10x+5)^3\,dx.$$

(13.4) If $f : [a, b] \longrightarrow \mathbf{R}$ is continuous and $m \le f(x) \le M$ for all $x \in [a, b]$, show that

$$m(b-a) \le \int_a^b f(x)dx \le M(b-a).$$

(13.5) If $f : [a, b] \longrightarrow \mathbf{R}$ is continuous, show that

$$\lim_{\epsilon > 0 \to 0} \int_{a+\epsilon}^{b} f(x)dx = \int_{a}^{b} f(x)dx.$$

(13.6) Let $P(x) = \sum_{j=1}^{n} a_j x^j$ for all $x \in \mathbf{R}$ where $a_j \in \mathbf{R}$ all j. If $\sum_{j=1}^{n} \frac{a_j}{j+1} = 0$, show that there exists c, $0 < c < 1$ such that $P(c) = 0$.

(13.7) Evaluate $\int_{1}^{n} (\log x)dx$. Using Riemann sums for this integral show that

$$(n - 1)! < n^n e^{-n} e < n!$$

and hence deduce that $\lim_{n \to \infty} \left(\frac{n^n}{n!} \right)^{1/n} = e$.

(13.8) Use Riemann sums to find

$$\lim_{n \to \infty} \frac{1}{n} \{ (n+1) \cdot (n+2) \cdots (n+n) \}^{1/n}.$$

Mention all results used to obtain this limit.

(13.9) Sketch the parabola $\{(x, y) : x^2 = y\}$. By integrating between the curves $y = \pm \sqrt{x}$ verify the result of Archimedes.

(13.10) Find indefinite integrals for $(\log x)^2$ and $2x \log x$.

(13.11) If $f : [0, 1] \longrightarrow \mathbf{R}$ is continuous and

$$\int_{0}^{t} f(x)dx = \int_{t}^{1} f(x)dx$$

for all $t \in [0, 1]$, show that $f(x) = 0$ for all $x \in [0, 1]$.

(13.12) Let $P, Q : (a, b) \longrightarrow \mathbf{R}$ denote continuous functions, suppose $x_0 \in (a, b)$ and $y_0 \in \mathbf{R}$ are arbitrary. If $\rho(x) = \exp \left(\int P(x)dx \right)$ show that

$$y(x) := \frac{1}{\rho(x)} \int Q(x)\rho(x)dx + C$$

is, for an appropriate choice of C, the unique solution $y : (a, b) \longrightarrow \mathbf{R}$ of the ordinary differential equation $y' + P(x)y = Q(x), y(x_0) = y_0$.

References [9; 15; 16; 24; 25; 26; 27; 30; 31; 34]

Chapter 14

APPLICATIONS OF
INTEGRATION THEORY

If I ever made any valuable discoveries,
it has been owing more to patient
attention, than any other talent.

Isaac Newton

The way mathematics as a whole progresses is
complicated, and often depends on unexpected
links between apparently very different areas.

Tim Gowers, 2000

Summary

We select a small number of applications from the large number available.
These include integral inequalities, infinite integrals, a test for series convergence, a definition of the *Gamma Function*, density functions and power series expansions. Our aim is to use integrals to shed further light on what we already know and, in this way, to emphasize the unity of mathematics.

14.1   Integral Inequalities

Inequalities are useful estimates that take the place of equalities when these are not available. In this section we consider two inequalities: *The Cauchy-Schwarz inequality* (Theorem 14.1) and *Jensen's inequality* (Theorem 14.2).[1]

Theorem 14.1. *If f and g are continuous real-valued functions on $[a, b]$, then*

$$\left| \int_a^b f(x)g(x)dx \right|^2 \le \left(\int_a^b f^2(x)dx \right) \cdot \left(\int_a^b g^2(x)dx \right). \qquad (14.1)$$

Proof.   For any real number λ and any $x \in [a, b]$,

$$(f(x) + \lambda g(x))^2 = f^2(x) + 2\lambda f(x)g(x) + \lambda^2 g^2(x) \ge 0.$$

By Theorem 13.3(a) and (b)

$$\left(\int_a^b f^2(x)dx \right) + 2\lambda \left(\int_a^b f(x)g(x)dx \right) + \lambda^2 \left(\int_a^b g^2(x)dx \right) \ge 0$$

for all $\lambda \in \mathbf{R}$. This is just a *quadratic* in λ of the form $A + 2B\lambda + C\lambda^2$ that is always non-negative. We know, from Chapters 1 and 2, that $(2B)^2 - 4AC \le 0$. Hence $AC \ge B^2$ and, translating back, we obtain the required result. $\square$

The Cauchy-Schwarz inequality is useful in analysis, statistics and probability theory. Our next result is known as *Jensen's Inequality*.[2] It has important applications in economics and probability theory. The $b - a$ factor that occurs in the proof is just a normalizing term.

---

[1]Hermann Amandus Schwarz, 1843-1921, from Hermsdorf, Silesia, was a student of Weierstrass in Berlin and held academic positions in Halle, Zurich, and Göttingen before returning to Berlin as successor to Weierstrass. He is responsible for a number of very important results in real analysis, geometry, complex analysis, and potential theory.

[2]Johan Jensen, 1859-1925, from Nakskov (Denmark) studied a range of sciences, including mathematics, at university but was, essentially, self taught as a research mathematician and never held an academic position. He had a successful professional career as a technical engineer with the Bell Telephone Company in Copenhagen from 1881 until 1924 and devoted his spare time to mathematics. He published high-quality research in complex and real analysis. Jensen's Inequality, first published in 1906, is regarded, along with the Cauchy-Schwarz inequality, as one of the fundamental inequalities in analysis.

Theorem 14.2. *If $f : [a,b] \longrightarrow [c,d]$ is continuous and $\phi : [c,d] \longrightarrow \mathbf{R}$ is convex then*

$$\phi\Big(\frac{1}{b-a}\int_a^b f(x)dx\Big) \leq \frac{1}{b-a}\Big(\int_a^b \phi \circ f(x)dx\Big).$$

Proof. We require the Riemann sums $\mathcal{R}(n, f, w)$, see Figure 13.3, where $x_{i,n} = w_{i,n} = a + \frac{i(b-a)}{2^n}$ for $0 \leq i \leq 2^n$. Since $x_{i+1,n} > x_{i,n}$ for all i and n and $\sum_{i=0}^{2^n-1} \frac{x_{i+1,n}-x_{i,n}}{b-a} = 1$, Exercise 4.18 implies

$$\phi\Big(\frac{\mathcal{R}(f,n,w)}{b-a}\Big) = \phi\Big(\sum_{i=0}^{2^n-1} \frac{x_{i+1,n} - x_{i,n}}{b-a}f(x_i)\Big)$$

$$\leq \sum_{i=0}^{2^n-1} \frac{x_{i+1,n} - x_{i,n}}{b-a}\phi(f(x_i))$$

$$= \frac{\mathcal{R}(\phi \circ f, n, w)}{b-a}.$$

By Example 9.15, ϕ is continuous and, by Lemma 9.2, $\phi \circ f$ is continuous. Applying the definition of continuity and (13.3) to both sides of the above inequality completes the proof. □

14.2   Integrals over Infinite Intervals

For many applications it is useful to have integrals over open intervals and over intervals of infinite length.

Definition 14.3. If $f : [a,b) \longrightarrow \mathbf{R}$ is continuous we let

$$\int_a^b f(x)dx = \lim_{y<b, y \longrightarrow b} \int_a^y f(x)dx \tag{14.2}$$

whenever the limit on the right-hand side of (14.2) is finite.

Definition 14.4. If $f : [a, \infty) \longrightarrow \mathbf{R}$ is continuous we let

$$\int_a^\infty f(x)dx = \lim_{n \to \infty} \int_a^n f(x)dx \tag{14.3}$$

whenever the limit on the right-hand side of (14.3) is finite.

When the limits in (14.2) and (14.3) are finite we say that f is Riemann integrable, or just integrable. It is now clear how to define integrals of continuous functions over any type of interval. The rules for integrating from the previous chapter extend in an obvious way and, in particular, we obtain the following lemma.

Lemma 14.5. *Let $a, b \in \mathbf{R} \cup \{\pm\infty\}$, $a < b$, and let $f, g : (a, b) \longrightarrow \mathbf{R}$ denote continuous functions. If $|f(x)| \leq g(x)$ for all $x \in (a, b)$ and g is integrable then f and $|f|$ are integrable and*

$$\left| \int_a^b f(x)dx \right| \leq \int_a^b |f(x)|dx \leq \int_a^b g(x)dx.$$

Example 14.6.

(a) If p denotes a positive real number, $p \neq 1$, then

$$\int_1^n \frac{1}{x^p}dx = \frac{x^{-p+1}}{-p+1}\Big]_1^n = \frac{1}{(-p+1)n^{p-1}} - \frac{1}{(-p+1)}.$$

When $p = 1$

$$\int_1^n \frac{dx}{x} = \log x\Big]_1^n = \log n$$

and hence $\int_1^\infty \frac{dx}{x^p} < \infty \iff p > 1$. When $p > 1$

$$\int_1^\infty \frac{dx}{x^p} = \frac{1}{p-1}.$$

(b) If $\alpha > 0$, then $\int_0^n e^{-\alpha x}dx = \frac{e^{-\alpha x}}{-\alpha}\Big]_0^n = (e^{-\alpha n} - 1)/(-\alpha) \longrightarrow \frac{1}{\alpha}$ as $n \longrightarrow \infty$. Hence

$$\int_0^\infty e^{-\alpha x}dx = \frac{1}{\alpha}.$$

(c) If $t \in \mathbf{R}$ and $x > 0$, then $e^x > \frac{x^m}{m!}$ for all $m \in \mathbf{N}$. Hence, if $t \in \mathbf{R}$ and we choose $m \in \mathbf{N}$ such that $m > t + 1$, then

$$0 \leq \frac{x^{t-1}e^{-x}}{x^{-2}} = \frac{x^{t+1}}{e^x} \leq \frac{m!x^{t+1}}{x^m} \longrightarrow 0 \text{ as } x \longrightarrow +\infty.$$

Hence, there exists $c > 0$ such that $x^{t-1}e^{-x} \leq cx^{-2}$ for all $x > 0$ and, by part (a) and Lemma 14.5, the integral $\int_\epsilon^\infty x^{t-1}e^{-x}dx$ is finite for any $\epsilon > 0$.

We have already used the numbers $n!$ in defining the exponential function. Euler was interested in defining a well behaved function on the set of positive real numbers which interpolated the function $f : \mathbf{N} \longrightarrow \mathbf{N}, f(n+1) = n!$. In 1729 he used integration to define such a function, see Exercise 14.7. Subsequently Legendre modified Euler's integral to give the now standard Definition 14.7 and introduced the Γ (Gamma) notation. This function has, over the last quarter millennium, proved its importance in mathematical analysis, probability theory and statistics.

Definition 14.7. If $t \in \mathbf{R}^+$ let

$$\Gamma(t) = \lim_{\epsilon > 0 \longrightarrow 0} \int_\epsilon^\infty x^{t-1} e^{-x} dx = \int_0^\infty x^{t-1} e^{-x} dx.$$

We have already noted in Example 14.6(c) that $\int_1^\infty x^{t-1} e^{-x} dx$ is finite. When $t \geq 1$, the function $x \longrightarrow x^{t-1}$ is continuous on $[0,1]$ and $\Gamma(t)$ is finite. For $t \geq 0$ and $\epsilon > 0$ we have

$$\int_\epsilon^\infty x^{t-1} e^{-x} dx = \int_\epsilon^\infty e^{-x} \frac{d}{dx}\left(\frac{x^t}{t}\right) dx = e^{-x} \frac{\epsilon^t}{t} + \int_\epsilon^\infty \frac{x^t}{t} e^{-x} dx$$

and, as the right-hand side tends to a finite limit as $\epsilon \longrightarrow 0$, we see that the limit used in the definition of the Gamma function[3] exists and, moreover, for all $t > 0$

$$t\Gamma(t) = \Gamma(t+1). \tag{14.4}$$

By Example 14.6(b), $\Gamma(1) = 1$, and hence, by (14.4), $\Gamma(n+1) = n!$.

Definition 14.8. A continuous positive real-valued function defined on a finite or infinite interval (a, b) is called a density function if

$$\int_a^b f(x) dx = 1.$$

---

[3] A highly non-trivial result which first appeared in a 1922 Danish textbook on probability theory by Harald Bohr, 1887-1951, and Johannes Mollerup, 1872-1937, states that if $f : \mathbf{R}^+ \longrightarrow \mathbf{R}^+$ satisfies $f(1) = 1$, $tf(t) = f(t+1)$ for all $t > 0$, and $\log f$ is convex, then $f = \Gamma$. Many properties of the Gamma function can be derived directly from this characterization. Bohr, from Copenhagen, was a younger brother of the Nobel prize winning physicist, Niels Bohr. He studied and later became professor at the University of Copenhagen. He made important contributions to complex analysis and founded the theory of *almost periodic functions*. He was also a noted footballer and was on the Danish team that were runners up at the London Olympics in 1908.

Density functions arise in probability theory and statistics. If a given population is suitably distributed then the probability that a randomly chosen member of the population will lie in the interval (c, d) is given by $\int_c^d f(x)dx$. The expected or average value and different parameters admitting significant statistical interpretation are found by evaluating $\int_a^b g(x)f(x)dx$ for appropriate g. A random sample with replacement of size n from a population is a set of independent identically distributed random variables. The *Central Limit Theorem*, first *mentioned* in 1732 by Abraham DeMoivre[4] and *proved* by Pierre Simon Laplace in 1801, states that, under fairly general conditions, the sample average for n large, is approximately normal.

Example 14.9. (a) The standard normal or Gaussian density function (see Figure 14.1) is given by $f(x) = \frac{1}{\sqrt{2\pi}}e^{-\frac{x^2}{2}}, x \in \mathbf{R}$. We assume the non-trivial result that this is a density function. Since

$$e^{\frac{1}{2}x^2} \geq \frac{x^{2n+2}}{(2n + 2)!2^{n+1}}$$

we have $x^{2n}e^{-\frac{1}{2}x^2} \leq (2n+2)!2^{2n+1}x^{-2}$ and $\int_{-\infty}^{+\infty} x^n e^{-\frac{1}{2}x^2} dx$ is finite for all n. By completing squares we can compute many other integrals associated with this density function. By symmetry $\int_{-\infty}^{+\infty} x^{2m+1}e^{-\frac{1}{2}x^2} dx = 0$ for all non-negative integers m and

---

[4]DeMoivre, 1667-1754, a Huguenot, fled France as a teenager to escape religious persecution after the revocation of the edict of Nantes and settled in London for the remainder of his life. He supported himself by private tutoring in mathematics and acting as a consultant on games of chance and annuities. Pierre Simon Laplace, 1749-1827, was born in Beaumont-en-Auge (Normandy) and educated in Caen and Paris. He made important contributions to maxima and minima problems, the integral calculus, differential equations, astronomy (he proved that the solar system was stable), probability theory and applied mathematics. He benefited greatly from his collaborations with Lagrange. Laplace's conservative nature and careful temperament, his attention to detail, his wide ranging interests in science, education, literature and philosophy helped him become the successful and politically astute leader of the French scientific community during difficult and changing times. He served on various government and academy committees and promoted many academic careers. However, his tendency to dominate proceedings and frequently remind his colleagues of his obvious scientific abilities irked some while others regarded his political gyrations as opportunistic.

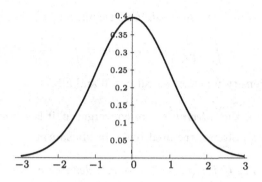

Fig. 14.1

$$\int_{-n}^{+n} x^2 e^{-\frac{1}{2}x^2}\,dx = \int_{-n}^{+n} x \cdot x e^{-\frac{1}{2}x^2}\,dx = \int_{-n}^{+n} (-x) \cdot \frac{d}{dx}\left(e^{-\frac{1}{2}x^2}\right)dx$$

$$= \left. -x e^{\frac{-1}{2}x^2} \right]_{-n}^{n} + \int_{-n}^{+n} e^{-\frac{1}{2}x^2}\,dx$$

$$\longrightarrow \int_{-\infty}^{+\infty} e^{-\frac{1}{2}x^2}\,dx = \sqrt{2\pi}.$$

Using $-\frac{1}{2}x^2 + 4x = -\frac{1}{2}(x^2 - 8x + 16) + \frac{1}{2}16 = -\frac{1}{2}(x-4)^2 + 8$ and the substitution $y = x - 4$, we obtain

$$\int_{-\infty}^{+\infty} e^{-\frac{1}{2}x^2+4x}\,dx = \int_{-\infty}^{+\infty} e^{-\frac{1}{2}(x-4)^2+8}\,dx = e^8 \int_{-\infty}^{+\infty} e^{-\frac{1}{2}y^2}\,dy = e^8\sqrt{2\pi}$$

and

$$\int_{-\infty}^{+\infty} x^2 e^{-\frac{1}{2}x^2+4x}\,dx = \int_{-\infty}^{+\infty} x^2 e^{-\frac{1}{2}(x-4)^2+8}\,dx$$

$$= e^8 \int_{-\infty}^{+\infty} (x - 4 + 4)^2 e^{-\frac{1}{2}(x-4)^2}\,dx$$

$$= e^8 \int_{-\infty}^{+\infty} (y^2 + 8y + 16) e^{-\frac{1}{2}y^2}\,dy$$

$$= e^8 \sqrt{2\pi}(1 + 16).$$

Example 14.10. The Gamma density function depends on two parameters $\alpha > 0$ and λ. We let

$$g_{\alpha,\lambda}(x) = \frac{\lambda^\alpha}{\Gamma(\alpha)} x^{\alpha-1} e^{-\lambda x}, \qquad x > 0.$$

The substitution $y = \lambda x$, $dy = \lambda dx$ and Definition 14.7 imply

$$\int_0^\infty g_{\alpha,\lambda}(x)dx = \int_0^\infty \frac{\lambda^\alpha}{\Gamma(\alpha)}x^{\alpha-1}e^{-\lambda x}dx = \int_0^\infty \frac{1}{\Gamma(\alpha)}y^{\alpha-1}e^{-y}dy = 1$$

and $g_{\alpha,\lambda}$ is a density function for all $\alpha > 0$ and all λ.

Our next result is the *integral test* for convergence, it is generally finer than the ratio test and often attributed to Colin Maclaurin.

Theorem 14.11. *If $f : (0, +\infty) \longrightarrow \mathbf{R}^+$ denotes a continuous decreasing positive function, then $\sum_{n=1}^\infty f(n)$ converges if and only if $\int_1^\infty f(x)dx$ converges.*

Proof.   It suffices to note (see Figure 14.2) that

$$\sum_{j=2}^n f(j) \le \int_1^n f(x)dx \le \sum_{j=1}^{n-1} f(j).$$

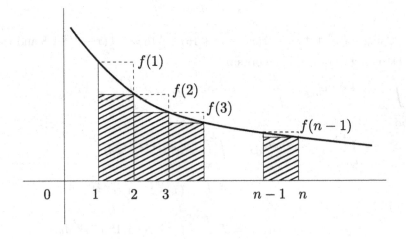

Fig. 14.2

$\square$

Example 14.12. (a) Let $f : \mathbf{R}^+ \longrightarrow \mathbf{R}^+, f(x) = x^{-p}$ where p is a positive real number. Then $f'(x) = -px^{-p-1}$ for all $x > 0$ and f is decreasing. By Example 14.6(a) and Theorem 14.11, $\sum_{n=1}^\infty \frac{1}{n^p} < \infty$ if and only if $p > 1$.

Alternative approaches are given in Example 8.11 and Exercise 11.24, when $p = 1$, and in Exercise 8.9 for arbitrary p.

(b) Consider $\sum_{n=2}^{\infty} \frac{1}{n \log n}$. Let $f(n) = \frac{1}{n \log n}$ for $n \geq 2$ and $f(x) = \frac{1}{x \log x}$ for $x \geq 2$. Since

$$f'(x) = -(x \log x)^{-2}(1 + \log x) < 0$$

we may apply Theorem 14.11. If $y = \log x$ then $dy = dx/x$ and

$$\int_2^n \frac{1}{x \log x} dx = \int \frac{1}{y} dy = \log y = \log(\log x)]_2^n$$
$$= \log(\log n) - \log(\log 2).$$

Hence

$$\int_2^\infty \frac{dx}{x \log x} = \sum_{n=2}^{\infty} \frac{1}{n \log n} = \infty.$$

(c) The following example, due to Ivan Niven, brings together in an interesting way, properties of the exponential function, prime numbers, convergent and divergent series, the log function, and via Theorem 14.11, integration theory (see also Exercise 11.24). A positive integer n is a square if $n = m^2$ for some positive integer m and it is *square free* if $n \neq m^2 r$ for positive integers m and r, $m > 1$. Clearly, a positive integer n is square free if and only if it can be written as a product of primes with no prime repeated and every positive integer n can be written in a unique way as a product of a square and a square free number. Let A_n denote the square free positive integers less than n, that is $A_n = \{j : j \in \mathbf{N}, j < n, j \neq k^2 m, k, m \in \mathbf{N}, k > 1\}$. Then

$$\left(\sum_{k<n} \frac{1}{k^2} \right) \cdot \left(\sum_{j \in A_n} \frac{1}{j} \right) \geq \sum_{m<n} \frac{1}{m}. \tag{14.5}$$

By part (a), $\sum_{m=1}^{\infty} \frac{1}{m} = \infty$ and $\sum_{k=1}^{\infty} \frac{1}{k^2} < \infty$ and hence, by (14.5), $\sum_{j \in A_n} \frac{1}{j} \longrightarrow \infty$ as $n \longrightarrow \infty$.

Let p_j denotes the jth prime in increasing order. By (7.3), $e^x > 1 + x$ for $x > 0$ and, by Theorem 7.17,[5]

$$\exp \left(\sum_{p_j < n} \frac{1}{p_j} \right) = \prod_{p_j < n} \exp \left(\frac{1}{p_j} \right) > \prod_{p_j < n} \left(1 + \frac{1}{p_j} \right) \geq \sum_{j \in A_n} \frac{1}{j}.$$

[5] $\prod_{j=1}^k a_j = a_1 a_2 \cdots a_k$ is the standard notation for finite products.

Hence $\exp\left(\sum_{p_j < n} \frac{1}{p_j}\right) \longrightarrow \infty$ as $n \longrightarrow \infty$ and, by continuity of the log function, we obtain a result first proved by Euler in 1737:

$$\sum_{j=1}^{\infty} \frac{1}{p_j} = \infty.$$

14.3   Taylor's Theorem

In Chapters 8 and 11 we saw that power series define differentiable functions. The approach used in Example 11.14 shows that, if the power series $\sum_{n=0}^{\infty} a_n x^n$ has radius of convergence $r > 0$, then $f(x) := \sum_{n=0}^{\infty} a_n x^n$ has derivatives of all orders at all points within the interval of convergence and, in particular, $n!a_n = f^{(n)}(0)$ for all n. We consider the converse question of how to recognize when a given function can be represented by a power series. This is known as Taylor's Theorem.[6] There are different formulae for the remainder term we just give one, and note also that the Mean

---

[6]Brook Taylor, 1685-1731, and Colin Maclaurin, 1698-1746, were Newton's chief supporters in the controversy with Leibniz's followers on the discovery of the calculus and, at one stage, Taylor was appointed to the committee of the Royal Society set up to settle the matter. Taylor was born in Edmonton, Middlesex, England and educated at Cambridge. He made contributions to probability theory, differential equations, linear perspective and the calculus. His writings were concise but difficult. A retrospective judgement of his contributions shows that he touched, but failed to substantially develop, many major concepts. Taylor had his own priority dispute with Johann Bernoulli. Maclaurin was born in Kilmodan, Argyllshire, Scotland and educated at Glasgow University, graduating with an M.A at age 15 and was appointed a professor of mathematics at Aberdeen when he was 19. He wrote a 700 page two volume text in 1742, *A Treatise of Fluxions*, defending the calculus against the criticism of Bishop Berkeley and in the preface diplomatically declared the misunderstanding of the concise manner of the usual presentations by a person of such abilities *to be sufficient proof that a fuller account of this was required*. Maclaurin later moved to Edinburgh where he was known as a excellent teacher and played a leading role in the defense of the city during the Jacobite rebelion of 1745. He made mathematical contributions to geometry, summation theory, and the calculus and is regarded as one of the founders of *actuarial science*. Nowadays we associate the names Taylor and Maclaurin with the power series $\sum_{n=0}^{\infty} a_n(x-a)^n$ and $\sum_{n=0}^{\infty} a_n x^n$, respectively. Taylor used power series expansions to approximate solutions to equations but the more substantial uses of power series are due to Euler and Lagrange.

Value Theorem is a special case of this result. Suppose the function f is defined and has derivatives of all orders up to n at all points on an interval containing the origin. Let

$$g(x) = f(x) - f(0) - f'(0)x - \frac{f^2(0)}{2!}x^2 - \cdots - \frac{f^{n-1}(0)}{(n-1)!}x^{n-1}$$

and let $h(x) = \frac{x^n}{n!}$. Then $g^i(0) = h^i(0) = 0$ for $i \le n-1$, $g^n(x) = f^n(x)$ and $h^n(x) = 1$ for all x. Repeated applications of the extended Mean Value Theorem, Theorem 11.2, imply that there exist $\theta_i, i \le n, 0 < \theta_n < \theta_{n-1} < \cdots \theta_1 < 1$, such that

$$\frac{g(x)}{h(x)} = \frac{g(x) - g(0)}{h(x) - h(0)} = \frac{g'(\theta_1 x)}{h'(\theta_1 x)} = \frac{g'(\theta_1 x) - g'(0)}{h'(\theta_1 x) - h'(0)}$$

$$= \frac{g^2(\theta_2 x) - g^2(0)}{h^2(\theta_2 x) - h^2(0)}$$

$$= \ldots \ldots$$

$$= \frac{g^{n-1}(\theta_{n-1} x) - g^{n-1}(0)}{h^{n-1}(\theta_{n-1} x) - h^{n-1}(0)}$$

$$= \frac{g^n(\theta_n x)}{h^n(\theta_n x)} = \frac{f^n(\theta_n x)}{1}$$

and we obtain the following theorem.

Theorem 14.13. *If f is defined and has n derivatives at all points in an open interval I about the origin, then, for all $x \in I$, there is a $\theta, 0 < \theta < 1$, such that*[7]

$$f(x) = f(0) + f'(0)x + \frac{f^2(0)}{2!}x^2 + \cdots + \frac{f^{n-1}(0)}{(n-1)!}x^{n-1} + \frac{f^n(\theta x)}{n!}x^n.$$

Corollary 14.14. *If f is defined and has derivatives of all orders at all points in the interval $(-r, r)$, $r > 0$, and if $\sup_{\{|x| < r\}} \{|r^n f^n(x)/n!|\} \longrightarrow 0$ as $n \longrightarrow \infty$, then*

$$f(x) = \sum_{n=0}^{\infty} \frac{f^n(0)}{n!}x^n \tag{14.6}$$

when $|x| < r$.

---

[7]The term $\frac{f^n(\theta x)}{n!}x^n$ is known as Lagrange's form of the remainder.

The series expansion in (14.6) was given in 1668 by James Gregory and was known to James Stirling. It was mentioned without proof in 1715 by Brook Taylor in *Methodus incrementorum directa et inversa* and its usefulness was demonstrated by Colin Maclaurin in 1742.

Example 14.15. If $f(x) = e^x$ then $f^n(x) = e^x$ for all n and x. If $r > 0$ then, by Definition 7.16 and Theorem 8.8,

$$\sup_{\{|x|<r\}} \{r^n |f^n(x)/n!|\} \le \frac{r^n e^r}{n!} \longrightarrow 0 \quad \text{as} \quad n \longrightarrow \infty.$$

Since r was arbitrary, this confirms that

$$e^x = \sum_{n=0}^{\infty} \frac{x^n}{n!}$$

for all $x \in \mathbf{R}$.

Example 14.16. If $\sum_{n=0}^{\infty} a_n x^n$ has radius of convergence $r > 0$ then, by Theorem 8.15, the power series $\sum_{n=0}^{\infty} \frac{a_n}{n+1} x^{n+1}$ converges absolutely when $|x| < r$. Example 11.14 implies that $f(x) := \sum_{n=0}^{\infty} a_n x^n$ is continuous, and hence integrable, over $[a, b]$ where $-r < a < b < r$. If

$$g(x) := \sum_{n=0}^{\infty} \frac{a_n}{n+1} x^{n+1}$$

for $|x| < r$ then, by Example 11.14, $g' = f$, and by the Fundamental Theorem of Calculus,

$$\int_a^b \left(\sum_{n=0}^{\infty} a_n x^n \right) dx = \int_a^b g'(x) dx = g(b) - g(a) = \sum_{n=0}^{\infty} \frac{a_n}{n+1} \left(b^{n+1} - a^{n+1} \right).$$

Thus, a power series may be integrated, term by term, over any closed bounded interval inside its interval of convergence. By integrating the geometric series

$$\frac{1}{1+x} = \sum_{n=0}^{\infty} (-1)^n x^n$$

and noting that $\log 1 = 0$, we obtain, for $|x| < 1$, the power series expansion

$$\log(1+x) = \sum_{n=0}^{\infty} \frac{(-1)^n x^{n+1}}{n+1} = x - \frac{x^2}{2} + \frac{x^3}{3} - \cdots \qquad (14.7)$$

already given in Example 11.14.[8]

---

[8] The power series expansion for $\log(1+x)$ was given in 1656, prior to the development of the differential calculus, by Johann van Waveren Hudde, 1628-1704. The renowned map

14.4   Exercises

(14.1) If $\sum_{n=1}^{\infty} |a_n|^2 < \infty$ and $\sum_{n=1}^{\infty} |b_n|^2 < \infty$ show that

$$\left(\sum_{n=1}^{\infty} |a_n b_n|\right)^2 \le \sum_{n=1}^{\infty} |a_n|^2 \cdot \sum_{n=1}^{\infty} |b_n|^2.$$

(14.2) For which values of α is the integral

$$\int_2^{\infty} \frac{dx}{x(\log x)^{\alpha}}$$

finite? Evaluate the integral when it is finite.

(14.3) Does $f : [a, b] \longrightarrow \mathbf{R}$ continuous and $\int_c^d f(x)dx = l$ for all rational $c, d, a < c < d < b$ and some fixed l imply $f(x) = 0$ for all $x \in [a, b]$. Justify your answer.

(14.4) Evaluate

$$\int_1^{\infty} xe^{-x^2}\,dx, \quad \int_1^{\infty} x^2 e^{-x^3}\,dx, \quad \int_2^{\infty} xe^{-3x}dx, \quad \int_1^{\infty} \frac{(2x+1)dx}{(x^2+x)^3}.$$

(14.5) Find all real numbers where the following power series converge:

$$\sum_{n=0}^{\infty} \frac{x^{2n}}{n2^n}, \quad \sum_{n=1}^{\infty} \frac{(\log n)x^n}{n3^n}, \quad \sum_{n=0}^{\infty} (-1)^n (3x^n)^n.$$

(14.6) If $f, g : \mathbf{R}^+ \longrightarrow \mathbf{R}^+$ and $\lim_{x \to +\infty} \frac{f(x)}{g(x)} \longrightarrow c > 0$, show that $\sum_{n=1}^{\infty} f(n) < \infty$ if and only if $\sum_{n=1}^{\infty} g(n) < \infty$. Test for convergence

$$\sum_{n=1}^{\infty} \frac{1}{n(1+\log n)^2 + \sqrt{n}}.$$

maker Nicolaus Mercator, 1620-1687, also obtained (14.7) in 1668. He used geometric series and infinite products. Hudde studied law at the University of Leiden and spent most of his working life in the services of the Amsterdam city council, serving 30 years as a burgomaster and directing, in 1657, the flooding of parts of Holland to impede the advance of the French army. His mathematical work, which included contributions to the theory of equations, maxima and minima problems and probability theory, influenced Newton and Leibniz. Hudde made important original contributions to actuarial science, worked on canal maintenance and optics, and was generally influential in promoting scholarship and higher education.

(14.7) Show that
$$\Gamma(t+1) = \lim_{\epsilon \to 0} \int_{\epsilon}^{1} \left(-\log x\right)^t dx =: \int_{0}^{1} \left(-\log x\right)^t dx$$
for all $t > 0$.

(14.8) What will be the amount accumulated if money is deposited continuously at a rate of $10,000 per year for seven years when the interest rate is 4.5%.

(14.9) If $f : [0, 1/2] \longrightarrow \mathbf{R}, f(x) = \frac{1}{2} + x + bx^2$ is a density function, find b.

(14.10) Test for convergence
$$\sum_{n=1}^{\infty} \frac{\log n}{n}, \quad \sum_{n=1}^{\infty} \frac{1}{n^4 + n^2}, \quad \sum_{n=1}^{\infty} \frac{n^2 + 1}{n^2(\log n + 4)}, \quad \sum_{n=1}^{\infty} \frac{n^2 - n}{\sqrt{n^5 + 2}}.$$

(14.11) Use Taylor's Theorem to find the power series expansion about the origin of
$$f(x) = (1 - x)^{-1}, \quad g(x) = \log\left(\frac{1}{1 - x^2}\right).$$

(14.12) Find $\int_{\mathbf{R}} e^{-q(x)} dx$ where $q : \mathbf{R} \longrightarrow \mathbf{R}$ is the quadratic function $q(x) = ax^2 + bx + c$ and $a > 0$.

(14.13) Let
$$f(x) = \frac{1}{\sqrt{2\pi}\sigma} \exp\left\{\frac{-(x - \mu)^2}{2\sigma^2}\right\}, \quad x \in \mathbf{R},$$
where $\sigma > 0$ and $\mu \in \mathbf{R}$. Show
$$\int_{\mathbf{R}} f(x)dx = 1, \quad \int_{\mathbf{R}} xf(x)dx = \mu, \quad \int_{\mathbf{R}} x^2 f(x)dx = \sigma^2 + \mu^2.$$

(14.14) If $\alpha > 0$, $\lambda \in \mathbf{R}$ and n is a positive integer, show that
$$\int_{0}^{\infty} x^n g_{\alpha, \lambda}(x)dx = \lambda^{-n} \int_{0}^{\infty} x^n g_{\alpha, 1} dx = \frac{\lambda^\alpha}{\Gamma(\alpha)} \int_{0}^{\infty} x^{\alpha + n - 1} e^{-x} dx$$
and evaluate the integral.

(14.15) Show that the Pareto[9] function p given, for $x > 0$, by the formula
$$p_{\alpha, \lambda}(x) = \frac{\alpha \lambda^\alpha}{(\lambda + x)^{\alpha + 1}}$$
is a density function where α and λ are positive parameters. If $\alpha > 1$, find $\int_{\mathbf{R}} x p_{\alpha, \lambda}(x) dx$ and, if $\alpha > 2$, find $\int_{\mathbf{R}} x^2 p_{\alpha, \lambda}(x) dx$.

---

[9]This density function is used to model income distribution and in the analysis of insurance claim distributions. It was introduced by Vilfredo Pareto, 1848-1923.

(14.16) Show, for any positive integer n, that

$$\int_0^\infty |x - n| x^{n-1} e^{-x} dx = 2n^n e^{-n}.$$

(14.17) If f and g have power series expansions in $|x| < r$, $r > 0$, show, using Exercise 10.19 and Corollary 14.14 or otherwise, that $f \cdot g$ has a power series expansion in $|x| < r$.

(14.18) Show, for any positive integer m, that

$$\int_{-\infty}^{+\infty} x^{2m} e^{-x^2/2} dx = \frac{(2m)! \sqrt{2\pi}}{m! \cdot 2^m}.$$

(14.19) Show that a positive integer m is square free if and only if for any positive integer n, $m|n$ if and only if $m|n^2$.

(14.20) Use (14.4) and the result in Example 14.9 to evaluate $\Gamma(\frac{1}{2})$ and $\Gamma(\frac{16}{3})\Gamma(\frac{5}{2})/\Gamma(\frac{4}{3})$.

(14.21) Use the formula $a^b = \exp(b \log a)$, the first two terms in the Taylor series expansion for $\log(1 + x)$ (Examples 11.14 and 14.16), and the first three terms in the Taylor series expansion of $\exp(x)$ (Definition 7.16) to estimate $(1.03)^{29}$.

(14.22) Using (a) the Cauchy-Schwarz inequality and (b) Jensen's inequality give two proofs, for f continuous, of the inequality

$$\int_0^1 |f(x)| dx \leq \left(\int_0^1 f^2(x) dx \right)^{1/2}.$$

References [9; 13; 16; 25; 30; 34; 36]

SOLUTIONS

Chapter 1

(1.1) $x^8 - 3x^4 - 5x^2 - 4x + 17 = (x^8 - 4x^4 + 4) + (x^4 - 6x^2 + 9) + (x^2 - 4x + 4) = (x^4 - 2)^2 + (x^2 - 3)^2 + (x - 2)^2 \geq 0$ and it is only 0 when all three terms are 0. If $x = 2$ then $x^2 = 4$ and $(x^2 - 3)^2 > 0$. Hence the expression is always positive.

(1.3) $-1, \sqrt{2} - 1, -\sqrt{2} - 1$. See also Exercises 4.24, 4.25, 10.15 and 11.28 and Example 11.6.

(1.4) 1st solution. If $Q(x, y) > 0$ when $(x, y) \neq (0, 0)$, then $Q(1, 0) = A > 0$ and

$$(*) \qquad Q(x, y) = Ax^2 + Bxy + Cy^2 = A\left(x + \frac{By}{2A}\right)^2 + \frac{y^2(4AC - B^2)}{4A}$$

269

implies $Q(-BA^{-1/2}, 2A^{1/2}) = 4AC - B^2 > 0$. Conversely, if $A > 0$ and $4AC - B^2 > 0$, then $(*)$ implies that $Q(x, y) > 0$ for $(x, y) \neq (0, 0)$.

2nd solution. $Q(x, y) > 0$ for $(x, y) \neq (0, 0) \iff A > 0$ and $Ax^2 + (By)x + Cy^2 = 0$ has no non-zero solutions for all $y \neq 0 \iff A > 0$ and $(By)^2 - 4A \cdot Cy^2 = y^2(B^2 - 4AC) < 0$ for all $y \neq 0 \iff A > 0$ and $4AC > B^2$.

(1.5) **Q $\Longrightarrow$ P.**

(1.6) $\pm\sqrt{\dfrac{10+\sqrt{124}}{4}}$.

(1.7) $f(x, y) = x^4 - 4x^2 + 4 + 2x^2 - 4x + 2 + y^4 - 2y^2 + 1 + y^2 + 2y + 1 + 1 = (x^2 - 2)^2 + 2(x - 1)^2 + (y^2 - 1)^2 + (y + 1)^2 + 1 > 0$.

(1.8) Complete squares to get $(x - y)^2 + (y - 2z)^2 + (x + 1)^2 = 16$. This implies $(x + 1)^2 \leq 16$, hence $|x + 1| \leq 4$ and $-4 \leq x + 1 \leq +4$ and, hence $-5 \leq x \leq 3$. Also $(x - y)^2 \leq 16$ and $|x - y| \leq 4$ and $-4 \leq x - y \leq +4$. This implies $y \leq x + 4 \leq 7$ and $y \geq x - 4 \geq -5 - 4 = -9$, etc. If $x = 3$, then $(x - y)^2 + (y - 2z)^2 = 0$, which implies $x = y = 2z$. Hence $x = y = 3$ and $z = 3/2$.

(1.9) $x^4 - 4x^3 + 8x^2 - 2x + 10 = x^4 - 4x^3 + 6x^2 - 4x + 1 + x^2 + 2x + 1 + x^2 + 8 = (x - 1)^4 + (x + 1)^2 + x^2 + 8 \geq 8$ for all real x.

(1.10) We do not define or use matrices in this book. This exercise is simply a diversion for any reader who has the background, temperament, time, and inclination to follow an unmarked path.

Chapter 2

(2.1) If $x \geq 1$ then $f(x) \geq x \geq 1$, if $0 < x < 1$ then $f(x) > (1/x) > 1$. If $x < 0$ then $f(-x) = -f(x)$ and apply the result for $x > 0$. $\dfrac{y^2}{1+y} \geq 0$ for $y > -1$ and when $y = 0$ we get 0.

(2.5) $|a - b|$ is the distance from a to b. Look for real numbers closer to 1 than 5. $\{x : x < 3\}$.

(2.7) $b^2 - 4ac = b^2 - (2a)(2c) > 0$, $a > 0$. The first diagram in Figure 2.8.

(2.9) The minimum would have to be 0 and, if this were achieved, it would imply $\sqrt{2}$ was rational (see Chapter 5 and also Section 12.1).

(2.11) $-4 < x < 2$, $a = 3$.

(2.12) Use $f(x) = \dfrac{x}{2} - \dfrac{1}{4} + \dfrac{9}{8x+4}$.

(2.13) Use the graphs of $2x$, $3/x$ and $f(x) = \frac{1}{2x+\frac{3}{x}}$ to draw the graph. The graph shows that the minimum and maximum occur when $\frac{x}{2x^2+3} = k$ has a unique solution. Maximum $1/2\sqrt{6}$. Minimum $-1/2\sqrt{6}$. These can also be found using the differential calculus.

(2.14) The number of solutions equals the number of times the graphs of $(x^2 - 10)^2$ and $2x + 50$ cross one another.

(2.16) 82.

Chapter 3

(3.1) $\{x \in \mathbf{R} : x > 5\}$, $im(f) = \mathbf{R}^+$.

(3.2) $f \circ g, g \circ f : \mathbf{R} \longrightarrow \mathbf{R}$, $f \circ g(x) = (x^2/2) - x + (3/2), g \circ f(x) = x^2$, $f \circ g(0) = 3/2 \neq 0 = g \circ f(0)$ so $f \circ g \neq g \circ f$.

(3.5) No solution if $a \leq 0$. If $a > 0$, $\log\left(\frac{e^a}{e^a-1}\right)$ is the unique solution.

Chapter 4

(4.1) If $f : A \longrightarrow B$ is injective, then the 4 arrows hit four different targets and hence all targets are hit and f is surjective and hence bijective. The result is also true for surjective functions.

(4.2) If $f(x) = f(y)$ then, since the exponential is inverse to the log function, $x^2 + 1 = \exp f(x) = \exp f(y) = y^2 + 1$ and $x^2 = y^2$. Hence $x^2 - y^2 = (x - y) \cdot (x + y) = 0$ and as $x > 0$ and $y > 0$ we have $x = y$ and f is injective. To prove surjectivity we need the Intermediate Value Theorem (Chapter 9) but from the graph it appears surjective. $f^{-1}(x) = \sqrt{e^x - 1}$.

(4.3) There are 4 arrows and 5 targets, no function can hit all targets, there are no surjective functions.

(4.4) $f : A = \mathbf{N} \longrightarrow 2\mathbf{N}$, (the set of all even integers), $f(n) = 2n$. $g : \mathbf{N} \longrightarrow \mathbf{Z}$, $g(2n) = n - 1$ and $g(2n - 1) = -n$ where $n \in \mathbf{N}$ in both cases.

(4.7) exp is convex but log is not.

(4.8) $\mathbf{1}_A \leq \mathbf{1}_B \Longleftrightarrow \mathbf{1}_A(x) = 1 \Longrightarrow \mathbf{1}_B(x) = 1 \Longleftrightarrow x \in A \Longrightarrow x \in B \Longleftrightarrow A \subset B$. If $x \in \Omega$, then $\min\{\mathbf{1}_A(x), \mathbf{1}_B(x)\} = 1 \Longleftrightarrow \mathbf{1}_A(x) = \mathbf{1}_B(x) = 1 \Longleftrightarrow x \in A \cap B \Longleftrightarrow \mathbf{1}_{A\cap B}(x) = 1$.

(4.11) If $a \neq 0$ and $b \neq 0$ then $a \cdot b \neq 0$. If $a \cdot b = 0$ then $a = 0$ or $b = 0$.

(4.12) If $6|p$ then $p = 6s$ and $p^2 = (6s)^2 = 6(6s^2)$. If $6 \nmid p$ then $p = 6s + r, r = 1, 2, 3, 4, 5$ and $p^2 = (6s+r)^2 = 6(6s^2+2rs)+r^2$. Since $r^2 = 1, 4, 9, 16, 25$ this proves the result. No, let $p = 4$.

(4.13) Since $1_\emptyset(x) = 0$ for all x, Exercise 4.8 shows that $1_A + 1_B = 1_{A \cup B}$ if $A \cap B = \emptyset$ and hence $1_{A \cup B} = 1_C$ if $A \cup B = C$. Conversely, suppose $1_A + 1_B = 1_C$. If $x \in A \cap B$ then $1_A(x) + 1_B(x) = 2 > 1_C(x)$ and $A \cap B = \emptyset$, if $x \in C \backslash (A \cup B)$ then $1_A(x) + 1_B(x) = 0 < 1 = 1_C(x)$ and $C \subset A \cup B$, and if $x \in (A \cup B) \backslash C$ then $1_A(x) + 1_B(x) \geq 1 > 0 = 1_C(x)$ and $A \cup B \subset C$.

(4.14) $x^2 - bxy + y^2 = (x - \frac{by}{2})^2 + y^2(1 - \frac{b^2}{4})$. $|b| < 2 \Longleftrightarrow b^2 < 4 \Longleftrightarrow \frac{b^2}{4} < 1 \Longleftrightarrow 1 - \frac{b^2}{4} > 0$. Hence $x^2 - bxy + y^2 \geq 0$ and $x^2 - bxy + y^2 = 0 \Longleftrightarrow y = 0$ and $x = \frac{by}{2} = 0$.

(4.15) $f^{-1}(x) = \sqrt{(x^2 - 4)} - 1$.

(4.16) $g(x) = x - 1$ and $h(x) = x^2$.

(4.17) If A had more than one element, then at least two arrows would be available, and one target could be hit more than once. This could be stated a number of ways. If B is nonempty and every mapping into B is surjective, then B has only one element.

(4.19) $21_{[-1,1]} + 31_{(1,3)} - 21_{(3,4)} - 31_{[4,5]}$.

(4.21) $x < y \Longrightarrow f(x) > f(y) \Longrightarrow g(f(x)) < g(f(y))$. The composition of p strictly increasing functions and q strictly decreasing functions is strictly increasing if q is even and strictly decreasing if q is odd.

(4.22) $\mathbf{Q} \Longrightarrow \mathbf{P}$ but the converse is not true. If f is injective, then $\mathbf{P} \Longleftrightarrow \mathbf{Q}$.

(4.24) $f(x) - f(y) = x^3 - y^3 + px - py = (x - y) \cdot (x^2 + xy + y^2 + p) = 0 \Longleftrightarrow x = y$ or $2x = -y \pm \sqrt{-4p - 3y^2}$. If $p > 0$, then, since $-4p - 3y^2 < 0$, we have $x = y$ and f is injective. If $p = 0$, then $f(x) = x^3$ is injective since $x^2 + xy + y^2 = ((x+y)^2 + x^2 + y^2)/2 = 0 \Longleftrightarrow x = y = 0$ (see also Figure 10.7). If $p < 0$, then $f((-p)^{1/2}) = f(-(-p)^{1/2}) = f(0) = 0$ and f is not injective.

(4.25) $p = -3\alpha^2$. Since $\alpha^3 - 3\alpha^2 \cdot \alpha + 2\alpha^3 = 0$, α is a solution. Then $x^3 + px + q = x^3 - 3\alpha^2 x + 2\alpha^3 = (x - \alpha)^2(x + 2\alpha)$ and $x^3 + px + q = 0$ has solutions α, α and -2α, $x^3 + px - q = 0$ has solutions $-\alpha, -\alpha$ and 2α. $p = -25/3, q = -250/27, 4p^3 + 27q^2 = 0, \alpha = -5/3$. Solutions $\frac{-5}{3}, \frac{-5}{3}, \frac{10}{3}$. See Example 11.6.

Chapter 5

(5.1) Proof by contradiction: suppose $p_1, \ldots, p_m$ is the set of all primes. If $n = p_1 p_2 \cdots p_m + 1$, then $n > p_i$ and $p_i \nmid n$ for all i. If n is prime, we have a contradiction. Otherwise, n is a product of positive integers all less than n. If one of these is prime it will be a prime factor of n and hence greater than each p_i, a contradiction. If none is prime, we continue the process but we can only continue a finite number of times and when the process stops we will have found a prime factor of n, a contradiction. *Wilson's Theorem* is another interesting connection between products and primes: $n + 1$ is prime $\Leftrightarrow (n + 1)|n! + 1$.

(5.2) If $n = 1$, $a^n - b^n = a - b$. If $n > 1$ then, by direct calculation, we have $a^n - b^n = (a - b) \cdot (a^{n-1} + a^{n-2}b + \cdots + a^{n-1-j}b^j + \cdots + b^{n-1})$.

(5.3) $n = a_m a_{m-1} \cdots a_0 = a_0 + 10a_1 + 10^2 a_2 + \cdots + 10^m a_m$, then $n = (a_0 + a_1 + \cdots + a_m) + 9a_1 + (9 \times 11)a_2 + \cdots + (9 \times 11 \times \cdots \times 11)a_m$ and $9|n$ if and only if 9 divides $a_0 + a_1 + \cdots + a_m$.

(5.4) Proof by contradiction in both cases. Use $\sqrt{2} + \sqrt{4} = \sqrt{2} + 2$ and $\sqrt{2} + \sqrt{4} + \sqrt{8} = 3\sqrt{2} + 2$, the fact that the sum and product of rational numbers is rational, and Theorem 5.9.

(5.5) $a, b \in A^+$. Let $c = an_0 + bm_0$. If $k \in \mathbf{N}$ and $kc < an + bm < (k+1)c$, then $0 < (an + bm) - kc < c$ and $(an + bm) - kc = a(n - kn_0) + b(m - km_0) =: e \in A \cap \mathbf{N}$. But $e < c$ is impossible and $A^+ = \{nc : n \in \mathbf{N}\}$. Since $a, b \in A^+$, c divides a and b. If $a = dn_1$ and $b = dm_1$, then $a, b \in A^+$, $c = (n_0 n_1 + m_0 m_1)d$ and $d \mid c$. If $r = (r, s)r_0$ and $s = (r, s)s_0$, then $a = (a, b)(r, s)r_0$ and $b = (a, b)(r, s)s_0$ and $(a, b)(r, s)$ divides both a and b and so $(a, b)(r, s) \mid (a, b)$. This implies $(r, s) = 1$. If $p \nmid a$, then $(a, p) = 1$ and $1 = an + pm$ for some integers n and m. Hence $b = abn + pmb$ and, as $p|ab$, $p|b$.

(5.6) If $\sqrt{2} + \sqrt{n} = r, r \in \mathbf{Q}$, then $\sqrt{2} = \frac{r^2 + 2 - n}{2r} \in \mathbf{Q}$. Consider $\sqrt{2}$ and $-\sqrt{2}$.

(5.7) This is equivalent to showing x rational implies x^2 is rational.

(5.8) All these numbers are prime but $P(40) = 41^2$.

(5.9) Let S denote the set of all integers which cannot be written as required. If S is non-empty it has a smallest element m_0. Clearly, m_0 in non-prime and so $m_0 = m_1 \cdot m_2$ where $m_i < m_0$. Hence both of these can be written as products of primes and m_0 is also a product of primes. If

$(p_i)_{i=1}^{\infty}$ are the primes in increasing order, $m_0 = p_1^{r_1} p_2^{r_2} \cdots = p_1^{s_1} p_2^{s_2} \cdots$ and i is the smallest integer for which $r_i > s_i$, then $p_i^{r_i - s_i} p_{i+1}^{r_{i+1}} \cdots = p_{i+1}^{s_{i+1}} p_{i+2}^{s_{i+2}} \cdots$. By Exercise 5.5, $p_i | p_i^{r_i - s_i} p_{i+1}^{r_{i+1}} \cdots$ and $p_i \nmid p_{i+1}^{s_{i+1}} p_{i+2}^{s_{i+2}} \cdots$. A contradiction unless $r_i \leq s_i$. Similarly $r_i \geq s_i$ and $r_i = s_i$.

(5.10) First, use the final part of Exercise 5.5. We may suppose, by the first part, that $(r, s) = 1$. Then $ns = mr$ and, if p is a prime divisor of s, say $s = ps_1$ where $s_1 < s$, then $p | mr$ and as $(r, s) = 1$, $p | m$ and $m = pm_1$. Then $\frac{n}{pm_1} = \frac{n}{m} = \frac{r}{s} = \frac{r}{ps_1}$ implies $\frac{n}{m_1} = \frac{r}{s_1}$. Clearly, $(n, m_1) = (r, s_1) = 1$ and $s = ps_1 \leq pm_1 = m \implies 1 \leq s_1 \leq m_1$. Repeating the process a finite number of times, we arrive at $s_j = 1$, $n = m_j r$ and $(n, m_j) = 1$ for some integer j. This implies $m_j = 1$ and $n = r$. Hence $m = s$.

(5.11) We may suppose $(r, s) = 1$. $n < \frac{r}{s} < n + 1 \implies ns < r < ns + s \implies 1 \leq r - ns < s$. The identity follows by multiplication. Right-hand side vanishes. Hence, $\sqrt{m} = \frac{ms - nr}{r - ns} = \frac{r}{s}$ (note that $\sqrt{m} > n \implies m > n\sqrt{m} = n(r/s) \implies ms > nr$). Now apply Exercise 5.10.

(5.12) $([1, 2])^2 = [1 + 4, 2 + 2] = [5, 4] = [2, 1] = 1$, $(-1)^2 = 1$. Let $x = [p, q]$ where p and q are positive rational numbers. If $p = q$, then $x^2 = [p, p]^2 = [2p^2, 2p^2] = 0$. If $p > q$, then $p = q + r$ where $r \in \mathbf{Q}^+$ and $x^2 = [p, q]^2 = [p^2 + q^2, 2pq] = [q^2 + 2qr + r^2 + q^2, 2(q + r)q] = [2q^2 + 2qr + r^2, 2q^2 + 2qr] = [r^2 + 1, 1] > 0$ since $r \in \mathbf{Q}^+$, if $q > p$ then $q = p + s$ where $s \in \mathbf{Q}^+$ and $x^2 = [s^2 + 1, 1] \geq 0$.

(5.14) By Exercise 5.5, we have $\sqrt{n} = \frac{p}{q}$ where $(p, q) = 1$. Then $nq^2 = p^2 \implies q | p^2 \implies q = 1$, since $(p, q) = 1$. Hence, $\sqrt{x} = p \in \mathbf{N}$.

(5.16) Since $(x + y)^n = (x + y) \cdot (x + y) \cdots (x + y)$, each term in the expansion has the form $x^a y^b$, where $a + b = n$, and we write $x^{n-j} y^j$. Collecting like terms, we get $\sum_{j=0}^{n} A_{j,n} x^{n-j} y^j$ where $A_{j,n}$ is the number of ways of choosing one letter from each $(x + y)$ and getting x precisely $(n - j)$ times. Consider a bowl with n balls labeled $1, 2, \ldots, n$ (these correspond to the $x's$). We can choose an ordered $(n - j)$-tuple in $n(n - 1) \cdots (j + 1) = n!/j!$ ways, since the first ball can be chosen in n ways, the second in $n - 1$ ways and so on. Alternatively we can first choose $(n - j)$ balls in $A_{j,n}$ ways and then arrange them in $(n - j)(n - j - 1) \cdots 2 \cdot 1 = (n - j)!$ ways. Hence $\frac{n!}{j!} = A_{j,n} \cdot (n - j)!$ as required.

(5.17) Let $a, b, c \in \mathbf{Q}^+$, $a = \left[\frac{n}{m}\right]$, $b = \left[\frac{n_1}{m_1}\right]$, $c = \left[\frac{u}{v}\right]$. Then $a + c = b + c \implies \left[\frac{n}{m}\right] + \left[\frac{u}{v}\right] = \left[\frac{n_1}{m_1}\right] + \left[\frac{u}{v}\right] \implies \left[\frac{nv + um}{mv}\right] = \left[\frac{n_1 v + um_1}{m_1 v}\right] \implies m_1 v(nv + um) =$

$$mv(n_1v + um_1) \implies m_1n = n_1m \implies a = \left[\frac{n}{m}\right] = \left[\frac{n_1}{m_1}\right] = b.$$

Chapter 6

(6.1) It suffices to apply Theorem 6.3 and to note that $\mathbf{R}$ is uncountable by Theorem 6.6.

(6.2) If $(n + 1)^{10}$ is bounded above by m, then $n \leq (n + 1)^{10} \leq m$ and the natural numbers are bounded above. This contradicts the result in Example 6.8. Since $2^n n^{-3}$ is always positive the sequence is bounded below by zero. If $2^n n^{-3} \leq m$, then $(2^{1/3})^n n^{-1} \leq m^{1/3}$. If $2^{1/3} = 1 + a$, then $a > 0$ and, by the Binomial Theorem (Exercise 5.16), $(2^{1/3})^n = (1 + a)^n \geq 1 + na + n(n+1)a^2/2 \geq n^2 a^2/2$ and so $\frac{n^2 a^2}{2} \cdot \frac{1}{n} \leq \frac{(2^{1/3})^n}{n} \leq m^{1/3}$. Hence, $n \leq 2m^{1/3}/a^2$ and this contradicts Example 6.8.

(6.3) Use a diagonal process: see Theorem 6.6.

(6.4) $\{x : x \leq 1\}$, $\{x : x < 1\}$.

(6.7) If m is an upper bound for B, then $y \leq m$ for all $y \in B$. If $x \in A$, then there exists $y \in B$ such that $x \leq y$. Then $x \leq y \leq m$ and m is an upper bound A. In particular $lub(B)$ is an upper bound for A. Hence $lub(A) \leq lub(B)$.

(6.8) If $y \in B$, then y is an upper bound for A, hence $lub(A) \leq y$ for all $y \in B$. Since y was an arbitrary element of B, this implies that $lub(A)$ is a lower bound for B and hence it is less than or equal to the greatest lower bound, that is $lub(A) \leq glb(B)$.

(6.9) If $r < s$ then, by Example 6.8, there exists a positive integer n_0 such that $n_0(s - r) > 1$. Since $n_0 r + 1 < n_0 s$ there is an $m \in \mathbf{Z}$ such that $n_0 r < m < n_0 s$. We have $r < \frac{m}{n_0} < s$.

(6.10) For (a) use Theorem 6.3(b). Let A_n be countable all n, let $f_n : A_n \longrightarrow \mathbf{N}$ be injective and let p_n denote the nth prime (increasing order). Define $f : \bigcup_{n=1}^{\infty} A_n \longrightarrow \mathbf{N}$ as follows: if $x_m \in A_n$ let $f(x_m) = p_n^{f_n(x_m)}$. Use Exercise 5.5, or the method in Example 6.4, to show that f is injective and apply Theorem 6.3(b).

(6.11) If $f : A \longrightarrow \mathbf{N}$ and $g : B \longrightarrow \mathbf{N}$ are injective functions (see Theorem 6.3), then $h : A \times B \longrightarrow \mathbf{N}, h(x, y) = 2^{f(x)} \cdot 3^{g(y)}$ is injective (by Exercise 5.9) and Theorem 6.3 shows that $A \times B$ is countable. If $A_i, i \leq n$, are countable with $f_i : A_i \longrightarrow \mathbf{N}$ injective and $(p_i)_{i=1}^n$ are n distinct primes, then $F : A_1 \times A_2 \times \cdots \times A_n \longrightarrow \mathbf{N}, F(x_1, \ldots, x_n) = p_1^{x_1} \cdots p_n^{x_n}$ is injective.

(6.12) Suppose $lub(A)$ is finite. Since $A_n \subset A$, Exercise 6.6 implies $x_n = lub(A_n) \leq lub(A)$ for all n. Hence, $lub(\{x_n\}_{n=1}^{\infty}) \leq lub(A)$. If $m < lub(A)$ then there is an $x \in A = \bigcup_n A_n$ such that $x > m$. Then $x \in A_{n_0}$ for some n_0 and $x_{n_0} = lub(A_{n_0}) \geq x > m$. An application of Example 6.10 completes the proof. If $lub(A)$ is not finite then for each $n \in \mathbf{N}$ we can find $x \in A$ such that $x > n$. If $x \in A_m$ then $x_m > n$ and $lub(\{x_n\})_{n=1}^{\infty})$ is not finite.

(6.13) By Exercise 6.9 we can choose, for all $x \in A$, rational numbers p_x and q_x such that $x - \delta_x < p_x < x < q_x < x + \delta_x$. Hence, $A \subset \bigcup_{x \in A}\{x\} \subset \bigcup_{x \in A}(p_x, q_x) \subset A$ and $A = \bigcup_{x \in A}(p_x, q_x)$. By Example 6.4, $\mathbf{Q}$ is countable and, by Exercise 6.11, $\mathbf{Q}^2$ is countable and hence, by Exercise 6.10(a), the set of all intervals with rational end points is countable.

Chapter 7

(7.2) No, consider $a_n = b_n = -1/n$. Note that by Theorem 7.3(c) the sequences could not be positive.

(7.3) The exponential function is increasing, hence $x < y \Longrightarrow -y < -x \Longrightarrow e^{-y} < e^{-x} \Longrightarrow -e^{-x} < -e^{-y} \Longrightarrow \exp(-e^{-x}) < \exp(-e^{-y})$ and f is increasing. See Exercise 4.21.

(7.4) (i) $\sqrt{n+1} - \sqrt{n} = \frac{(\sqrt{n+1}-\sqrt{n})\cdot(\sqrt{n+1}+\sqrt{n})}{\sqrt{n+1}+\sqrt{n}} = \frac{1}{\sqrt{n+1}+\sqrt{n}}$. Use Exercise 6.7 and Theorem 7.6 to show that the limit is 0 (see Example 7.7). (ii) $\frac{n^3}{3^n} \geq 0$, $\frac{(n+1)^3}{3^{n+1}} \leq \frac{n^3}{3^n} \Longleftrightarrow (1 + \frac{1}{n})^3 \leq 3$, true for $n \geq 4$. Next show $(\frac{3^n}{n^3})_{n=1}^{\infty}$ is not bounded above (see Exercise 6.2) and apply Theorem 7.6.

(7.5) First show, by induction, that $a_n < 2$ for all n. If $a_n < 2$, then $a_{n+1} = +\sqrt{2 + a_n} < \sqrt{2 + 2} = 2$. Next show $(a_n)_{n=1}^{\infty}$ is increasing: $a_{n+1} \geq a_n \Longleftrightarrow 2 + a_n \geq 2 + a_{n-1} \Longleftrightarrow a_n \geq a_{n-1} \Longleftrightarrow a_{n-1} \geq a_{n-2} \cdots \Longleftrightarrow a_2 = \sqrt{3} \geq 1$. By Definition 7.5 the sequence converges to a positive limit x. Then $x^2 = \lim_{n\to\infty} a_{n+1}^2 = \lim_{n\to\infty}(2 + a_n) = 2 + x$ and $x^2 - x - 2 = (x - 2)(x + 1) = 0$. Since $x \geq 0$, $\lim_{n\to\infty} a_n = 2$.

(7.6) $x < 1/2$.

(7.8) $x = e^4, x = 1/e$.

(7.9) $\{x : 0 < x < \frac{1+\sqrt{5}}{2}\}$, $\frac{x^2}{1+x-x^2}$.

(7.10) All positive real numbers, $x^8 e^{2x} - 1$.

(7.11) $1 < x < e$.

(7.12) $0 < e - \sum_{i=0}^{n} \frac{1}{i!} = \sum_{i=n+1}^{\infty} \frac{1}{i!} < \frac{1}{n!} \sum_{i=1}^{\infty} \frac{1}{(n+1)^i} = \frac{1}{n!} \frac{1/(n+1)}{1-(1/(n+1))} = \frac{1}{n!} \frac{1}{n}$.
If $e = p/q$ where p, q are positive integers, then $q!(e - \sum_{n=0}^{q} \frac{1}{n!}) = (q-1)!p - \sum_{n=0}^{q}((n+1) \cdot (n+2) \cdots q)$ is an integer. Since $0 < q!(e - \sum_{n=0}^{q} \frac{1}{n!}) < \frac{1}{q}$, this is impossible.

(7.13) Series sums to $\frac{(1/\sqrt{2})}{1-\frac{1}{\sqrt{2}}} = \frac{1}{\sqrt{2}-1}$. If $\frac{1}{\sqrt{2}-1} = \frac{p}{q}$ where p, q are positive integers, then $\sqrt{2} = \frac{p+q}{p}$ is rational and we know that this is not true.

(7.14) $\frac{n^4+1}{n^4+2} = 1 - \frac{1}{n^4+2}$ is increasing and bounded above by 1, hence it converges. The sequence $(\frac{1}{n^4+2})_{n=1}^{\infty}$ is decreasing and positive. Since $(n^4 + 2)_{n=1}^{\infty}$ is not bounded above (see Example 6.9 or Exercise 6.2). Theorem 7.6 implies $\lim_{n\to\infty} \frac{1}{n^4+2} = 0$ and $\lim_{n\to\infty} \frac{n^4+1}{n^4+2} = 1 = lub\left(\{\frac{n^4+1}{n^4+2}\}_{n=1}^{\infty}\right)$.

(7.15) $\sum_{n=0}^{\infty} \frac{x^{2n}}{n!e}$.

(7.16) Suppose $.x_1 x_2 \cdots$ and $.y_1 y_2 \cdots$ are two different decimal expansions of a. Then for some n_0 we have $x_{n_0} \neq y_{n_0}$ and $x_i = y_i$ for $i < n_0$. Suppose $x_{n_0} > y_{n_0}$ (which implies $x_{n_0} > 0$). We have $a \geq .x_1 x_2 \cdots x_{n_0} = \sum_{i=1}^{n_0} \frac{x_i}{10^i} = \sum_{i=1}^{n_0-1} \frac{x_i}{10^i} + \frac{x_{n_0}-1}{10^{n_0}} + \sum_{i=n_0+1}^{\infty} \frac{9}{10^i} \geq \sum_{i=0}^{\infty} \frac{y_i}{10^i} = a$. This implies that we have equality at all stages and hence if one of the $9's$ is replaced by $k, 0 \leq k \leq 8$, then we get an inequality, which is impossible. Hence $y_i = 9$ for $i > n_0$.

(7.17) -1.

(7.18) Let $A_n := \{\alpha \in \Gamma : a_\alpha \geq 1/n\}$. Then, $|A_n|/n \leq \sum_{\alpha \in A_n} a_\alpha \leq \sum_{\alpha \in \Gamma} a_\alpha =: M$, where $|A_n|$ denotes the number of elements in A_n. Hence, $|A_n| \leq nM < \infty$ and A_n is finite and $\{\alpha : a_\alpha > 0\} = \bigcup_{n=1}^{\infty} A_n$ is countable by Exercise 6.10.

(7.19) The set of all subsets of Ω with n elements, Ω_n, can be identified with $\Omega \times \Omega \times \cdots \times \Omega$ (n-times), and the set of all *finite subsets* of Ω with $\cup_n \Omega_n$. Now use Exercise 6.10. If $A \subset \Omega$ define $f_A : \Omega \longrightarrow \{0, 1\}, f(x) = 0 \iff x \notin A$ (the indicator function of A). The set of all such functions corresponds to the set of all subsets of Ω. Now use a diagonal process.

(7.20) See the solution to Exercise 6.12.

(7.22) It suffices to choose, using Exercise 6.9, for each positive integer n rational numbers p_n and q_n such that $r - \frac{1}{n} < p_n < r - \frac{1}{n+1} < r + \frac{1}{n+1} < q_n < r + \frac{1}{n}$.

(7.23) If $\alpha := lub(A) \in A$ let $x_n = \alpha$ all n. Otherwise, choose $x_1 \in A \cap (\alpha-1, \alpha)$, $x_2 \in A \cap (\alpha - \frac{1}{2}, \alpha) \cap (x_1, \alpha), \ldots, x_n \in A \cap (\alpha - \frac{1}{n}, \alpha) \cap (x_{n-1}, \alpha)$ and so on.

Chapter 8

(8.2) Suppose $\lim_{n \to \infty} x_n = 0$. If $y_n \le x_n \le z_n$ with $(y_n)_{n=1}^{\infty}$ increasing to 0 and $(z_n)_{n=1}^{\infty}$ decreasing to 0, then $0 \le |x_n| \le z_n$ and $\lim_{n \to \infty} |x_n| = 0$. Conversely, if $\lim_{n \to \infty} |x_n| = 0$, then there exists $(z_n)_{n=}^{\infty}$ decreasing to 0 such that $0 \le |x_n| \le z_n$ for all n and, as $-z_n \le x_n \le z_n$ and $(-z_n)_{n=1}^{\infty}$ is increasing to 0 we have $\lim_{n \to \infty} x_n = 0$.

(8.3) If n and m are positive integers, then $a_n \le a_{n+m} \le b_{n+m} \le b_m$ and $a_n \le b_m$ for all n and m. Hence $(a_n)_{n=1}^{\infty}$ is bounded above and $(b_n)_{n=1}^{\infty}$ is bounded below. By Exercise 6.8, $\lim_{n \to \infty} a_n := lub(\{a_n\}_{n=1}^{\infty}) \le glb(\{b_n\}_{n=1}^{\infty}) = \lim_{n \to \infty} b_n$. Now use $b_n = a_n + (b_n - a_n)$.

(8.5) If $(x_n)_{n=1}^{\infty}$ converges to x, then any subsequence of $(x_n)_{n=1}^{\infty}$ also converges to x, in particular any two subsequences contain subsequences which converge to the same limit. Conversely suppose any two subsequences of $(x_n)_{n=1}^{\infty}$ contain subsequences which converge to the same limit. Let x denote one such limit. If $(x_n)_{n=1}^{\infty}$ does not converge to x then, by Theorem 8.4, there is an open interval about x, (r, s), such that for no positive integer k do we have all $r < x_n < s$ for all $n \ge k$. Hence we can choose a strictly increasing sequence of positive integers $(n_j)_{j=1}^{\infty}$ such that $x_{n_j} \notin (r, s)$ for all j. A further application of Theorem 8.4 shows that no subsequence of $(x_{n_j})_{j=1}^{\infty}$ converges to x.

(8.6) $2e^{\sqrt{n}} > 2(1 + \sqrt{n} + (n/2)) > n$ and $\frac{\sqrt{n} + \log 2}{n} > \frac{\log n}{n} > 0$. Apply Definition 8.1 (see also Exercise 11.26).

(8.7) If $s_n = r + 2r^2 + \cdots + nr^n$, then $rs_n = r^2 + \cdots + nr^{n+1}$ and $(1 - r)s_n = r + r^2 + \cdots + r^n - nr^{n+1} = r(1 + r + \cdots + r^{n-1}) - nr^{n+1} = \frac{r(1-r^n)}{1-r} - nr^{n+1} \le \frac{r}{1-r}$. Hence $s_n \le \frac{r}{(1-r)^2}$ and the series is bounded above and converges. By Theorem 8.8, $nr^n \longrightarrow 0$ as $n \longrightarrow \infty$ and the limit s satisfies $(1-r)s = r(\sum_{n=0}^{\infty} r^n) = r/(1 - r)$. Hence, $s = \sum_{j=1}^{\infty} jr^j = \frac{r}{(1-r)^2}$. Alternatively $\frac{d}{dr}(1 + r + \cdots + r^n) = \frac{d}{dr}[(r^{n+1} - 1)/(r - 1)]$ and $1 + 2r + \cdots + nr^{n-1} = \frac{nr^{n+1} - (n+1)r^n + 1}{(1-r)^2}$ which implies $s_n = \frac{nr^{n+2} - (n+1)r^{n+1} + r}{(1-r)^2}$. We are well aware that differentiation has not yet been discussed in the text, in mathematics you have to be prepared for the unexpected.

(8.8) $(p_n)_{n=1}^{\infty}$ is increasing (by definition) and, as $p_n > n$, the sequence is not bounded (see Example 6.9 and Exercise 6.8), $(1/p_n)_{n=1}^{\infty}$ is decreasing and converges to 0. An application of Theorem 8.10 completes the proof.

(8.9) $p > 1$. This is known as the condensation test. See Example 14.12.

(8.10) $\left|\frac{a_{n+1}}{a_n}\right| = \left|\left(\frac{\log(n+1)}{\log n}\right)^a \left(\frac{n+1}{n}\right)^b 2^c \left(\frac{(n+1)!}{n!}\right)^d\right| = \left|\left(1 + \frac{\log(1+\frac{1}{n})}{\log n}\right)^a (1 + \frac{1}{n})^b 2^c (n+1)^d\right|$. By the ratio test the series converges if (a) $d < 0$, (b) $d = 0, c < 0$, and diverges if $d > 0$ or $d = 0, c > 0$. Using Exercise 8.9, the series converges (c) if $d = 0, c = 0, b < -1$, (d) $d = 0, c = 0, b = -1, a < -1$ and diverges if $d = 0, c = 0, b \geq -1$ or $d = 0, c = 0, b = -1, a \geq -1$. One can also use the integral test (Chapter 14) in place of Exercise 8.9. See also Exercise 14.6.

(8.11) Working backwards one arrives at the following solution. By Definition 8.1, it suffices to consider an increasing sequence, $(a_n)_{n=1}^{\infty}$, and we may also suppose that the sequence is positive. The sequence $(b_n)_{n=1}^{\infty}$ is increasing, since $b_n \leq b_{n+1} \iff (n+1)\sum_{i=1}^{n} a_i \leq n\sum_{i=1}^{n+1} a_i \iff \sum_{i=1}^{n} a_i \leq na_{n+1}$ and this is true. Moreover, $b_n \leq a_n$ for all n. Hence $(b_n)_{n=1}^{\infty}$ converges and $a' := \lim_{n\to\infty} b_n \leq \lim_{n\to\infty} a_n =: a$. Let $a' < \alpha < \beta < a$ be arbitrary. Then there exists a positive integer n_0 such that $a_n > \beta$ for all $n \geq n_0$ and hence $\frac{1}{n-n_0}\sum_{i=n_0+1}^{n} a_i > \beta$. Now $\frac{n-n_0}{n} = 1 - \frac{n_0}{n} \longrightarrow 1$ as $n \longrightarrow \infty$ and there exists $n_1 > n_0$ such that $\frac{n-n_0}{n} > \frac{\alpha}{\beta}$ for all $n > n_1$. Hence $\frac{1}{n}\sum_{i=1}^{n} a_i \geq \frac{1}{n}\sum_{i=n_0+1}^{n} a_i = \frac{n-n_0}{n} \cdot \frac{1}{n-n_0}\sum_{i=n_0+1}^{n} a_i > \frac{\alpha}{\beta} \cdot \beta = \alpha$ for all $n > n_1$. Now apply Example 6.10 and Definition 7.5.

(8.12) For $0 < x < 1$, $e^x = \sum_{n=0}^{\infty} x^n/n! < \sum_{n=0}^{\infty} x^n = 1/(1-x)$. Hence $x < \log(\frac{1}{1-x})$ (see Lemma 9.4). $a_{n+1} - a_n = \frac{1}{n+1} - \log(n+1) + \log(n) = \frac{1}{n+1} - \log(\frac{n+1}{n}) = \frac{1}{n+1} - \log(\frac{1}{1-\frac{1}{n+1}}) < 0$ and $(a_n)_{n=1}^{\infty}$ is strictly decreasing. Since $e^{1/j} > (j+1)/j$, $\exp(a_n) = \frac{e^1}{2} \cdot \frac{e^{1/2}}{3/2} \cdots \frac{e^{1/(n-1)}}{n/(n-1)} \cdot e^{1/n} > \frac{e}{2}$ and, as $a_1 = 1$, $(a_n)_{n=1}^{\infty}$ converges to γ where $(1 - \log 2) \leq \gamma < 1$. Moreover, $b_{2n} = \sum_{j=1}^{2n} \frac{(-1)^{j+1}}{j} = \sum_{j=1}^{2n} \frac{1}{j} - 2(\sum_{j=1}^{n} \frac{1}{2j}) = (a_{2n} + \log 2n) - (a_n + \log n) = a_{2n} - a_n + \log 2 \longrightarrow \log 2$ as $n \longrightarrow \infty$. See also Examples 11.14, 11.15 and 14.16.

(8.15) If $n = 2^m, n + p_n = 2^m + 2$ and $(-1)^{n+p_n} = 1$. If $n = 1, n + p_n = 1$. If $n \neq 2^m, n > 1$, then p_n is odd and $(-1)^{n+p_n} = (-1)^{n+1}$. By Exercise 8.12, $\sum_{n=1}^{\infty} \frac{(-1)^{n+p_n}}{n} = -1 + \sum_{n=2}^{\infty} \frac{(-1)^{n+1}}{n} + 2\sum_{m=1}^{\infty} \frac{1}{2^m} = -1 + ((\log 2) - 1) + \frac{2(1/2)}{1-\frac{1}{2}} = \log 2$.

(8.16) The first 3 converge by the Ratio Test. Since $n^n \geq n!$ the final sequence diverges by Theorem 8.8.

(8.17) $\sqrt{2}, +\infty$.

(8.18) $\{x : x \leq -\log 2\}$.

(8.19) $63/256$.

(8.20) $(-1, 1), [-4^{1/3}, 4^{1/3}], [-1, 1], (-1, 1), [-1, 1], [-1, 1]$.

(8.24) Only the first series converges.

(8.25) $1, \lambda, \lambda^2 + \lambda$.

(8.26) Let B denote the set of all $x \in [a, b]$ such that $[a, x]$ can be covered by a finite number of intervals from $\mathcal{A}$. B is non-empty, since $a \in B$, and b is an upper bound for B. Let $c = lub(B)$ and let $(x_n)_{n=1}^{\infty}$ denote a sequence in B which converges to c. Then $c \in (u, v)$ for some $(u, v) \in \mathcal{A}$. By Theorem 8.4, there exists a positive integer n such that $u < x_n < v$. Since $[a, x_n]$ can be covered by a finite number of elements in $\mathcal{A}$, $[a, c] = [a, x_n] \cup [x_n, c]$ can also be covered by a finite number of elements in $\mathcal{A}$ and $c \in B$. If $c < b$ we can choose $c_1, c < c_1 < v$,. Then $c_1 \in B$, this contradicts the definition of c. Hence $c = b$ and $B = [a, b]$.

Chapter 9

(9.2) Let $\delta f(c) := \lim_{x \to c, x > c} f(x) - \lim_{x \to c, x < c} f(x)$ for all $c \in (a, b)$. Then f is not continuous at $c \iff \delta f(c) > 0$. Let $A_n = \{c \in (a, b) : \delta f(c) \geq \frac{1}{n}\}$ and let $A = \cup_n A_n$. A is the set of all points where f is non-continuous. If $x_1 < x_2 < \cdots < x_m \in A_n$, then $f(b) - f(a) \geq \sum_{j=1}^{m} \delta f(x_j)$ and, if $|A_n|$ is the number of elements in A_n, then $|A_n| \leq n(f(b) - f(a))$. Hence each A_n is finite and, by Exercise 6.10, A is countable.

(9.3) By Theorem 9.10, $im(f) = [m, M]$. If $m = M$, that is if f is a constant function, then $im(f)$ has one element. If $m < M$, then $im(f)$ is uncountable by Theorem 6.6.

(9.4) $n^{1/n} = \exp(\log(n^{1/n})) = \exp(\frac{\log n}{n})$.

(9.5) $P(-2) > 0, P(-1) < 0, P(0) > 0$.

(9.6) Let $P(x) = x(2x^3 - 9x^2 + 7x + 1) =: xQ(x)$. Then $P(0) = 0$, $Q(0) \neq 0$, $Q(1) = 1$ and $Q(2) = -5 < 0$. By the Intermediate Value Theorem Q, and hence P, has a zero in $(1, 2)$. Now $Q(x) = x^3(2 \pm$ something small$)$ for x large. Choose $a > 2$ such that $Q(a) > 0$, then Q and P has a zero in $(2, a)$ by the Intermediate Value Theorem. Choose $b < 0$ such that

$Q(b) < 0$, then Q and P has a zero in $(b, 0)$ by the Intermediate Value Theorem. Since P has degree 4 it has no more than 4 zeros.

(9.7) Proof by contradiction. Suppose there exist $x_1 < x_2, f(x_1) < f(x_2)$, and $y_1 < y_2, f(y_1) > f(y_2)$. Now consider all possible combinations (diagrams may help). For example, suppose $y_2 < x_2$ and $f(y_2) \leq f(x_2)$. If $f(y_2) \leq z \leq \min\{f(y_1), f(x_2)\}$, then by the Intermediate Value Theorem there exists $w_1 \in (y_1, y_2)$ and $w_2 \in (y_2, x_2)$ such that $f(w_1) = f(w_2) = z$, a contradiction.

(9.8) Apply the Intermediate Value Theorem to $f - g$.

(9.9) If x is rational, then $x_n := x + \frac{\sqrt{2}}{n}$ is irrational, $\lim_{n \to \infty} x_n = x$, and $0 = \lim_{n \to \infty} f(x_n) \neq 1 = f(x)$, and f is not continuous at any rational point. If y is irrational with decimal expansion $\sum_{n=1}^{\infty} \frac{a_n}{10^n}$ let $y_k = \sum_{n=1}^{k} \frac{a_n}{10^n}$ for all $k \in \mathbf{N}$. Then $y_k \in \mathbf{Q}, y_k \longrightarrow y$ as $k \longrightarrow \infty$, but $f(y_k) = 1$ and $f(y) = 0$ imply $1 = \lim_{k \to \infty} f(y_k) \neq 0 = f(y)$.

(9.10) If not, then for all sufficiently large n, we can find $x_n, c - \frac{1}{n} < x_n < c + \frac{1}{n}$, $f(x_n) = 0$. Then, $\lim_{n \to \infty} x_n = c$, $\lim_{n \to \infty} f(x_n) = 0 \neq f(c)$. A contradiction.

(9.11) We have $0 < f(x) \leq 1$, $f(0) = 1$ and $f(x) \longrightarrow 0$ as $x \longrightarrow \pm\infty$. By the Intermediate Value Theorem, $im(f) = (0, 1]$. Using Definition 3.7 we see that $f \circ f, f \circ g$ and $g \circ g$ are defined and $g \circ f$ is not defined. Since $g(x) = 1 - x^4(x^4 + x^2 + 1)^{-1} = 1 - (1 + x^{-2} + x^{-4})^{-1}$, g is decreasing and $im(g) = [g(1), g(0)] = [2/3, 1]$.

(9.12) 0.

(9.13) Fix $x \in [a, b]$ and let $x_n \in [a, b] \longrightarrow x$ as $n \longrightarrow \infty$. For each m choose an increasing sequence $(a_{n,m})_{n=1}^{\infty}$ and a decreasing sequence $(b_{n,m})_{n=1}^{\infty}$ such that $a_{n,m} \leq f_m(x_n) \leq b_{n,m}$ and $\lim_{n \to \infty} a_{n,m} = \lim_{n \to \infty} b_{n,m} = f_m(x)$ for all m. By Theorem 8.4, we may suppose $|a_{n,m}| < M_m$ and $|b_{n,m}| < M_m$ for all n and m. For each n, the series $A_n := \sum_{m=1}^{\infty} a_{n,m}$ and $B_n := \sum_{m=1}^{\infty} b_{n,m}$ converge. Moreover, $(A_n)_{n=1}^{\infty}$ is increasing, $(B_n)_{n=1}^{\infty}$ is decreasing, $A_n \leq f(x_n) \leq B_n$, and, for every positive integer k, $\sum_{m=1}^{k} f_m(x) = \lim_{n \to \infty} \sum_{m=1}^{k} a_{n,m} \leq \lim_{n \to \infty} A_n$. Hence $\sum_{m=1}^{\infty} f_m(x) = f(x) \leq \lim_{n \to \infty} A_n$. Similarly $\sum_{m=1}^{\infty} f_m(x) = f(x) \geq \lim_{n \to \infty} B_n$ and, as $A_n \leq f(x_n) \leq B_n$, $\lim_{n \to \infty} A_n = \lim_{n \to \infty} B_n = \lim_{n \to \infty} f(x_n) = f(x)$. If $g_n(x) = \frac{1}{n^2 + x^2}$ for all n, then g_n is continuous. On any interval $[-a, a]$, $|g_n(x)| < \frac{2}{n^2}$ and $\sum_{n=1}^{\infty} \frac{2}{n^2} < \infty$. Since $g = \sum_{n=1}^{\infty} g_n$, the first part gives the required result.

(9.14) Since $A \subset [a,b]$, $\alpha := lub(A) \in [a,b]$. If $x \in A$, then $x \leq \alpha$, and, as f is increasing, $f(x) \leq f(\alpha)$ and $f(\alpha)$ is an upper bound for $f(A)$. Hence $lub(f(A)) \leq f(\alpha) = f(lub(A))$. If $\alpha \in A$, then $f(\alpha) \in f(A)$ and $lub(f(A)) \geq f(\alpha) = f(lub(A)$. If $\alpha \notin A$, then there exists, by Exercise 7.23, an increasing sequence $(x_n)_{n=1}^{\infty} \subset A$ such that $\lim_{n \to \infty} x_n = \alpha$. Then $(f(x_n))_{n=1}^{\infty}$ is an increasing in $f(A)$ and $\lim_{n \to \infty} f(x_n) = f(\alpha) = f(lub(A))$. By Exercise 6.6, $lub(f(A)) \geq lub((f(x_n))_{n=1}^{\infty}) = f(lub(A))$.

(9.15) If $(x_n)_{n=1}^{\infty} \subset A$ converges to $x \neq 0$ in A, then $x_n = x$ for all sufficiently large n and $\lim_{n \to \infty} f(x_n) = f(x)$. Hence, continuity at the origin is the only question to be considered. If f is continuous then, since $\frac{1}{n} \to 0$ as $n \to \infty$, $\lim_{n \to \infty} f(\frac{1}{n}) = f(0)$. Conversely, suppose $\lim_{n \to \infty} f(\frac{1}{n}) = f(0)$. If I denotes an arbitrary open interval which contains $f(0)$ then, by Theorem 8.4, there is a positive integer n_1 such that $f(\frac{1}{n}) \in I$ all $n \geq n_1$. If $(x_n)_{n=1}^{\infty} \subset A$ and $\lim_{n \to \infty} x_n = 0$ then, by Theorem 8.4, we can choose a positive integer n_2 such that $0 \leq x_n < \frac{1}{n_1}$ for all $n \geq n_2$. If $n \geq n_2$, then $x_n = \frac{1}{k}$ for some $k \geq n_1$ and $f(x_n) \in I$. By Theorem 8.4, $\lim_{n \to \infty} f(x_n) = f(0)$. If $x = t\frac{1}{n+1} + (1-t)\frac{1}{n}, 0 \leq t \leq 1$, let $g(x) = tf(\frac{1}{n+1}) + (1-t)f(\frac{1}{n})$, if $x \leq 0$ let $g(x) = f(0)$, and if $x \geq 1$ let $g(x) = f(1)$.

(9.16) If $a < f(x) < b$ and $g(x) = f(x) - a$, then $g(x) > 0$ and there exists, by Exercise 9.10, $\alpha_x > 0$ such that $g(y) > 0$, that is $f(y) > a$, for all $y \in (x - \alpha_x, x + \alpha_x)$. Similarly for some $\beta_x > 0$, $f(y) < b$ for all $y \in (x - \beta_x, x + \beta_x)$. If $\delta_x = \min(\alpha_x, \beta_x)$, then $a < f(y) < b$ for all $y \in (x - \delta_x, x + \delta_x)$. Hence $A := \{x \in \mathbf{R} : a < f(x) < b\} = \bigcup_{x \in A}(x - \delta_x, x + \delta_x)$ and an application of Exercise 6.13 completes the proof in one direction. Conversely, suppose $\{x \in \mathbf{R} : a < f(x) < b\}$ is a countable union of open intervals for all $a, b \in \mathbf{R}$. Let $\lim_{n \to \infty} x_n = c$ and let $j \in \mathbf{N}$ be arbitrary. Then, $c \in A := \{x \in \mathbf{R} : f(c) - \frac{1}{j} < f(x) < f(c) + \frac{1}{j}\}$ and, by Theorem 8.4, there exists a positive integer n_j such that $x_n \in A$ for all $n \geq n_j$, that is $f(c) - \frac{1}{j} < f(x_n) < f(c) + \frac{1}{j}$ for all $n \geq n_j$. By Theorem 8.4, $\lim_{n \to \infty} f(x_n) = f(c)$ and f is continuous.

(9.17) Suppose (a) holds. Let $\lim_{n \to \infty} y_n = 0, y_n \neq 0$ and let x be any point in (a, b). By Theorem 9.10 there exists for each n, $z_n, |z_n| \leq |y_n|$, such that $\max\{|f(x + y) - f(x)| : |y| \leq |y_n|\} = |f(x + z_n) - f(x)|$. Since $|y_n| \longrightarrow 0$ as $n \longrightarrow \infty$ and $|z_n| \leq |y_n|$, Theorem 8.5 implies $|z_n| \to 0$ as $n \to \infty$ and, by Definition 9.1, (b) holds. Clearly (b) implies (c). Now suppose (c) holds for $(y_n)_{n=1}^{\infty}$. Let $x \in (a, b)$ and suppose $x_n \longrightarrow x$ as

$n \longrightarrow \infty$. Then, $x_n = x + w_n$ where $w_n \longrightarrow 0$ as $n \longrightarrow \infty$. For each positive integer k choose a positive integer n_k such that $|w_n| \leq |y_k|$ for all $n \geq n_k$. Then, $|f(x+w_n)-f(x)| \leq \max\{|f(x+y)-f(x)| : |y| \leq |y_k|\}$ for all $n \geq n_k$ and, without loss of generality, $(n_k)_{k=1}^{\infty}$ is strictly increasing. By (c), $\lim_{n\to\infty} |f(x + w_n) - f(x)| \leq \lim_{k\to\infty} \max\{|f(x + y) - f(x)| : |y| \leq |y_k|\} = 0$ and, by Definition 9.1, f is continuous.

(9.18) Suppose x is irrational and $\lim_{n\to\infty} x_n = x$. Then, $f(x) = 0$. If $0 \in (r, s)$, there are only a finite number of points, A, of the form $\frac{p}{q}$ in $(0,1)$ with p and q having no common factor and $\frac{1}{q} \geq s$. Hence we can choose $n_s \in \mathbf{N}$ such that $x_n \notin A$ for all $n \geq n_s$, otherwise we would have $x_n = a$ for some $a \in A$ and an infinite number of n and this would imply that $(x_n)_{n=1}^{\infty}$ has a subsequence which converge to a and hence $x = a$, a contradiction, since x is irrational and all points in A are rational. Hence $r < f(x_n) < s$ for all n sufficiently large and, by Theorem 8.4, $\lim_{n\to\infty} f(x_n) = 0 = f(x)$. If x is rational, then $x + \frac{\sqrt{2}}{n} \longrightarrow x$ as $n \longrightarrow \infty$ but $f(x + \frac{\sqrt{2}}{n}) = 0 \nrightarrow f(x) \neq 0$. Hence, f is not continuous at rational points.

Chapter 10

(10.2) Fix $x > 0$. Let $f(y) = x^y$ for $y \in \mathbf{R}$. Then, $f(y) = \exp(y \log x)$ and $f'(y) = \exp(y \log x) \cdot \log x$. Hence, $f'(0) = \log x = \lim_{n\to\infty} \frac{f(\frac{1}{n})-f(0)}{\frac{1}{n}} = \lim_{n\to\infty} n(x^{1/n} - 1)$.

(10.3) $f'(2/\sqrt{3}) = 3$, $g'(\sqrt{\frac{-3}{2} + \sqrt{15.45}}) = 106$.

(10.4) First sketch $(x^2 - 1)^2$. See Chapter 2.

(10.5) $x = 0$, $\{x \in \mathbf{R} : x > \log \sqrt{3}\}$, 0 and $-\infty$.

(10.6) $\mathbf{R}$, $f'(x) = -2x(1 + x^2)^{-1}$, $\{x > 1\}$, $g'(x) = (\log x)^{\log x} \cdot \frac{1 + \log(\log x)}{x}$.

(10.7) (a) $f^{-1}(x) = ((\log x) - 1)^{1/2}$, $(f^{-1})'(x) = (2x((\log x) - 1)^{1/2})^{-1}$. (b) $f'(x) = 2x \exp(x^2 + 1)$, $f'(f^{-1}(x)) = 2f^{-1}(x) \exp((f^{-1}(x))^2 + 1) = 2f^{-1}(x) \exp(((\log x) - 1) + 1) = 2f^{-1}(x) \exp(\log x) = 2x f^{-1}(x) = 2x(\log(x)-1)^{1/2}$ and $(f^{-1})'(x) = \frac{1}{f'(f^{-1}(x))} = (2x((\log x)-1)^{1/2})^{-1}$. (c) $f^{-1} = f_1 \circ f_2 \circ f_3$ where $f_3(x) = \log x$, $f_3'(x) = 1/x$, $f_2(x) = x - 1$, $f_2'(x) = 1$ and $f_1(x) = \sqrt{x}$, $f_1'(x) = 1/2\sqrt{x}$ and $f_2 \circ f_3(x) = f_2(\log x) = (\log x) - 1$. Hence $(f^{-1})'(x) = f_1'(f_2 \circ f_3(x)) \cdot f_2'(f_3(x)) \cdot f_3'(x) = \frac{1}{2\sqrt{(\log x) - 1}} \cdot 1 \cdot \frac{1}{x} = \frac{1}{2x\sqrt{(\log x) - 1}}$.

(10.8) See the proof of Corollary 10.9.

(10.9) $f'(x) = 8(4 - 3x^2)(3x^2 + 4)^{-2}$, f increasing only in $[-2/\sqrt{3}, 2/\sqrt{3}]$, maximum $2/\sqrt{3}$, minimum $-2/\sqrt{3}$.

(10.10) $f'(0) > 0, f'(1) < 0, f'(2) > 0$.

(10.11) f is strictly increasing and hence injective. The Mean Value Theorem applied to f over $[0, x]$ and $[-x, 0]$ shows that $f(x) \geq f(0) + cx \longrightarrow +\infty$ as $x \longrightarrow +\infty$ and that $f(x) \leq f(0) + cx \longrightarrow -\infty$ as $x \longrightarrow -\infty$ and f is surjective.

(10.12) Clearly, f is continuous and, as the composition of continuous functions $x \longrightarrow f(2^n x)$ is continuous. By Exercise 9.13, g is continuous. Fix $x \in \mathbf{R}$. For each positive integer n there exists a unique integer j such that $u_n := j2^{-n} \leq x < (j+1)2^{-n} =: v_n$. It suffices to show that $\lim_{n \to \infty} \frac{g(v_n) - g(u_n)}{v_n - u_n}$ does not exist. Then $v_n - u_n = 2^{-n}$, $\frac{g(v_n) - g(u_n)}{v_n - u_n} = \sum_{m=0}^{n-1} 2^{n-m}[f(2^{m-n}(j+1)) - f(2^{m-n}j)] = \sum_{m=0}^{n-1} \pm 1$ since $f(2^{m-n}k) = 0$ for $m \geq n$ and $[f(\frac{j}{2^{n-m}} + \frac{1}{2^{n-m}})) - f(\frac{j}{2^{n-m}})] = \frac{1}{2^{n-m}}$, if $\frac{j+1}{2^{n-m}}$ lies in an interval $(s, s + \frac{1}{2}]$ and equals $\frac{-1}{2^{n-m}}$ if $\frac{j+1}{2^{n-m}}$ lies in an interval $(s + \frac{1}{2}, s]$ for some integer s.

(10.13) If $f'(x) \geq 0$ all $x \in (a, b)$ use the Mean Value Theorem, as in the proof of Corollary 10.7(a). If $f(x) \geq f(y)$ for all $x \geq y$, then (10.3) implies $f'(x) \geq 0$ for all $x \in (a, b)$.

(10.14) $f'(x) = e^x - x = 1 + \sum_{n=2}^{\infty} \frac{x^n}{n!} \geq 1$ for $x \geq 0$, if $x < 0$, $e^x > 0$ and $-x > 0$ imply $f'(x) > 0$. Hence, f is strictly increasing and injective. If $x > 0$, then $f(x) > \frac{x^3}{6} \longrightarrow +\infty$ as $x \longrightarrow +\infty$. Since $e^x \longrightarrow 0$ as $x \longrightarrow -\infty$, $f(x) \approx -x^2/2 \longrightarrow -\infty$ as $x \longrightarrow -\infty$. By the Intermediate Value Theorem f is surjective and hence bijective.

(10.15) See Exercises 1.3 and 4.24 and Example 10.10(d). Since $f(x) \longrightarrow \pm\infty$ as $x \longrightarrow \pm\infty$, the Intermediate Value Theorem implies that f is always surjective, and hence f is bijective if and only if it is injective. Since $f(x) = f(y) \Longleftrightarrow x^3 + px = y^3 + py$ we may suppose $q = 0$. If $p > 0$, then $f'(x) = 3x^2 + p > 0$ and f is strictly increasing and injective. The cases $p = 0$ and $p < 0$ are handled as in Exercise 4.24.

(10.16) $\mathbf{R}^+$, $\mathbf{R} \backslash \{0\}$, $(1, \infty)$. $\frac{2 \log x}{x}$, $\frac{2}{x}$, $\frac{1}{x \log x}$.

(10.17) Apply the Mean Value Theorem to the exponential function on $[a, b]$.

(10.18) If $e^x \geq 1 + mx$, then $\lim_{x > 0 \to 0} \frac{e^x - e^0}{x - 0} = e^0 = 1 \geq m$ and $\lim_{x < 0 \to 0} \frac{e^x - e^0}{x - 0} = e^0 = 1 \leq m$. Hence, 1 is the only possible value for m. Using power series expansions, we have $e^x \geq 1 + x$ for $x \geq 0$,

$e^x = 1 + x + \sum_{n=1}^{\infty} \frac{x^{2n}}{2n!}\left(1 - \frac{x}{2n+1}\right) \geq 1 + x$ for $-1 \leq x \leq 0$, and $e^x > 0 > 1 + x$ for $x < -1$.

Chapter 11

(11.1) $\frac{n(n+1)(2n+1)}{6}$ (this was known to the Babylonians and Archimedes), $\left(\frac{n(n+1)}{2}\right)^2$.

(11.2) $f'(x) = 27x^8 + 3(x-2)^2 \geq 0$. $f'(x) = 0 \iff x^8 = 0$ and $(x-2)^2 = 0$ and $f'(x) \neq 0$ all x. Hence, f is strictly increasing and injective. For x large $f(x) \approx 3x^9 \longrightarrow \pm\infty$ as $x \longrightarrow \pm\infty$. Hence, $f : \mathbf{R} \longrightarrow \mathbf{R}$ is bijective (see Corollary 9.7). Now $f(0) = -8$ and $f'(0) = 12$ hence $f^{-1}(-8) = 0$. By Theorem 10.8, $(f^{-1})'(-8) = 1/12$.

(11.3) Let $f(x) = x^2 e^{-x}$ and $g(x) = x^4 e^{-x^2}$. f in increasing on $[0,2]$, decreasing on $(-\infty, 0]$ and $[2, \infty)$, convex on $(-\infty, 2 - \sqrt{2})$ and $(2 + \sqrt{2}, +\infty)$. On $[0, 4]$ maximum $4e^{-2}$, minimum 0. g in increasing on $(-\infty, -\sqrt{2}]$ and $[0, \sqrt{2}]$, decreasing on $[-\sqrt{2}, 0]$ and $[\sqrt{2}, \infty)$. $g''(x) = 2x^2(6 - 9x^2 + 2x^4)e^{-x^2}$, g is convex on $(-\infty, \frac{-9-\sqrt{33}}{4})$, $(\frac{-9+\sqrt{33}}{4}, \frac{9-\sqrt{33}}{4})$ and $(\frac{9+\sqrt{33}}{4}, \infty)$. On $[0, 4]$, g has maximum $4e^{-2}$ and minimum 0.

(11.4) U and g convex imply $(g \circ U)''(x) = g''(U(x)) \cdot (U'(x))^2 + g'(U(x)) \cdot U''(x) \geq 0$ since $g'(x) \geq 0$ if g is increasing. If $g(x) = U(x) = e^{-x}$ for all x, then g and U are both convex and, as $(g \circ U)''(x) = (e^{-x} - 1)e^{-(x + e^{-x})}$, $g \circ U$ is not convex on $\mathbf{R}^+$.

(11.6) $f(x) = \frac{x^2}{x^2+1}$, $f'(x) = \frac{2x}{(x^2+1)^2} > 0$ for $x > 0$.

(11.7) 1678, 2.95 days.

(11.8) \$1041 if no interest was charged. \$1719.

(11.9) (a) 1/2 (b) 1/32 (c) e.

(11.10) f is even, maximum 1 over both intervals, occurs at ± 1, Since f is never negative and $f(0) = 0$ a minimum occurs at 0. Note that $f(\frac{1}{x}) = f(x)$ when $x \neq 0$.

(11.11) $(f(2) - f(1))/(2 - 1) = 13$, $f'(2) = 27$, $f'(1) = 4$. Since f' is continuous, the Intermediate Value Theorem implies it takes the value 13 between 1 and 2.

(11.12) $g(x) = \exp(c - e^{-x} - xe^{-x})$.

(11.13) $(1 + 2x^2)e^{x^2+1}$, $\frac{1+2x^2}{x}$, $1 + 2x^2$.

(11.14) $\sum_{n=0}^{\infty} r^n = (1-r)^{-1}$. On differentiating twice, $r \cdot \sum_{n=1}^{\infty} nr^{n-1} = \sum_{n=1}^{\infty} nr^n = r(1-r)^{-2}$ and $r \cdot \sum_{n=1}^{\infty} n^2 r^{n-1} = \sum_{n=1}^{\infty} n^2 r^n = \frac{r+r^2}{(1-r)^3}$, $0 \le r < 1$, $\sum_{n=1}^{\infty} \frac{n^2}{2^n} = 6$, $r \cdot \sum_{n=1}^{\infty} n^3 r^{n-1} = \sum_{n=1}^{\infty} n^3 r^n = \frac{r+4r^2+r^3}{(1-r)^4}$. Let $r = x^3/4$, $|x| < 4^{1/3}$, $\sum_{n=1}^{\infty} \frac{n^3 x^{3n}}{4^n} = \frac{64x^3+64x^6+4x^9}{(4-x^3)^4}$.

(11.15) $[-2, +2]$, $g''(x) = (1 - e^{-2x})^{-3} e^{-3x} ((e^{2x} - 1)^2 + 8)$, g is convex $\Longleftrightarrow$ $1 - e^{-2x} > 0 \Longleftrightarrow x > 0$.

(11.17) Level of Production $(c - d)/2d$. Absorb half the tax.

(11.18) Price of gasoline used, $P(x) = 3 \cdot 120 \cdot \frac{1}{100} \left(\frac{3200}{x} + x \right), P'(x) = \frac{36}{10}(-3200x^{-2} + 1) = 0$ if $x^2 = 3200$, $x = 56.6$mph. Driver included $C(x) = \frac{36}{10}\left(\frac{3200}{x} + x \right) + 15\left(\frac{120}{x} \right)$ and $C'(x) = \frac{36}{10}(-3200x^{-2} + 1) - 15(120)x^{-2} = 0$ if $x^2 = 3700$, that is $x \approx 61$mph (to see that these are minima see Figure 2.13).

(11.19) $f'(x) = \frac{a}{ax} - \frac{1}{x} = 0$. $f(1) = \log a$. Hence $f(x) = \log a$ all x and $f(b) = \log(ab) - \log b = \log a$.

(11.20) $f''(x) = 60((x^2 - 1)^2 + (x - 1)^2 + 1) > 0$.

(11.21) $x = 16$ and $y = 40$.

(11.22) We may suppose, without loss of generality, that $(x_n)_{n=1}^{\infty}$ is strictly decreasing to 0. By Example 9.14, f and g are continuous and $a_0 = f(0) = g(0) = b_0$. By Rolle's Theorem or the Mean Value Theorem applied to $f - g$ we see that for each n there exists y_n, $x_{n+1} < y_n < x_n$ such that $f'(y_n) = g'(y_n)$. By Example 11.14 this implies that f' and g' satisfy the original hypotheses and hence $a_1 = f'(0) = g'(0) = b_1$. Continuing in this way implies $a_n = b_n$ all n. It is very worthwhile to write out a complete proof of this result, use either proof by contradiction or proof by induction.

(11.23) A charge of $100 + 10x$ yields $\Pi(x) = (100 + 10x)(24 - x) - 20x = 2400 + 240x - 100x - 10x^2 - 20x = 2400 + 120x - 10x^2$. This is a quadratic with negative coefficient of x^2 so it has a unique maximum at its critical point. Since $\Pi'(x) = 120 - 20x = 0$ when $x = 6$ the price charged should be \$160.

(11.24) $f(0) = 0$, $f'(x) = -x/(1 + x) < 0$. Hence f is strictly decreasing and $\log(1 + x) < x$ for $x > 0$. $\sum_{k=1}^{n} \log(1 + \frac{1}{k}) = \sum_{k=1}^{n}(\log(1 + k) - \log k) = \log(1 + n) < \sum_{k=1}^{n} \frac{1}{k}$. Now use $\lim_{n \to \infty} \log n = \infty$. (See also Examples 8.11 and 14.12(c) and Exercise 8.12.)

(11.25) If $y = f(x)$, then $x = f^{-1}(y)$ and $(f^{-1})''(y) = -f''(x) \cdot (f'(x))^{-3}$.

(11.26) Simple modifications to Theorem 11.2 prove this version of L'Hôpital's Rule. Let $f_1(x) = f(\frac{1}{x})$ and $g_1(x) = g(\frac{1}{x})$ for $x > 0$ and let $f_1(0) = g_1(0) = 0$. Then f_1 and g_1 are continuous on $\mathbf{R}_0^+$ and differentiable on $\mathbf{R}^+$. By the first part $\lim_{x \to +\infty} \frac{f(x)}{g(x)} = \lim_{y>0, y \to 0} \frac{f_1(y)}{g_1(y)} = \lim_{y>0, y \to 0} \frac{f_1'(y)}{g_1'(y)} = \lim_{y>0, y \to 0} \frac{-\frac{1}{y^2} f'(y)}{-\frac{1}{y^2} g'(y)} = \lim_{y>0, y \to 0} \frac{f'(y)}{g'(y)} = \lim_{x \to +\infty} \frac{f'(x)}{g'(x)}$. 0 and 1 (see Exercise 8.6 and 9.4).

(11.27) $e^x(1 + sx) \leq 1$ all $x \in \mathbf{R} \iff e^{-x}(1 - sx) \leq 1$ all $x \in \mathbf{R} \iff (1 - sx) \leq \frac{1}{e^{-x}} = e^x$ all $x \in \mathbf{R} \iff e^x \geq 1 + mx$ where $m = -s \iff m = 1 = -s \iff s = -1$ (by Exercise 10.18).

(11.28) Use $\left(\frac{-q}{2} \pm \sqrt{\frac{p^3}{27} + \frac{q^2}{4}}\right)^{-1/3} = \frac{-3}{p}\left(\frac{-q}{2} \mp \sqrt{\frac{p^3}{27} + \frac{q^2}{4}}\right)^{1/3}$. See Exercise 1.3. Solutions $0, 1, -1$ and $3/2, -5/4, -9/4$.

(11.29) Sketch the graph of f, it is in two parts, local maximum at 2, local minimum at 6. Using this information, show that the graph of $g(x) = \frac{5}{x}$ for positive x lies in between the two parts of f.

(11.30) $A = 4/3, B = -16/9$. $Ce^x + De^{-x} + (Ax + B)e^{2x}$ where C, D are arbitrary real numbers.

(11.31) Differentiate the power series term by term. If $h = f^2 + g^2$, then $h' = 0$. Apply Corollary 10.7.

(11.32) $y(x) = a(1 - x^2) + b \sum_{k=0}^{\infty} \frac{x^{2k+1}}{2^k k! (4k^2 - 1)}$, $a, b \in \mathbf{R}$, $x \in \mathbf{R}$.
$y = a(1 - 2x^2) + b\{x + \sum_{k=1}^{\infty} \frac{(2k-3)(2k-5)\cdots(-1)}{2^k k!} x^{2k+1}\}$, $a, b \in \mathbf{R}, |x| < 1$.

(11.33) (i) $Ae^{2x} + Be^{-6x} - \frac{1}{16}e^{-2x}$, (ii) $Ae^{2x} + Be^{3x} - 3xe^{2x}$, (iii) $-2e^{3x} + Ae^{(2+\sqrt{3})x} + Be^{(2-\sqrt{3})x}$, (iv) $4 + x - 4e^{2x} + Ae^{(2+\sqrt{3})x} + Be^{(2-\sqrt{3})x}$, (v) $Ae^{2x} + Bxe^{2x} + \frac{x^2}{4} - \frac{1}{8}$, (vi) $r^3 - 2r^2 - r + 2 = (r-1)(r+1)(r-2)$, $Ae^x + Be^{-x} + Ce^{2x} + \frac{1}{18}e^{5x}$, where A, B and C are arbitrary real numbers.

Chapter 12

(12.1) $x^4 - 10x^2 + 1$.

(12.2) Let A denote the set of algebraic real numbers and let A_n denote the real numbers x which satisfy $P(x) = 0$ for some polynomial of degree n. Since $A = \bigcup_{n=1}^{\infty} A_n$ it suffices, by Exercise 6.10(b), to show that each A_n is countable. Let P_n denote the set of polynomials of degree n with integer coefficients. Each $P \in P_n$ is determined by its coefficients and as there are $n + 1$ of these the number of elements in P_n is less that the number of elements in $\mathbf{Z}^{n+1}$. By Exercises 6.11 and 6.10(a) this set is countable. By Exercise 1.2, each polynomial of degree n has at most n solutions, and the total number of solutions is less than the number of elements in $(P_n)^n$ and this is countable by Exercises 6.10 and 6.11.

(12.3) Let $\alpha = S((a_n)_{n=0}^\infty)$ where $(a_n)_{n=0}^\infty$ is an increasing sequence of positive dyadic numbers, let the dyadic number m be an upper bound for β. By Theorem 5.12 we can choose a positive integer n such that $na_0 > m$. If $b \in \beta$ then $b < m < na_0$. Hence $\beta \le n\alpha < (n+1)\alpha$.

(12.7) The mapping $\varphi : \mathbf{Z} \times \mathbf{N} \longrightarrow \mathbf{D}, \varphi(p,n) = \frac{p}{2^n}$ is surjective. Now $\mathbf{N}$ is countable and $\mathbf{N} \cup -\mathbf{N} \cup \{0\} = \mathbf{Z}$ is countable by Exercise 6.10(b), $\mathbf{Z} \times \mathbf{N}$ is countable by Exercise 6.11. By Theorem 6.3 there is a surjective mapping from $\mathbf{N}$ onto $\mathbf{Z} \times \mathbf{N}$ and hence from $\mathbf{N}$ onto $\mathbf{D}$. By Theorem 6.3, $\mathbf{D}$ is countable.

(12.8) We have $\alpha \subset \beta$ and the inclusion is proper. Choose an $a \in \beta$ such that $a \notin \alpha$. Next choose $n \in \mathbf{N}$ such that $a + \frac{1}{2^n} \in \beta$. Then $\alpha \subset S(a + \frac{1}{2^{n+1}}) \subset \beta$ and both inclusions are proper. Then $a + \frac{1}{2^{n+1}} \in \mathbf{D}$ and $\alpha < S(a + \frac{1}{2^{n+1}}) < \beta$.

(12.10) Clearly $\beta := \bigcup_{j=0}^\infty \{x \in \mathbf{D} : x \le a_{n_j}\} \subset \bigcup_{n=0}^\infty \{x \in \mathbf{D} : x \le a_n\} = \alpha$. If $m \in \mathbf{N}$ is fixed choose j such that $n_j > m$. Then $S(a_m) \subset S(a_{n_j}) \subset \beta$ and $\alpha \subset \beta$. If $c \in \mathbf{D}^+$, then $c^{-1} \in \mathbf{D}^+$. Hence it suffices to show that for any positive dyadic number c we can choose n_c such that $a_{n_c} + c \notin \alpha$. Otherwise, for all n, $a_n + c \in \alpha$. Then $a_0 + c \in \alpha, (a_0 + c) + c \in \alpha = a_0 + 2c \in \alpha, \ldots, a_0 + nc \in \alpha$ for all n. But the sequence $(a_0 + nc)_{n=1}^\infty$ is easily seen, by Theorem 5.12, not to be bounded above. Hence α is not bounded above, and this contradicts the fact that α is a dyadic section (see Definition 12.2(b)).

Chapter 13

(13.1) Find indefinite integrals by trial and error or use integration by parts.
(a) $x 10^x (\log 10)^{-1} - 10^x (\log 10)^{-2}$, $10^4 (46 \log 10 - 9)(\log 10)^{-2}$, (b) $(2x^2 \log x - x^2)/4$, $(72 \log 12) - (143/4)$, (c) $xe^x - e^x$, $11e^{12}$, (d) $2 \log(3/2)$, (e) 4.

(13.2) $3x^2 \cdot 2^{x^3}$. Let $y = x^2 + 1$ and then use Exercise 13.1(b). Let $y = \log x$.

(13.3) $e^{x^4}/4, \frac{-(1 + \log x)}{x}$, (c) $3(1 + x^2)^{2/3}/4$, (d) $-e^{-x^3}/3$, (e) $(x^2 + 10x + 5)^4$.

(13.6) If $Q(x) = \sum_{j=1}^n a_j (j+1)^{-1} x^{j+1}$ then $Q' = P$ and $Q(0) = Q(1) = 0$. Apply the Mean Value Theorem to Q.

(13.7) $(x \log x - x)' = \log x$. Hence $\int_1^n (\log x) dx = n(\log n) - (n-1) = \log(n^n) - \log(e^{n-1}) = \log\left(\frac{n^n}{e^{n-1}}\right)$. Using intervals of length one on $[1, n]$ gives $\underline{\mathcal{R}}(\log x, n) = \sum_{j=1}^{n-1} \log j = \log((n-1)!)$ and $\overline{\mathcal{R}}(\log x, n) = \sum_{j=2}^{n} \log j = \log(n!)$. Taking exponentials $(n-1)! \leq n^n e^{-n+1} \leq n!$, hence $\frac{e^n}{en} \leq \frac{n^n}{n!} \leq \frac{e^n}{e}$.

(13.8) $\frac{1}{n}\{(n+1) \cdot (n+2)\cdots(n+n)\}^{1/n} = \left\{\frac{n+1}{n}\frac{n+2}{n}\cdots\frac{n+n}{n}\right\}^{1/n} = \exp\left\{\frac{1}{n}\sum_{j=1}^{n} \log(1 + \frac{j}{n})\right\} \longrightarrow \exp\left\{\int_0^1 \log(1+x) dx\right\} = \exp\left\{\int_0^1 \frac{d}{dx}[(1+x)\log(1+x) - x] dx\right\} = \exp\left\{(1+x)\log(1+x) - x\right]_0^1\} = \exp\{(2\log 2) - 1\} = e^{\log 4} \cdot e^{-1} = 4/e$.

(13.9) $2\int_0^a \sqrt{x}\,dx = 4x^{3/2}/3]_0^a = 4a^{3/2}/3 = \frac{4}{3}(2 \times \frac{1}{2} \times a \times \sqrt{a})$.

(13.10) $x(\log x)^2 - 2x \log x + 2x$, $x^2(\log x) - \frac{x^2}{2}$.

(13.11) If $F(t) := \int_0^t f(x) dx - \int_t^1 f(x) dx + \int_0^1 f(x) dx = 2\int_0^t f(x) dx$, then F is constant and $F'(t) = 2f(t) = 0$.

Chapter 14

(14.2) Example 14.12(b) implies that the integral diverges if $\alpha \leq 1$. If $\alpha > 1$ and $y = \log x$, then $dy = dx/x$ and $\int_2^n \frac{dx}{x(\log x)^\alpha} = \int \frac{dy}{y^\alpha} = \frac{y^{-\alpha+1}}{-\alpha+1} = \frac{(\log x)^{-\alpha+1}}{-\alpha+1}\Big]_2^n = \frac{(\log n)^{-\alpha+1}}{-\alpha+1} - \frac{(\log 2)^{-\alpha+1}}{-\alpha+1} \longrightarrow \frac{1}{(\log 2)^{\alpha-1}(\alpha-1)} < \infty$.

(14.3) If $a < c < d < e < b$ and c, d, e are rational, then, by Theorem 13.3(c), $l = \int_c^e f(x) dx = \int_c^d f(x) dx + \int_d^e f(x) dx = 2l$ and $l = 0$. If $f(x_0) \neq 0$, $a < x_0 < b$ then, by Exercise 9.10, there exists $\delta > 0$ such that $a < x_0 - \delta < x_0 + \delta < b$ and $f(x) \neq 0$ for all $x \in [x_0 - \delta, x_0 + \delta]$. By Exercise 12.8, we can choose $c, d \in \mathbf{Q}$ such that $x_0 - \delta < c < d < x_0 + \delta$. By the Intermediate Value Theorem, there exists $m > 0$ such that either $f(x) \geq m$ on $[c, d]$ or $f(x) \leq -m$ on $[c, d]$. By (13.5), we have either $\int_c^d f(x) dx \geq m(d - c) > 0$ or $\int_c^d f(x) dx \leq -m(d - c) < 0$, both impossible.

(14.4) $1/2e, 1/3e, 7e^{-6}/9, 1/8$.

(14.5) $(-\sqrt{2}, \sqrt{2})$, $[-3, 3)$, $(-1, 1)$.

(14.6) Series converges, use Exercise 14.2 with $\alpha = 2$.

(14.8) \$82,280.

(14.9) $b = 15$.

(14.10) Only the second series converges.

(14.11) $\sum_{n=0}^{\infty} x^n$, $\sum_{n=1}^{\infty} \frac{x^{2n}}{n}$.

(14.12) $\sqrt{\pi/a} \cdot e^{(b^2-4ac)/4a}$.

(14.15) $\frac{\lambda}{\alpha-1}$, $\frac{2\lambda^2}{(\alpha-1)(\alpha-2)}$.

(14.17) Fix $s, t, 0 < s < t < r$ and let $A(n) = \sup_{\{|x|<t\}}\{t^n|f^n(x)/n!|\}$ and let $B(n) = \sup_{\{|x|<t\}}\{|t^n g^n(x)/n!|\}$. By Corollary 14.14, $A(n), B(n) \longrightarrow 0$ as $n \longrightarrow \infty$. If $|x| < s$, Exercise 10.19 implies that for some constant c

$$\frac{|s^n(f \cdot g)^n(x)|}{n!} \leq \sum_{k=0}^{n}\left|\frac{s^k f^k(x)}{k!}\right| \cdot \left|\frac{s^{n-k}g^{n-k}(x)}{(n-k)!}\right|$$

$$\leq \sum_{k=0}^{n}\left(\frac{s}{t}\right)^k A(k) \cdot \left(\frac{s}{t}\right)^{n-k} B(n-k) \leq c(n+1)\left(\frac{s}{t}\right)^n.$$

By Examples 7.13 and 11.13, $\sup_{|x|<s}\{|s^n(f \cdot g)^n(x)/n!|\} \longrightarrow 0$ as $n \longrightarrow \infty$ and, since s and t were arbitrary, an application of Corollary 14.14 proves the result.

(14.18) Use integration by parts, induction and the result assumed in Example 14.9.

(14.19) Fix $m \in \mathbf{N}$. Let $\mathbf{P}$ be the statement that m is square free, $\mathbf{Q}$ be the statement that $m|n^2$, let $\mathbf{R}$ be the statement that $m|n$. Since $\mathbf{R} \implies \mathbf{Q}$ we need to show $\mathbf{P} \iff \{\mathbf{Q} \implies \mathbf{R}\}$. If $\mathbf{P}$ holds, then $m = p_1 \cdots p_k$ where $p_1, \ldots, p_k$ are distinct primes. If $m|n^2$ then, for each i, $p_i|n^2$ and hence $p_i|n$ and $m|n$, that is $\mathbf{Q} \implies \mathbf{R}$. If $\mathbf{P}$ does not hold, then $m = q_1^{a_1}q_2^{a_2}\cdots q_t^{a_t}$ for distinct primes $q_1, \ldots, q_t$ and $a_1, \ldots, a_t$ in $\mathbf{N}$, $a_1 > 1$. If $b_1 = a_1 - 1$ and $n := q_1^{b_1}q_2^{a_2}\cdots q_t^{a_t}$, then $m|n^2$ but $m \nmid n$.

(14.20) $\Gamma(\frac{1}{2}) = \sqrt{\pi}$, $\Gamma(\frac{16}{3})\Gamma(\frac{5}{2})/\Gamma(\frac{4}{3}) = \frac{910\sqrt{\pi}}{27}$.

BIBLIOGRAPHY

[1] D.N. Arnold, K.K. Fowler, Nefarious Numbers, Newsletter of the European Mathematical Society, 80, 2011, 34-37.

[2] C. Babbage, *Passages from the Life of a Philosopher*, Longman, London, 1864. Republished in 1969 by Gregg International Publishers.

[3] P. Billingsley, Van der Waerden's continuous nowhere differentiable function, Amer. Math. Monthly, 89, 9, 1982, 691.

[4] G. Birkhoff, S. MacLane, *A Survey of Modern Algebra*, McMillan, New York, 1953.

[5] L.J. Boya, Another Relation Between π, e, γ and $\zeta(n)$, Rev. R. Acad. Cien. (Madrid), Serie A, Mat., Vol. 102, 2008, 199-202.

[6] C.B. Boyer, *The History of the Calculus and its Conceptual Development*, Dover Publications, New York, 1949.

[7] F. Ó Cairbre, The Importance of Being Beautiful in Mathematics, Newsletter, Irish Mathematics Teachers Association, 109, 2009, 29-45.

[8] E. Chiera, *They Wrote on Clay, The Babylonion Tablets Speak To-day*, Cambridge University Press, 1939.

[9] J.J. O'Conner, E.F. Robertson, http://www-history.mcs.st-and.ac.uk .

[10] A.D.D. Craik, *Mr. Hopkins' Men: Cambridge Reform and British Mathematics in the 19th Century*, Springer, Berlin, 2007.

[11] T. Dantzig, *Number, the Language of Science*, Penguin Group, USA, 2007 (First Edition, 1930).

[12] P.J. Davis, Fidelity in Mathematical Discourse: Is one and one really true? Amer. Math. Monthly, 79, 1978, 252-263.

[13] P.J. Davis, Leonhard Euler's Integral: A Historical Profile of the Gamma Function. *The Chauvenet Papers*, Volume II, J.C. Abbot (Ed.), The Mathematical Association of America, 1978, 330-351.

[14] P.J. Davis, R. Hersh, *The Mathematical Experience*, Houghton Mifflin Com-

pany, Boston, 1981.

[15] J.M. Dubbey, *The Mathematical Work of Charles Babbage*, Cambridge University Press, 1978.

[16] W. Dunham, *The Calculus Gallery, Masterpieces from Newton to Lebesgue*, Princeton University Press, Princeton and Oxford, 2008.

[17] E.G. Effros, Mathematics as Language, *Truth in Mathematics*, H.G. Dales, G. Oliveri (Eds), Clarendon Press, Oxford, 1998, 131-145.

[18] H. Eves, *Great Moments in Mathematics (Before 1650)*, Dolciani Mathematical Expositions, 5, The Mathematical Association of America, 1980.

[19] H. Eves, *Great Moments in Mathematics (After 1650)*, Dolciani Mathematical Expositions, 7, The Mathematical Association of America, 1981.

[20] H. Eves, *Foundations and Fundamental Concepts in Mathematics*, Third Edition, Dover Books, 1990.

[21] R.L. Fernandes, Evaluation of Faculty at IST-a Case Study, European Math. Soc., Newsletter 82, 2011, 13-17.

[22] J. Friberg, *A Remarkable Collection of Babylonian Mathematical Texts*, Sources and Studies in the History of Mathematical and Physical Sciences, Springer, Berlin, 2007.

[23] I.M. Gelfand, E.G. Glagoleva, E.E. Shnol, *Functions and Graphs*, Birkhäuser, Boston, 1990.

[24] W.T. Gowers, The Importance of Mathematics, Keynote Address at the Millenium meeting in Paris. Typed Manuscript.

[25] I. Grattan-Guinness, *The Development of the Foundations of Mathematical Analysis from Euler to Riemann*, MIT Press, Cambridge, Mass., 1970.

[26] J. Hadamard, *The Psychology of Invention in the Mathematical Field*. Princeton University Press, 1945.

[27] T.L. Heath, *The Work of Archimedes*, Cambridge University Press, 1897.

[28] L.A. Henkin, Are Logic and Mathematics Identical, Address to the 5th Canadian Mathematical Congress, September 5, 1961.

[29] J.E. Hofmann, *The History of Mathematics*, Philosophical Library, New York, 1957.

[30] S. Hollingdale, *Makers of Mathematics*, Dover Publications, New York, 1989.

[31] M. Kline, *Mathematical Thought From Ancient to Modern Times*, Vol.1, Oxford University Press, 1972.

[32] S.N. Kramer, Schooldays: A Sumerian Composition Relating to the Education of a Scribe, Journal of the American Oriental Society, 69, 4, 1949, 199-215.

[33] N. Levinson, A Motivated Account of an Elementary Proof of the Prime

Number Theorem, *The Chauvenet Papers*, Volume II, J.C. Abbot (Ed.), The Mathematical Association of America, 1978, 490-510.

[34] P. Maritz, James Stirling: Mathematician and Mine Manager, The Mathematical Intelligencer, 33, 3, 2011, 141-147.

[35] O. Neugebauer, *The Exact Sciences in Antiquity*, Princeton University Press. 1952.

[36] I. Niven, A Proof of the Divergence of $\sum 1/p$, Amer. Math. Monthly, 78, 3, 1971, 272-272.

[37] M. Poovey, Can Numbers Ensure Honesty? Unrealistic Expectations and the U.S. Accounting Scandal, Notices of the AMS, 50, 1, 27-35, 2003.

[38] L. Roth, Old Cambridge Days, Amer. Math. Monthly, 78, 1971, 223-236.

[39] G. Zapata, On a Dyadic Foundation for Analysis, Notes in Preparation.

INDEX